Environmental and Natural Resources Economics

Xiangzheng Deng · Malin Song · Zhihui Li ·
Fan Zhang · Yuexian Liu

Environmental and Natural Resources Economics

Springer

Xiangzheng Deng
Institute of Geographic Sciences
and Natural Resources Research
Chinese Academy of Sciences
Beijing, China

University of Chinese Academy of Sciences
Beijing, China

Zhihui Li
Institute of Geographic Sciences
and Natural Resources Research
Chinese Academy of Sciences
Beijing, China

University of Chinese Academy of Sciences
Beijing, China

Yuexian Liu
University of Chinese Academy of Sciences
Beijing, China

Malin Song
School of Statistics and Applied
Mathematics
Anhui University of Finance
and Economics
Bengbu, China

Fan Zhang
Institute of Geographic Sciences
and Natural Resources Research
Chinese Academy of Sciences
Beijing, China

University of Chinese Academy of Sciences
Beijing, China

ISBN 978-981-99-9922-4 ISBN 978-981-99-9923-1 (eBook)
https://doi.org/10.1007/978-981-99-9923-1

This Springer imprint is published by the registered company Springer Nature Singapore Pte Ltd.
The registered company address is: 152 Beach Road, #21-01/04 Gateway East, Singapore 189721, Singapore

Paper in this product is recyclable.

Acknowledgements

This book was supported by the Textbook Publishing Center of University of Chinese Academy of Sciences, the Science Fund for Creative Research Groups of the National Natural Science Foundation of China (Grant No. 72221002), and Strategic Priority Research Program of the Chinese Academy of Sciences (Grant No. XDA23070400).

Environmental and Natural Resources Economics

Textbook Series of University of Chinese Academy of Sciences (Graduate Level) (YJC0705009).

Contents

Chapter 1
Environment and Natural Resources Economics: Overview

1.1 Introduction

1.1.1 Background

Previous economic studies that considered the particularities of natural resources and environmental factors in economic activities did not have a particularly substantial impact on economic analysis. Therefore, in the general theoretical framework of economic analysis, it is usually classified as the factors under "other conditions remain unchanged" and is abstracted away from the analysis object. However, under new historical conditions, the relationship between economic activities and resources and the environment has undergone substantial significant changes. Resources and the environment have objectively become internal factors in economic decision-making. This should be considered as much as possible in the theoretical framework of economic analysis of resources and environmental factors, especially in the field of applied economics. Resource and environmental factors must be introduced in a general manner, as one research object, or one basic component of the research object.

Under certain conditions, good environmental factors and rational utilization of natural resources are conducive to the development of productive forces, thereby having a positive role in promoting social and economic development (Beder, 2011). However, some undesirable products of economic development, such as sewage and air pollutants, have been reflected in the environment, seriously endangering environmental quality. In the second half of the twentieth century, human development threatened its own living environment on a global scale. In China, the problem of environmental pollution created by sustained and rapid economic development cannot be ignored.

China has recently increased its emphasis on the environment. The report of the 18th National Congress of the Communist Party of China clearly stated that the

X. Deng et al., *Environmental and Natural Resources Economics*,
https://doi.org/10.1007/978-981-99-9923-1_1

construction of an ecological civilization should focus on the prominent contradictions in China. Without violating the laws of natural ecological development and while considering actual social conditions, ecological civilization projects in China should be constructed with a policy of prioritizing conservation, protection, and natural restoration. This policy not only indicates the direction of ecological civilization construction in China but also can become an important guiding ideology (Chen & Shi, 2022). In 2019, the National Natural Resources Work Conference was held in Beijing, which fully discussed ecological civilization conceptions and natural resource management strategies.

In the new era, ecological economics is guided by the idea of ecological civilization and uses socialist economic theory and ecological principles. From the industrial ecology and ecological industrialization perspectives, it is a comprehensive discipline that studies the structure, function, behavior, and laws of an integrated system, which includes both the ecological and economic systems. As China entered a new era of socialist ecological civilization, ecological economics became increasingly prominent. However, the ecological environment is a matter of people's livelihoods and is related to economic development and serving the people. Identifying effective communication links between economic development and ecological and environmental protection is an important issue for regional economic development.

China's basic national policy insists on saving resources and protecting the environment. Environmental protection is related to economic development and people's livelihoods, and is everyone's common responsibility and obligation. China has faced many challenges since the reform and opening up. Environmental impact was ignored, and a rough-emission economic development method was adopted to catch up with other developing countries. Waste has results in quite serious consequences for the ecological environment of the economic development area and surrounding areas. The "industrialized" economic model in eastern China is a success for local economic development, but a failure for the local ecological environment. The industrialization of eastern coastal areas has produced large amounts of industrial solid waste and atmospheric pollutants. Rough treatment and random discharge of these undesirable "industrial products" have caused serious and even irreversible water, soil, and heavy metal pollution, further affecting the quality of cultivated land, food safety, and national health (Jacobs, 2013). To enable us and our descendants to live with green water, green mountains, and blue sky, the central committee of the Communist Party of China (CPC) has proposed the development concept that "lucid waters and lush mountains are invaluable assets."

1.1.2 Relationship Between Economic Development and Environmental Protection

The resource depletion problem in China is becoming increasingly prominent, including in relation to freshwater, minerals, forests, and land resources, which are closely related to environmental pollution and damage. The greater the pollution and damage, the faster the resource depletion. Further, resource depletion is related to the economic development of cities. The faster the economic development of a city, the more resources are required. Thus, rapid economic development leads to excessive resource exploitation.

China's economic development cannot be separated from the support and help of environmental resources. However, many resources have not been properly developed, utilized, and protected. In the long run, this destroys the balance of the environment, and in turn, leads to a lack of resources and deteriorating environmental problems, hindering further economic development. Recently, the demand for environmental resources, such as land, oil, and other energy resources, has increased significantly (Meng et al., 2021). The supply of some important energy sources in China is far less than the demand, thus significantly increasing the dependence on foreign countries for these resources. Some scholars have found that, to alleviate the constraints of important energy resources on economic development, people should conserve energy and resources, while also promoting high utilization efficiency and recycling.

Dialectical materialism holds that the relationships between things should be uniformly viewed dialectically; that is, we should see both the unified and opposite side of things. Economic development and environmental protection are dialectically unified from the perspective of their content and purpose. Environmental pollution caused by economic development hinders the smooth progress of environmental protection work, and environmental protection also restricts economic development to a certain extent; therefore, the two have an opposing relationship. However, under certain conditions, economic development and environmental protection can achieve harmony and promote each other. Economic development should follow the basic principles of environmental protection and not involve pollution before treatment. Conversely, environmental protection can, to some extent, also promote economic development in a sound and sustainable direction. These two principles complement, promote, struggle with, and restrict each other. Therefore, their relationship is one of dialectical unity. Ultimately, both of them aim to wholeheartedly serve the people and meet their needs for a better life (Chen & Zhao, 2019). In the new era of socialism with Chinese characteristics, China's economic development is transforming from high-speed to high-quality development, and from "high-speed competition" to "quality comparison and sustainable development." That means high-quality economic development must be green and sustainable.

At present, some conventional and unconventional problems in the economic development process must be solved during the high-quality development stage. Environmental protection issues should be addressed to achieve economic development. In practice (Nilsen & Ellingsen, 2015), the first need is to address the relationship between economic development and environmental protection. Appropriate help should be provided to enterprises that meet environmental protection standards, while successfully implementing environmental protection and governance. Thus, enterprises can give full play to the "exemplary role" of environmental protection and encourage more industrial enterprises to take the same direction.

Two principles should be followed to understand and address the relationship between economic development and environmental protection. First, the ecological environment cannot be sacrificed in favor of economic development. Second, no "one size fits all" formalism exists. Specific issues require specific analyses according to local conditions.

A modern economic development system must be built to achieve high-quality development. Coordinating with the government and the market through scientific management and green scheduling is necessary. Creating a convenient environment and encouraging enterprises with markets and profits can help achieve better and faster development (Shafik, 1994). While focusing on the development of traditional industries, China should accelerate industrial transformation and actively foster and develop emerging industries with promising market prospects and low resource and energy consumption. Development is the CCP's top priority in governing and rejuvenating the country. Innovation is the core of the five development concepts of "innovative, coordinated, green, open and shared." Development without innovation is only following the old path, which cannot keep up with modern needs and is a manifestation of backwardness. It is necessary to cultivate scientific and technological innovation talent, build scientific and technological innovation platforms, and improve enterprises' independent innovation abilities and competitiveness. While innovating in development, opening up and development are also necessary. China requires both reform and development to introduce it and go global. China needs to seize major strategic opportunities, such as the Belt and Road Initiative, strengthen foreign exchange and cooperation, and increase the trade-based economy.

In terms of environmental protection, China should practice the development concept that "lucid waters and lush mountains are gold and silver mountains." Comprehensively rectifying "scattered and polluted" enterprises by shutting down enterprises with incomplete and being difficult to transform procedures, and developing clean energy sources, such as nuclear power and thermal energy, and gradually reducing high-polluting energy sources. The coverage area for urban greening should be increased by implementing dust-control measures on urban roads and construction sites. Five major battles that urgently need to be fought are the defense of water sources, urban black and odorous water bodies, Yangtze River protection and restoration, Bohai Sea comprehensive treatment, and agricultural and rural pollution control. A positive list system for ecological environment supervision should be established to implement precise and scientific management, advance judgment, highlight key points, and respond actively. Implementing important ecosystem protection and restoration projects can help achieve a good ecological balance with a blue sky, green land, and clean water.

1.2 Relationship Between Human Beings and Environmental Resources

1.2.1 Human Behavior and the Environment

Environmental impact refers to the environmental changes caused by human activities (e.g., economic and social activities) and the resulting effects on human society and the economy. Human activities constantly affect the natural environment, causing changes in environmental quality, which in turn affect normal human life and health. In the production activities is relatively simple and the scale is small, human activities have little impact on the environment. However, with industrialization, coal, steel, oil, and other raw production materials have discharged a large amount of industrial waste into the natural environment; the acceleration of urbanization, the population around the city into urban life, work, domestic waste is also increasing. Economic development has enriched people's material lives; however, the unexpected products it has brought about, such as waste water, gas, and residue, are also constantly polluting and affecting people's living environment. This section explains the impact of human behavior on the environment from the perspectives of population growth, economic growth, and technological progress.

1.2.1.1 Population Growth

Population growth is a change in a country or region over a certain period of time and is closely related to economic development and the merits of the environment. Population status: the birth rate is greater than the mortality rate; the large population base, the number of biological populations change law is not fully applicable to the population growth situation.

The pressures population growth has placed on the environment include impacts on food, water resources, forests, and grasslands and other natural resources, living spaces, and spiritual needs. Large population growth has increased human demand for the environment, and people have placed increasing value on the environmental problems this has caused. Under the premise of sustainable development, energy is obtained and new energy is developed while population growth is controlled (Fig. 1.1).

Land is an essential material condition for people's survival. People's food, clothing, shelter, and transportation are inseparable from that gifts the land provides to people. Most of the food required for human survival is derived from land-based crops. Although China's land area ranks third worldwide, the per capita land area is relatively low because of factors such as the large population base and topographic characteristics. With rapid economic development, in the industrialization and urbanization process, cultivated land area is increasingly used for transportation, water conservation, and industrial and mining construction. This reduces the

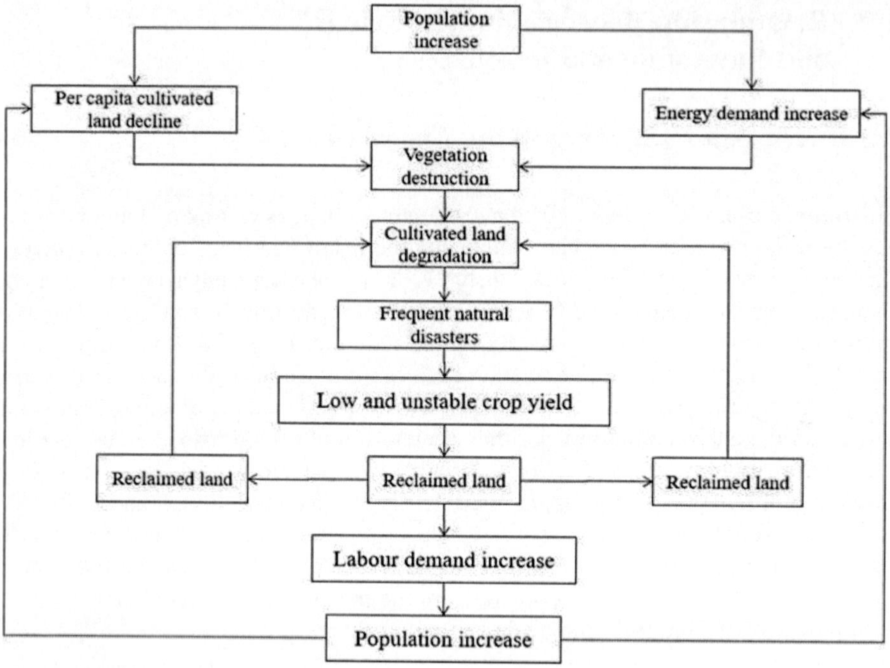

Fig. 1.1 Relationship between population growth and the environment

amount of excellent cultivated land, leading to the declining overall quality of cultivated land and a lack of reserve resources. China is a large agricultural producer, and land resources are facing many problems, such as unreasonable land use, land quality degradation, serious soil erosion, soil desertification, and serious soil pollution. Population increases and human production activities are under increasing pressure from land resources. Human activities affect land resource development and utilization. Reasonable human land reclamation activities can transform natural land ecosystems into farmland ecosystems, and facilitate agricultural production. However, unreasonable human reclamation activities will aggravate the trend in soil erosion, land desertification, and secondary salt salinity.

Water is the source of life. Water resources are of great significance to humans and other natural beings because of their indispensable and irreplaceable nature. Under the influence of geographical factors, water distribution are mainly distributed in hills and plains. The distribution of water resources in China has been studied mainly in terms of precipitation, surface water resources, underground water resources, and total water resources. In general, rivers and lakes are the primary freshwater resources in China. The per capita runoff in China is 2,200 m^3, which is 24.7% of the world's per capita runoff. The distribution of water resources in China is higher in the south and lower in the north.

The effects of human activities on water resources can be divided into reservoir, agricultural, and forest hydrological effects. Regarding the hydrological effect of reservoirs, people build reservoirs and other artificial lakes to intercept surface runoff, reducing the flow into the sea, such as the South-to-North Water Diversion Project in China. Reservoirs are constructed in relation to the surrounding temperature changes, becoming warm in the winter and cool in the summer, and the evaporation of the water surface of the reservoir area increases, changing the water vapor content of the attached atmosphere and reducing the airborne water on the surface, thereby leading to an increase in precipitation around the reservoir. The construction of reservoirs such as dams along some river basins will reduce soil fertility and even increase the saline content of the soil. Water quality deteriorates downstream after dams and reservoirs intercept the water and affect the flow rate. Agricultural hydrological effects are mainly reflected in water intake projects, such as farmland drainage projects, which reduce diving levels and increase river runoff. Mountainous terraces can effectively prevent soil erosion by changing the slope and river slope, intercepting and reducing surface runoff, increasing surface water infiltration, and delaying flood processes. Forest hydrological effects include high water vapor content, high humidity, and atmospheric horizontal gas flow through the forest; therefore, horizontal precipitation in the forest has some influence on groundwater. Mountain forest seepage water is beneficial to groundwater recharge, and the influence of plain forest flow on groundwater varies with different climatic conditions.

Daily human life, industrial and agricultural production, and a series of activities will affect water resource quality. The toxic and harmful gases and dust emitted by fossil fuel combustion and industrial production enter the atmosphere, and then enter the groundwater as precipitation, causing the water quality to deteriorate. Urban sewage, industrial wastewater, industrial waste residue, and municipal waste dissolved water enter the groundwater with surface water infiltration, and groundwater affects water quality. Aside from being absorbed, volatile, and decomposed by crops, most remain in farmland soil and water with farmland drainage and surface runoff, causing harm.

Energy is an important material basis for human survival and production. Energy resources are the driving force for the development of human production activities, and the utilization and development of energy play a substantial role in promoting human society. Given the need for economic development, increasing amounts of energy are required for economic production. However, some unreasonable energy consumption structures include coal and oil as the main energy that are consumed, while natural gas, hydropower, wind power, nuclear power, and other clean energy sources are consumed less. This causes environmental problems, such as increased air pollution and serious soil and water pollution. With the increases in the population and economic development, the human demand for energy is also increasing.

With economic development and scientific and technological progress, human beings have displayed greed toward nature for their own interests and overcut the forests. The tree growth rate is far lower than the cutting rate, and combined with the relevant forestry departments not having reasonable planting and cutting plans, to

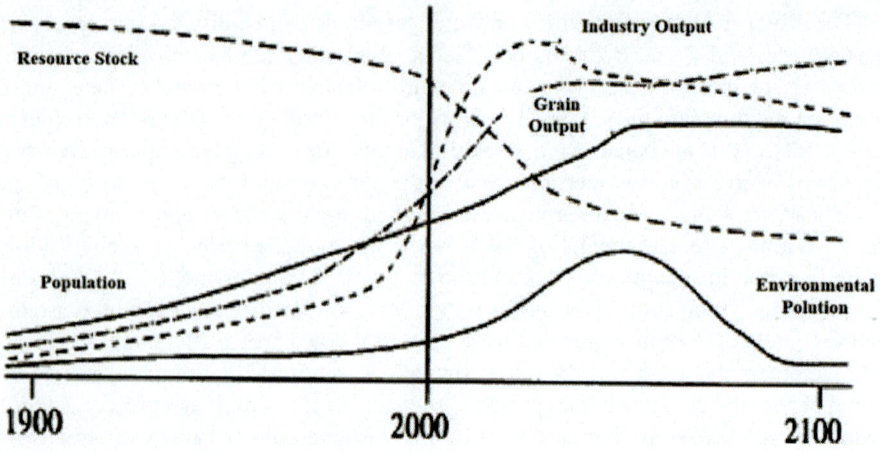

Fig. 1.2 Environmental impact of population growth

meet the needs of timber, many years of timber are classified as logging plan, deforestation of frequent phenomenon (Hepburn, 2010), made the original is not poor forest resources increasingly exhausted, seriously damaging the ecological environment. The economic development of China cannot be achieved without the support of forest resources, but in recent years, China's forest resources have suffered major damage, and the size of forest exploitation has reached its limit. China has needed to import wood from abroad to promote economic development, visible the importance of forest resources for economic development to be reckoned with, the influence of human activities on forest resources is also very important.

Environmental pollution has intensified over time. Accelerating urbanization processes have produced substantial amounts of domestic garbage and sewage discharge, industrial production and industrial waste residue, wastewater, and waste, which are constantly discharged into the atmosphere, rivers, and soil, leading to air pollution, river water quality deterioration, soil pollution, and other serious problems. These exceptions are in the vulnerable environment constantly suffering from pollution (Fig. 1.2).

1.2.1.2 Economic Growth

Especially in China, the rapid economic development of the entire country is promoted through resource utilization. However, excessive natural resource utilization will inevitably lead to environmental changes, such as a large amount of mining, resulting in resource depletion. The use of resources has also caused serious environmental pollution. Economic development in China is accompanied by various problems, the biggest of which being that economic development depends on environmental costs. Therefore, environmental protection has become an obstacle to economic development (Hossain et al., 2019). Developed countries, which previously focused on economic development and neglected environmental protection, now have to sacrifice economic development to reduce environmental damage. Although the environment has been improved, the environmental chain has broken and requires considerable human and material resources for repair. This not only wastes the environment but also creates economic damage (Paavola & Adger, 2005). Drawing on the experiences of developed countries, China strengthened its environmental protections and implemented a sustainable development strategy for economic development. This is conducive to not only economic development but also improved quality of life for future generations. Currently, all countries worldwide are attaching importance to strategic sustainable development research and vigorously developing green and pollution-free industries.

From a historical perspective, the relationship between economic development and environmental protection goes through three stages: first, economic development is emphasized and environmental protection is ignored; second, economic development and environmental protection play a game with each other; and finally, economic development and environmental protection are considered together.

The industrial economy replaced the agricultural economy in the dominant position during the industrial revolution, and human demand for natural resources increased. The extensive economic development model has led to a surge in pollutant emissions, and industrial wastewater and gas have caused unprecedented ecological damage. However, industrial development has brought great changes to human life, coupled with people's lack of awareness of environmental protection, making environmental damage insignificant. Therefore, at this stage, people pay more attention to economic development than ecological environment protection.

The ecological damage caused by blind economic development has awakened some people with insight. People are beginning to realize that to achieve long-term development, we must protect the ecological environment and the earth (Sun & Wang, 2022). Long-term development is even more important, especially to a country's economy. However, some small groups have paid more attention to their immediate interests. Therefore, at this stage, some countries are vigorously promoting sustainable economic development, while small groups still seek short-term interests at the expense of the environment.

The need for environmental protection has recently reached a global consensus. An increasing number of countries have joined carbon–neutral target actions. Active measures have been formulated to control and alleviate environmental pollution problems caused by economic development and vigorously develop renewable technologies to balance the relationship between environmental protection and economic development, while promoting sustainable development.

1.2.1.3 The Environmental Impact of Technological Progress

Scientific and technological development and utilization have both positive and negative effects on social ecology. Science and technology play an inestimable role in the progress and development of human society. The rapid scientific and technological development has strongly affected every part of society and changed the socio-ecological environment (Vasileva & Ivanova, 2014). Scientific and technological development constantly changes the conditions and environment of human life, which is conducive to socioecological development. However, the development and use of some science and technology have also led to environmental pollution and huge ecological disasters, endangering socioecological development.

People have different answers to the problems of technological progress and the ecological environment. Although technological progress promotes societal development, it may also damage the environment. Technological development and progress are both necessary and inseparable from societal development. Technology used in the development process can also negatively impact the environment. Everything has two sides; thus, to advance technology and avoid environmental damage to the environment, it must form the view of Marxist nature and dialectical materialism of nature, with the dialectical view of technological development, using dialectical thinking to guide technological development and application. By correctly understanding the relationship between human beings and nature, and finding a balance between technological progress and the ecological environment, technological progress, social development, and the ecological environment are protected.

1.2.2 *Asset Attributes of the Environment and Natural Resources*

1.2.2.1 Natural Resources and Their Basic Attributes

Natural resources are material resources that humans can use under certain conditions, changing or unchanging their natural form, and combined with human resources and capital to generate economic value, such as mineral, forest, water, and biological resources.

There are many kinds of natural resources, whose basic attributes mainly include "Usefulness" "Regionality" "Systematicness" and "Timeliness". Regarding "usefulness," the resources in nature are very rich, but not all can be called "natural resources." At present, from the perspective of economics, the temporary definition of natural resources is the useful resources in nature that can be part of economic production. Because of scientific and technological limitations, many resources in nature are unknown, temporarily not included in economic production, and not classified as natural resources. "Regionality" means that the geographical distribution of natural resources. According to economic zones, China can be divided into eastern, western, and central regions, where the central region is rich in mineral resources and has numerous rivers and lakes, such as Henan, Hubei, and Hunan Provinces. Forests and other vegetation resources have very high soil requirements, and are therefore zonal. For example, the forest resources in Northeast China are mainly concentrated in the Greater Khingan Mountains, Lesser Khingan Mountains, and Changbai Mountains. The forest resources in Southwest China are mainly concentrated in the Hengduan Mountains, most of which are located in high-altitude areas (Fang et al., 2021a, 2021b). Systematicness means that natural resources are interconnected. For example, in the process of agricultural production, only the suitable soil and good quality water are conducive to smooth progress. The "systematization" of natural resources requires the systematization between natural resources in people's resource protection to be considered (Kemp & Long, 1984). Timeliness refers to the fact that the understanding scope of natural resources is not fixed, but keeps pace with the times. This will constantly change with the levels of economic, technical, and scientific development. In general, its connotation is constantly enriched.

1.2.2.2 The Basic Connotation of Natural Resource Assets

Material assets existing in the form of natural resources are known as natural-resource assets. Although the total amount of stored natural resources is large, it has not been sufficiently developed to support the development and utilization of all natural resources, and only a small portion of the resources can be exploited and utilized. The vast majority of natural resources are still stored in nature by protection and storage. Therefore, the natural resources that are extracted are very precious and scarce. This also determines the scarcity of natural resource assets. Further, natural resource assets have important economic value. With economic development, natural resource assets are also attributed to important production input factors, such as mineral resources and petroleum resources. As production factors, their economic value can be assessed and circulated in the market. Finally, the property rights of natural resource assets must be discussed (Zuo et al., 2021). The property rights of assets can only be realized once those rights are clarified. The property rights of natural resource assets are also important, and the economy and scarcity of these assets make it particularly important to know who owns the property rights. In general, the state owns most natural resource assets, as well as the benefits obtained from their development and utilization.

In general, natural resource assets refer to material assets that have clear scarcity, economy, and property rights and exist in the form of natural resources. They are conducive to economic development and benefit the development of human ecological civilization.

1.2.2.3 Basic Attributes of Natural Resource Assets

Natural resource assets exist in their natural state (and are accompanied by the artificial state), such as the water flow, forests, mountains, grassland, wasteland, and beaches. First, natural resource assets are assets owned by the state. If the reasonable use of natural resource assets within a controllable range will be conducive to national economic development, for the purposes of national security, we should protect national resource security. Second, regarding the exclusivity of natural resource assets, the property rights of natural resource assets determine the right to use them as a special subject. Once a subject is determined for use, other subjects can no longer obtain the corresponding right to use that resource. Third, regarding regional natural resource assets, the heterogeneity of the distribution of natural resources determines the region of constituent assets.

1.2.3 Economic Explanation of the Relationship Between Human Beings and the Environment and Resources

1.2.3.1 Scarcity of Resources

The first principle of economics is resource scarcity. Compared with unlimited human needs, resources are limited, and to meet the infinite growth of human desire, more resources are needed to produce more goods and services. However, the relative shortage of resources and absolute growth of human demand cause resource scarcity. This scarcity is both absolute and relatively scarce. Absolute scarcity refers to the scarcity of aggregate materials. Thus, the stock of resources in nature is certain, not inexhaustible. If human beings are too greedy, the speed of resource exploitation and utilization will be greater than the regeneration rate of the resource itself, which will certainly lead to resource exhaustion. Relative scarcity refers to the scarcity as understood in mainstream economics, in which resources are compared to unlimited human desires and needs. Resource scarcity leads to social production activities facing selection and competition, while the task of economics is to study how to realize the effective allocation of resources based on the basic proposition of "resource scarcity."

Scarcity leads to society, the collective, and individuals needing to choose what to produce, how to produce, and for whom to produce using limited resources. Because of resource scarcity, resources can only be used for a certain purpose; therefore, effective allocation of resources must be realized among many choices. As Demsetz (2013) argues, the cause of a conflict of interest is the need for scarce resources. Conflicts arise when people attempt to possess and utilize limited resources. To resolve this contradiction, human beings need to restrict each other. Laws are an important way to implement effective restrictions. It is also widely respected by people for its universality, standardization, and stability beyond time and space. Environmental power came into being because of excessive industrial economic growth and neglect of the environmental crisis by later developed countries blindly seeking economic benefits (Zhang & Jiang, 2019). The basic premise of environmental rights is resource scarcity. Natural resources provide social production factors for human beings, such as water, land, and natural scenery, which are both economic and ecological natural resources. Both the absolute scarcity and the relative scarcity in relation to human desire indicate the rare expenditure of natural resources. Therefore, it is important to formulate environmental rights with the purpose of protecting the environment.

1.2.3.2 Economic Man Hypothesis

The term "economic man" has long been debated in economic theory circles, and scholars at different times have had different understandings. In Adam Smith's seminal work "*The Wealth of Nations*" (1776), *The Wealth of Nations*, the "economic man" hypothesis is the basic foothold of his theoretical system. According to Smith (1776), the essence of people engaged in any activity is to maximize their own interests, and self-interest is human nature. In *The Wealth of Nations*, Smith notes that self-interest is the fundamental motivation that controls all human actions and the starting point and purpose of all economic activities. The "economic man" has the dual attributes of self-interest and altruism, which is embodied in the nature of self-interest, which constitutes the basic value system of the "economic man," while altruism is realized through division of labor and exchange, resulting in common social interests. "Economic people" are self-beneficial, and all economic activities are determined by their self-interest; however, their economic activities also enhance the common interests of others and society in the unconscious state. Therefore, in Alfred Marshall's work *Principles of Economics*, (1890) the argument was made that "Economic motives are not all self-interest. The desire for money does not exclude influenced by factors other than money, desire itself may be noble motivation. The scope of economic measures can be gradually expanded to include many altruistic activities."

With the application of mathematical models and empirical tests, the value judgment in economics is decreasing, and the "economic man" gradually transforms into a synonym of "rational choice." Modern economics has provided a simplified and narrow treatment of human behavior motivation, and regards "self-interest maximization" as a basic feature of rational behavior. Especially after the emergence of theoretical methods, such as econometrics and game theory, economics seems to be increasingly unrelated (Heltberg, 2002). Ammatya-sen (1977) is critical of this phenomenon, noting that modern economics lacks not only normative theoretical analysis but also complex and diverse ethical considerations about people. Human behavioral motivations are diverse, and self-interest is only one aspect. Self-interest maximization alone has been unable to reasonably explain practical problems.

1.2.3.3 The Environmental Kuznets Curve

The Environmental Kuznets Curve (EKC), named after a hypothesis proposed by Kuznets (1955), argues that the relationship between unequal income distribution and income level has an inverted U-shape. If the EKC assumption is universal, all environmental degradation will eventually decrease when the national economy develops and the per capita income rises from low to high levels.

When the economy develops at a low level, the speed of economic recession is related to the utilization of basic resources through production activities. The higher the utilization of basic resources, the slower the speed of economic recession, and vice versa. When the economy develops at a medium speed, the consumption of resources caused by industrialization and agriculture rapidly begins to increase, leading to a regeneration rate of resources that is much lower than the increasing rate of resource consumption (Xiao et al., 2022). When the economy is developing rapidly, human environmental awareness begins to increase, science and technology become more advanced, relevant environmental protection laws and regulations are perfected, and the environmental degradation caused by early economic development is gradually resolved.

1.3 Concept Explanation

1.3.1 Environment

In a broad sense, the environment is the place where human beings live, the inorganic and organic combination space of human production and living activities, and the basis for human life and development. In a narrow sense, according to the requirements of the research content, the research object is artificially distinguished from the entire research body. This artificially selected research object is called the system, or the central thing, and the relevant factors around the central thing are called the environment of the central thing. Different central things have different environments, in that the environment changes around the changes of the central thing. In the research fields of natural and environmental sciences, the central thing are human beings. The environment refers to the surrounding things related to human survival, and can be defined as the sum of all external things and forces acting on the central thing of human beings, directly or indirectly affecting human life and development.

From a legal perspective, countries worldwide have clear provisions on environmental protection policies (Shafik, 1994); however, these provisions are mostly different from the national legal interpretation of the environment. China issued the environmental protection law of the People's Republic of China: "environment, refers to the human survival and development of various natural and artificial transformation of natural factors, including atmosphere, water, ocean, land, mineral, forest, grassland, protists, natural relics, cultural relics, nature reserves, scenic spots, cities and countryside, etc."

From the perspective of natural and environmental science, the environment refers to the space around a population and the total natural factors that directly or indirectly affect human production and life. The concept of the environment can be divided into the natural and social environment, based on their own attributes. The natural environment is the necessary basis for human life and production, including natural or artificially modified substances such as air, soil, water, animals, plants, microorganisms, rocks, and minerals; resources containing energy such as light energy, temperature, gravity, and magnetic fields; and natural phenomena such as typhoons, tsunamis, atmospheric flow, the global water cycle, and soil evolution. The social environment has a wider scope, and refers to new achievements formed by processing and transforming existing natural materials through human-conscious social labor. The social environment includes material achievements such as green gardens and urban and rural areas, as well as superstructural conditions such as productive forces, language, customs, culture, systems, technology, art, and law; the economic foundation of society; and non-material achievements adapted to various social and economic systems. These intangible achievements are created by human beings, and contain the humanistic spirit. The formation and development of the social environment are constrained by natural, economic, and social laws, and its environmental quality is a symbol of the construction of human material and spiritual civilization. The development and change of the social environment directly affect the development and change of the natural environment, which plays a role in promoting the development of human society; however, it will become a factor restricting the development of human society because of its rapid development.

The environment can be classified in several other ways. The scope of the environment includes the physical, chemical, atmospheric, water, geological, soil, biological, living, block, urban, regional, and global environments. The natural and social environments form the basis of human survival and development. Human behavior in nature has a comprehensive effect on the environment; through labor activities, humans use and dominate the natural environment, and affect it to a certain extent, and then the natural environment becomes a new living environment, and the new living environment and reflection on humans, embodies the natural environment's ability to adjust. Therefore, the environment in which people live today is part of the process of human evolution, and its structure has developed from simple to complex, from low to high, and through artificial transformations. The role of human beings in the environment has caused environmental problems together in transforming the environment. Environmental problems refer to the uncoordinated relationship between human economic and social development and the environment, which can generally be divided into two categories. "One is the unreasonable development and utilization of natural resources, beyond the carrying capacity of the ecological environment quality or natural resources. The other is the environmental pollution and damage caused by the expansion of population, industrialization, and urbanization. Since its birth, mankind has been in a unity of opposites with nature. With the development of human society, environmental problems are also developing. Human beings should respect the environment, develop production based on ensuring environmental protection, and achieve unity between themselves and the environment.

1.3.2 Natural Resources

In a broad sense, natural resources refer to substances naturally formed in nature that can be used for human survival and development. They are the source and locations of raw materials required for human production and life (Pindyck, 2020). The United Nations Environment Programme considers "natural resources" to be a general term for natural environmental factors that can generate economic value and create welfare for humankind within a certain period of time and under certain technical conditions.

1.3.2.1 Classification of Natural Resources

Many types of natural resource classifications are available. From the perspective of the physical form of natural resources, natural resources can be divided into tangible and intangible. Tangible natural resources include land, water, animal, plant, and minerals. Intangible natural resources are not physical entities but still play crucial roles in ecosystems and human well-being, including light, heat, and air. From the perspective of the attributes of natural resources themselves, they can be divided into environmental resources, biological resources, forest resources, marine resources, land resources, and mineral resources. Species in different natural environments have different characteristics. Environmental resources include light resources, heat resources, and air, which are characterized by stability and cannot be reduced by human use. If properly developed, humans can use these resources continuously. Biological resources refer to organisms that can be used for human production and life, and include animals, green spaces, forests, and grasslands. After they are used by human beings, such resources can reappear through their own production and reproduction. If they can be reasonably used and operated, humans can still use them continuously.

Biological resources play an important role in human production and life. Since ancient times, human beings have survived through rich animal and plant resources, and they have also cooperated with animals in life and development. Plants are renewable resources that provide raw materials for human production and life. Further, plants play an important role in optimizing air quality, resisting natural disasters, and providing animal habitats. Marine resources include marine life, energy, minerals and chemicals. The vast majority of Earth's area is ocean, which is an important part of the environment. Oceans contain extremely rich resources, and the exploration and excavation of marine resources remains the focus of research.

Land resources can be both broadly and narrowly classified. In a broad sense, land resources refer to the general terms of resources, including territory, airspace, continental shelves, and exclusive economic zones under the jurisdiction of a country. In a narrow sense, land resources refer to land under state jurisdiction. Land is not only where humans live but also the means of labor owned by human beings. China has a large total amount of land resources, but its per capita occupation is small, with more land types, and the utilization situation is more complicated. A current issue that needs attention is how land development and protection can be improved. Mineral resources, have accumulated over a long period of time, with limited reserves, and are non-renewable. The uneven distribution of mineral resources in China needs to build a transportation network and strengthen the transportation capacity to facilitate effective resource allocation.

From the perspective of quantitative changes in natural resources, they can be divided into exhausted natural resources, stable natural resources and liquid natural resources. Exhausted natural resources have limited reserves and decrease as people use them. Mineral resources are a typical exhausted natural resource. Stable natural resources have the characteristics of stable reserves and fixed places, and land resources are typical stable natural resources. Liquid natural resources regenerate and disappear at a certain rate. Water is a typical liquid natural resource.

1.3.2.2 Characteristics of Natural Resources

Multipurpose means that most resources have multiple functions. Integrity refers to the natural resource elements of each region being ecologically related and forming a whole. Therefore, more attention should be paid to the comprehensive research, development, and utilization of natural resources. Finiteness refers to the number of required resources. Human production and life will constantly increase the demand for natural resources, and the rapid regeneration of natural resources will lead to resource depletion. Therefore, attention should be paid to the reasonable development and protection of these resources. Regional refers to an imbalance in the distribution of resources. The quantity and quality of natural resources can show significant regional differences, with certain distribution rules. Natural resources are important for measuring sustainable development levels, the material basis for human survival and development, and the source of social material wealth. Humans should focus on the rational use of natural resources to achieve the goal of sustainable development.

1.3.3 Environmental Economics

Environmental economics is an interdisciplinary discipline of environmental science and economics, and applied simultaneously in the economic and environmental sciences, to analyze the relationship between the economy and the environment. By finding the best way to optimize the economy and the environment, humans can improve their quality of life and living standards at a minimum cost (Shogren & Taylor, 2020). The learning process of environmental economics requires not only economic theory but also knowledge of environmental science and engineering. The interdisciplinary integration of these two categories of knowledge is the key to the rational use of environmental economics. Environmental economics arises on the basis of overturning traditional neoclassical economics. Neoclassical economics holds that the environment is infinite. However, in the real world, various crises caused by environmental damage has led economists to reflect on the neoclassical view. Economists have gradually established and developed the discipline of environmental economics by realizing the relative scarcity of the environment through absolute scarcity perception.

The rapid development, wide application scope, and deep research level of environmental economics are not only far beyond the expectations of economic circles and society but also that which the first group of environmental economics researchers expected. Environmental economics research in China can be traced back to 1978. From 1978 to 1985 was an eight-year development planning period of environmental economics and environmental protection technology and economy, during which environmental economics began to sprout and develop in China (Xie et al., 2021). Early scholars found that environmental pollution and destruction were caused by human beings producing and living, which was caused by human failure to recognize natural laws and link social economic development with environmental protection, or by only weighing short-term economic benefits while ignoring future sustainable development. Early human beings regarded natural resources as inexhaustible, and regarded nature as a place to purify waste. With the increase in population and acceleration of industrialization and urbanization, this mode of production has gradually revealed its disadvantages. In the middle of the twentieth century, the economic density of human beings continued to improve, and humans continued to make demands on the environment due to the demand for development, making natural resources greatly exceed the current environmental carrying and regeneration capacity, and the global environmental pollution problem followed (Farzin, 1996). The importance of environmental economics has been recognized, leading to the research on countermeasures for pollution prevention and control and environmental protection. Environmental economics research mainly follows the production purposes of socialist society. Under the premise of public ownership of the means of production, and planned and proportional development of the national economy, it is possible to correctly adjust the material transformation between humans and nature.

Environmental economics research includes the basic theory of environmental economics, reasonable organization of social productivity, economic effect of environmental protection, and use of economic means for environmental management. The basic theory of environmental economics aims to determine the internal connection between human social systems and development law and environmental protection, as well as the application of economics in environmental protection. The relationship between economic development and environmental protection can be weighed to promote recycling and sustainable use of the environment (Jiang et al., 2021). Environmental quality is closely related to the quality of human life; therefore, the current environmental quality should be considered when considering the level of economic development. All departments should ensure the principle of coordination between production and the environment, rationally develop all types of resources, and rationally distribute the industrial structure and social productive forces. The economic effect of environmental protection includes the valuation theory and method of the economic loss of environmental pollution, optimal treatment method of waste (Pindyck, 2020), and establishment of various environmental mathematical models. The use of economic means for environmental management refers to combining policies and laws with environmental protection, and adjusting the relationship between economic activities and environmental protection through economic levers, such as taxation. At present, environmental economics research focuses on environmental value assessment, global environmental and economic analysis, ecological tax reform, and the application of general balance analysis.

Environmental economics has developed rapidly as an emerging discipline, and scholas widely value its application scope. Currently, environmental economics is gradually developing into a discipline with a mature system, integrating the theoretical research methods of mainstream economics and relevant policies and strategic objectives of environmental management, and providing an impetus for the continuous development of environmental economics.

1.3.4 Natural Resource Economics

Natural resource economics is a new discipline with great vitality and development potential. The origin of natural resource economics can be traced back to the seventeenth century economist and statistician William Petty, whose statement "land is the mother of wealth, labor is the father of wealth," is the embryonic idea of providing resources with value. After the eighteenth century, economists continued to study the relationship between economy and nature, concluding that resource scarcity could be alleviated by controlling resource prices. In the twentieth century, natural resource economics has become a systematic discipline branch, which can be divided into two directions. One direction is to study the discipline integration of natural resources and economics, and the other is to study the optimal allocation of natural resources with the theory of pure economics. During this period, many scholars have compiled related works on natural resource economics. Scholars' contributions have enriched the discipline and exposed more researchers to this emerging field.

Research on natural resource economics in China started late and was limited to geographical research and comprehensive investigation of natural resources. In the 1980s, Chinese scholars began to study price theory and the property rights system of natural resources. In recent years, natural resource economics in China has focused on resource integration and environmental, and economic models to realize the value measurement and institutional policies while solving the sustainable use-scheduling problem of natural resources. Chinese scholars learn from foreign research results (Sandmo, 2020), associate the economic growth model with the reality of natural resources and the environment, and design new models. The research process of resource economics in China reveals that resource development in China has gradually shifted toward to an intensive type, and China's economics are constantly adapting to Western economics, reflecting the rational characteristics of the transition from a planned to a market economy.

Classical economics holds that natural resources are infinite and free; however, with the continuous development of human production and life, natural resource consumption has attracted scholarly attention. Natural resource economics considers the paid use of natural resources as a theoretical basis for studying the economic attributes of resources. Natural resource economics research includes the value attributes of natural resources, pricing theory research on natural resources, and natural resource accounting.

1.3.4.1 The Value Attributes of Natural Resources

Chinese scholars have different views on the price attributes of natural resources. Therefore, there are four general types. The first is that natural resources have no real value but a price. The second view is that natural resources have value that depends on the role natural resources serve for human beings. Their value can be determined according to the theory of land rent or production price. The third view concurs with the theory of labor value that natural resources were not valued in the early stages of society, but have value in modern society. The fourth view involves inter-generational compensation of natural resources, that land rent is the value of natural resources. Chinese scholars' research on the value attributes of natural resources and the utility value theory of Western economics differ in the value and price of natural resources. Current theoretical research in resource and environmental economics focuses on the comprehensive application of both theories.

1.3.4.2 Research on the Pricing Theory of Natural Resources

Natural resource value and pricing theory aims to examine and determine the prices of natural resources. Current mainstream price theory includes the Marx attention price and market economy price theories. The Marxist theory of price measures the value of necessary social labor time, and then takes the price as the expression form of value. Several price models of market economy price theory have been proposed, including the equilibrium price model and marginal opportunity cost model.

In recent years, natural resource accounting has become a focus of research in the field of resource environmental economics. China has made some achievements in research on the introduction of natural resource accounting into the national economic accounting system. Therefore, studying accounting systems for natural resources has great significance. First, it can reflect increases and decreases in various resources by accounting for natural resources. Second, it can reflect an increase or decrease in the total amount of natural resources in the form of value. Third, it can fully reflect increases and decreases in national wealth. Value accounting of natural resources is a key but difficult aspect of natural resource accounting. The key to solving this problem lies in the scientific nature of natural resource accounting.

1.4 Environmental and Resource Problems and Economic Benefits

1.4.1 Efficient Government System: Government Failure and Government Role

The government plays a critical role in the operation of the market economy. Government intervention in the market is often intended to safeguard national security, promote social equality, conduct macro-control, and has a role in the normal operation of the market economy to a certain extent. However, if the government's decision does not serve its original role and instead brings unexpected harm, then this phenomenon is called government failure.

The problem of government failure is related to government operation mechanisms, and could be explained by several reasons, which can be summarized as deliberately destroying the market to seek its own interests and destroying market interests by too much or too little government intervention. The former type of government failure occurs because the government or a group formed by government officials engages in bribery, market monopoly, or collusion between government and business to seek their own interests, which seriously damages normal market operations. The latter type of government failure is caused by the government's improper understanding of the current economic market, leading the government to make the wrong judgment or decision: excessive intervention when no intervention is needed in the market economy or inaction when the market requires intervention. This type of market failure is an unintentional act of the government that can be improved by adjusting taxes and increasing subsidies after gaining a full understanding of the market.

Government failure is mostly reflected in the government's destruction of the price mechanism and rent-seeking. Government destruction of the price mechanism is the most common, and it means that the original price system is disturbed by the government's intervention in the market. Changes in the price mechanism directly affect residents' living standards and quality of life. Once the government destroys the price mechanism, the consequences become more severe. At present, the government has realized the importance of controlling the price mechanism and issued several documents, such as *the Government sets the Rules on Price Behavior*, to ensure the normal operation of the government and stability of the price mechanism. Rent-seeking is the act of achieving purposes and obtaining benefits through improper means, which mostly involves bribery and corruption. Rent-seeking is unlike profiting from normal production, and the government plays an important role in it. Many companies seek huge profits from engineering projects at the cost of destroying natural resources, disrupting the market balance, and bribing the government to ensure the development and operation of their projects.

China is currently experiencing high-quality economic development. Owing to the excessive pursuit of economic development speed and neglect of resources and environmental problems, the government began to play an intervening role to solve the problems related to resources and the environment. The failure of the market to allocate resources is a necessary condition for government intervention. The government must give full play to its functions, actively respond to the call for environmental protection, and intervene to a certain degree in resource and environmental issues. First, the government should formulate and implement laws related to environmental protection to maintain economic and market stability in a normal environment. For laws and regulations on resource protection and environmental governance, the pollutant emission standards of enterprises and factories should be strictly controlled, projects that create environmental damage should be strictly prohibited, and different degrees of punishment should be imposed for acts that cause serious environmental damage. By placing legal constraints on environmental damage, the government can effectively stop enterprises or individuals from harming the environment. Second, the government should strictly investigate internal corruption to purify the law enforcement environment and improve the efficiency of government intervention. The boundaries between government officials and environmentally polluting enterprises should be clearly defined and officials should not use their power for personal gain. Third, the government can improve officials' environmental protection abilities to improve their efficiency in solving ecological and environmental problems. At present, many local governments in China have included environmental protection capacity in annual employee assessments and increased support for environmental protection departments. Environmental protection departments themselves should also have the courage to take responsibility and pursue their responsibilities for improper employees according to law. The performance assessment of government officials is closely linked with environmental protection, and the relevant departments will pay more attention to environmental protection work at a higher level to improve the efficiency of government officials on resource and environmental protection and achieve the requirements of resource conservation and environmental protection. Fourth, the government can integrate various departments' functions and powers to form a joint force for environmental protection work. Owing to the different professional and technical knowledge and talent reserves in different departments, the ecological and environmental departments can give full play to their basic guiding role, take the lead in contacting various departments to perform ecological and environmental protection work, and provide talent and technical support when needed.

The government has played a vital role in protecting resources and the environment. The measures the government has taken can effectively stop enterprises and residents from destroying resources and the environment to a certain extent; however, if not handled properly, these measures could cause market failure, and the gain would not outweigh the loss. Therefore, government intervention should be conducted on the premise of fully understanding market mechanisms to improve processing efficiency and maximize the solutions to current resource and environmental problems in China.

1.4.2 *Effective Market Allocation: Externality and Imperfect Market Structure*

Market intervention is another important means of managing resource allocation. Resource allocation quality is determined by whether the market can operate effectively. Effective market interventions can make resource allocation sufficient for reducing environmental deterioration. Effective market allocation often requires many strict conditions, such as a fully competitive price market, a clear division of resource and property rights, and a price determined only by the supply and demand relationship. If these conditions are not met, the allocation of natural resources will be suboptimal. This phenomenon is known as market failure.

Market failure refers to the inability of market mechanisms to transfer resources, resulting in market price imbalance, inefficient resource allocation, and environmental deterioration. Market failure occurs for many reasons; however. there are generally three aspects: incomplete competitive market, the existence of public goods, and externality.

1.4.2.1 Incompletely Competitive Market

Perfectly and imperfectly competitive markets reflect two different market structures. In a perfectly competitive market, different economic entities cannot affect the price, and the price can only change with a change in the supply and demand relationship. A fully competitive market is the ideal model of the market economy, while most real markets are in a state of incomplete competition, that is, there is a monopoly phenomenon. This market structure is referred to as the incomplete competitive market. A monopoly will cause many adverse effects. For example, monopolistic enterprises will control their own product prices, making their prices higher than marginal costs, thereby resulting in low resource allocation efficiency. In addition, a monopolies have other social costs. For example, a complete monopoly reduces competition, removing the incentives for technological innovation.

1.4.2.2 Public Goods

Public goods have both non-exclusive and non-competitive features. Non-exclusivity means that members of society are not excluded from using the item, and any consumer can try it for free. Non-competitive means that the product does not affect the use of other consumers by the use of any consumer. Public goods weaken market competitiveness. Because public goods are non-exclusive and non-competitive, they cannot be supplied like other products, and generally can only be provided for free.

1.4.2.3 Externality

Externality refers to the economic impact imposed by one economic activity on another economic subject without trading. Externality is not transmitted through the price mechanism, in that the economic subject does not bear the corresponding decision cost or receive the corresponding decision remuneration, which makes the difference between the social and personal costs. The effects of externality can be good or bad, and are divided into positive and negative externalities according to their different effects. Positive externality refers to the income of the affected subject of economic activities without paying remuneration. Negative externality means that the affected subject of the economic activities is damaged, but no one bears the corresponding cost. In general, positive externalities can lead to insufficient production, and negative externalities can lead to excess production.

Externality is a non-market connection, and the existence of externalities is usually the result of purposes outside of economic activity, which is difficult to reflect in market transactions. Several scholars have shown that externality is an important reason for the disorder of market mechanisms and inefficient resource allocation. In the field of resources and the environment, it is often difficult for the market to consider environmental costs; thus, when externalities are present, the market price will not reflect pollution emissions and their impact on the environment, resulting in undervalued resources or further serious environmental damage. In addition to the above three reasons, market failure can be caused by other reasons, such as an unreasonable market structure or information asymmetry. When externalities exist, relying only on market mechanisms often cannot ensure optimal resource and social welfare allocation, thereby causing serious economic problems and environmental damage that should be solved by government intervention.

To resolve market failure, different measures should be taken according to the different causes of market failure. To eliminate the influence of monopoly factors, the government must take anti-monopoly measures such as enterprise recombination and punishment. The key to addressing the impact of public goods is to solve the supply of public goods, which must rely on the two collective decision-making methods of the market and non-market. For externalities, the government needs make direct adjustments through taxes or subsidies, or improve the situation by clarifying property rights and coordinating regional economic development.

An imperfect market structure is caused by monopolies, externalities, and many other factors. Most environmental pollution and the lack of resources are due to imperfect market mechanisms or the low efficiency of resources allocation. To solve the problem of market failure, resource utilization rates must be improved and environmental pollution must be reduced by improving the market mechanisms.

1.4.3 Economic Benefits of Environmental and Resource Utilization

Human survival and development cannot be separated from environmental resource utilization. Before the industrial revolution, the productivity of human society was low, and the environmental pollution caused by the exploitation of resources was soon purified by the natural system. Early humans believed that the stock of natural resources was unlimited and that it could be exploited and utilized without restraint. With the advent of the industrial revolution and rapid economic development, uncontrolled human exploitation of resources and excessive discharge of household waste seriously exceed the capacity of natural environmental systems, resulting in a severe lack of natural resources and high levels of environmental pollution.

Resources and the environment affect a country's economic development, and the level of national economic development also affects resource and environmental quality to a certain extent. Rational utilization of environmental resources can benefit human development in many ways. Over the past 40 years of reform and opening up, China has made many remarkable achievements, and people's living standards have greatly been improved. China has begun to pay more attention to families' health, living standards, and living environment. Ecological progress has gradually become the focus of the Chinese dream of national rejuvenation. The quality of the ecological environment directly affects people's yearning for and pursuit of a better life. To improve living standards, improving the ecological environment must be prioritized.

A good ecological environment is the basis for the survival of human society. The quality of the ecological environment directly affects the undertaking and development of civilization. Improving the ecological environment is necessary for consolidating the foundations of human survival. The new model of economic development needs to adapt to the specific resource and environmental situations and fully consider the carrying capacity of the environmental system. China is a developing country with a large population and few resources per capita. The insufficient supply of natural resources has become the main factor restricting economic development. China used to mainly rely on high investment in resources to achieve economic growth, and this extensive growth model led to serious resource waste. China's economic growth model has long been inefficient. It is estimated that every $1 of output generated in China consumes resources 4.3 times that of the United States, 7.7 times that of Germany and France, and 11.5 times that of Japan (Hossain et al., 2019). Economic growth at the expense of high resource consumption is unsustainable. Faced with increasingly high resource costs for its economic development, China has begun to look for efficient ways to use its resources. At the 2004 Central Economic Work Conference, the concept of a circular economy was proposed, with efficient utilization and recycling of resources as the core and changing the traditional economic growth model. During the 14th Five-Year Plan period, China has positioned sustainable development strategy with greater importance, taking promoting green and low-carbon development and solving ecological and environmental problems in the process of development. China has launched several of strategies that aim to correctly handle the relationship between economic development, resource utilization, and environmental protection, transforming the Chinese economy from high-speed to high-quality growth, and developing in on the premise of protecting the environment.

The current research on the economic benefits of environmental resource utilization in China has inherited the findings of Western scholars and made some breakthroughs in the theory of optimal resource consumption. Some scholars have used this theory to research and establish a model of optimal of mineral resource utilization to maximize the interests of society as a whole and determine the law of the coordinated development of resources, the environment, and the economy. Some scholars have conducted quantitative analyses of the optimal natural resource consumption process according to the optimal utilization and stock conditions of natural resources. Studying the economic benefits of environmental resource utilization is helpful in determining the optimal method for resource utilization. The macroeconomic benefits of natural resource utilization are a combination of social, economic, and ecological benefits.

1.5 Summary

Environmental factors are closely associated with human survival. These issues are not only confined to enterprises or economies, but also are universal challenges experienced worldwide. Today, the world faces complex situations, such as a lack of resources, wasted resources, and environmental pollution. Taking ecological civilization construction as an important cause of socialism with Chinese characteristics, China has issued several major decisions and arrangements that show the importance China attaches to resources and the environment. Research on resources and the environment affects the well-being of the people, future of the nation, and realization of the Chinese dream of the great rejuvenation of the Chinese nation. China has made significant progress and achieved positive results in promoting ecological progress.

Resources and the environment are linked to almost all disciplines, and environmental and natural resource economics are formally such a discipline. Environmental and natural resource economics are committed to addressing people's concern for environmental problems and solving the negative impact of resources and the environment on economic growth, and have been rapidly developed and widely applied since the disciple was established. Environmental and natural resource economics has become a systematic and perfect discipline, and the related research content and methods have been constantly innovated and developed.

Environmental and natural resource economics uses classical methods and means of economics to solve resource consumption, environmental pollution, climate change, and other phenomena existing in modern society, while simultaneously constantly promoting the improvement and development of its own discipline system on the basis of classical economics. For example, in the past ten years, this discipline has applied new growth theory to analyze methods for sustainable development and property rights theory to analyze environmental system policies, regulations, and pricing issues. Notably, environment and natural resource economics will apply finance theory to environmental protection, use environmental protection theory in two-way development, promote green finance as a new field of environmental and natural resource economics, and have a substantial role in the world's financial green and environmental protection practices.

Resource and environmental economics is a new subject with great vitality and development prospects. The world has emphasized sustainable development, and China has proposed green development. In fact, environmental and natural resource economics provide theoretical support, including the optimal allocation of environmental and natural resources within the scope of economic growth and national macro-control. The seriousness and uniqueness of China's environmental problems and theoretical need to support green development promote professional research development and talent training in resources and environmental economics. At present, low per-capita resources and an insufficient supply of natural resources have become the main factors restricting economic development. Reasonable natural resource allocation and improved resource utilization efficiency are premises for protecting the environment and realizing high-quality economic development.

References

Beder, S. (2011). Environmental economics and ecological economics: The contribution of interdisciplinarity to understanding, influence and effectiveness. *Environmental Conservation, 38*(2), 140–150.

Chen, P., & Shi, X. (2022). Dynamic evaluation of China's ecological civilization construction based on target correlation degree and coupling coordination degree. *Environmental Impact Assessment Review, 93*, 106734.

Chen, Y., & Zhao, L. (2019). Exploring the relation between the industrial structure and the eco-environment based on an integrated approach: A case study of Beijing, China. *Ecological Indicators, 103*, 83–93.

Demsetz, H. (2013). Toward a theory of property rights. In *Modern understandings of liberty and property* (pp. 125–137). Routledge.

Fang, C., Liu, H., & Wang, S. (2021a). The coupling curve between urbanization and the eco-environment: China's urban agglomeration as a case study. *Ecological Indicators, 130*, 108107.

Fang, X., Shi, X., Phillips, T. K., Du, P., & Gao, W. (2021b). The coupling coordinated development of urban environment towards sustainable urbanization: An empirical study of Shandong Peninsula, China. *Ecological Indicators, 129*, 107864.

Farzin, Y. H. (1996). Optimal pricing of environmental and natural resource use with stock externalities. *Journal of Public Economics, 62*(1–2), 31–57.

Heltberg, R. (2002). Property rights and natural resource management in developing countries. *Journal of Economic Surveys, 16*(2), 189–214.

Hepburn, C. (2010). Environmental policy, government, and the market. *Oxford Review of Economic Policy, 26*(2), 117–136.

Hossain, M., Malek, M. A., Hossain, M. A., Reza, M. H., & Ahmed, M. S. (2019). Agricultural microcredit for tenant farmers: Evidence from a field experiment in Bangladesh. *American Journal of Agricultural Economics, 101*(3), 692–709.

Jacobs, M. (2013). The limits to neoclassicism: Towards an institutional environmental economics. In *Social theory and the global environment* (pp. 75–99). Routledge.

Jiang, L., Zuo, Q., Ma, J., & Zhang, Z. (2021). Evaluation and prediction of the level of high-quality development: A case study of the Yellow River Basin, China. *Ecological Indicators, 129*, 107994.

Kemp, M. C., & Van Long, N. (1984). The role of natural resources in trade models. *Handbook of International Economics, 1*, 367–417.

Kuznets, S. (1955). Economic growth and income inequality. *American Economic Review, 45*(1), 1–28.

Marshall, A. (1890). *Principles of Economics*. London: Macmillan and Co.

Meng, F., Guo, J., Guo, Z., Lee, J. C., Liu, G., & Wang, N. (2021). Urban ecological transition: The practice of ecological civilization construction in China. *Science of the Total Environment, 755*, 142633.

Nilsen, H. R., & Ellingsen, M.-B. (2015). The power of environmental indifference. A critical discourse analysis of a collaboration of tourism firms. *Ecological Economics, 109*, 26–33.

Paavola, J., & Adger, W. N. (2005). Institutional ecological economics. *Ecological Economics, 53*(3), 353–368.

Pindyck, R. S. (2020). Uncertainty in environmental economics. *Review of Environmental Economics and Policy.*

Sandmo, A. (2020). The early history of environmental economics. *Review of Environmental Economics and Policy.*

Sen, A. K. (1977). Rational fools: A critique of the behavioral foundations of economic theory. *Philosophy & public affairs*, 317–344.

Shafik, N. (1994). Economic development and environmental quality: An econometric analysis. *Oxford Economic Papers*, 757–773.

Shogren, J. F., & Taylor, L. O. (2020). On behavioral-environmental economics. *Review of Environmental Economics and Policy.*

Smith, A. (1776). *An Inquiry into the Nature and Causes of the Wealth of Nations*. London: W. Strahan and T. Cadell.

Sun, Y., & Wang, N. (2022). Sustainable urban development of the π-shaped curve area in the Yellow River basin under ecological constraints: A study based on the improved ecological footprint model. *Journal of Cleaner Production, 337*, 130452.

Vasileva, E., & Ivanova, D. (2014). Towards a sustainable consumer model: The case study of Bulgarian recyclers. *International Journal of Consumer Studies, 38*(5), 475–484.

Xiao, Y., Wang, R., Wang, F., Huang, H., & Wang, J. (2022). Investigation on spatial and temporal variation of coupling coordination between socioeconomic and ecological environment: A case study of the Loess Plateau, China. *Ecological Indicators, 136*, 108667.

Xie, X., Fang, B., Xu, H., He, S., & Li, X. (2021). Study on the coordinated relationship between Urban Land use efficiency and ecosystem health in China. *Land Use Policy, 102*, 105235.

Zhang, H., & Jiang, N. (2019). Discussion on environmental protection and enterprise economic development. *IOP Conference Series: Earth and Environmental Science.*

Zuo, Z., Guo, H., Cheng, J., & Li, Y. (2021). How to achieve new progress in ecological civilization construction?—Based on cloud model and coupling coordination degree model. *Ecological Indicators, 127*, 107789.

Chapter 2
Environmental Resource Benefit Assessment

2.1 Introduction

In the wake of increasing environmental pollution, the international community has been actively exploring mechanisms to protect the environment and reduce emissions through continuous communication and cooperation. Scholars from various countries are also conducting extensive research on the optimal allocation of environmental resources. Considering environmental pollution control as the basis, this chapter introduces two concepts for controlling environmental pollution—direct control and market incentives—and uses the Environmental Kuznets Curve to explain the link between environmental protection and economic development, thereby leading to the necessity of estimating the value of environmental resources. The second section of this chapter focuses on environmental resource benefit assessment methods, including the classification of methods in detail, benefit valuation analysis, and cost–benefit analysis, which provide some effective tools for measuring environmental resource benefits. The practical case at the conclusion of this chapter describes how economic, ecological, and social benefits of natural resources can be achieved through ecotourism.

2.2 Decision-Making Criteria

2.2.1 Pollution Control

2.2.1.1 Two Ideas of Pollution Control

The previous chapter elucidated the externalities of environmental resources, where pollution is considered a manifestation of negative externalities and government failures. Thereby, how can pollution be controlled? There are two main ideas: direct

X. Deng et al., *Environmental and Natural Resources Economics*,
https://doi.org/10.1007/978-981-99-9923-1_2

control and market incentives. Direct control seeks to establish environmental quality standards and enforce them through legislative means, while market incentives aims to protect the environment through the modification of market mechanisms.

In economics, prices and market mechanisms can achieve Pareto optimality under normal circumstances. However, for non-marketed environmental goods and services, the general market mechanism cannot achieve Pareto optimality because the value of all the resources consumed to produce goods and services is often difficult to reflect in its price, leading to a deviation between social and private costs. Therefore, it is challenging to achieve efficient resource allocation, although it cannot be denied that markets cannot improve environmental quality. There are two ways to use markets to improve environmental quality: first, to establish new markets for certain environmental products for which there are no markets and establish markets for environmental resources by charging admission fees or changing property rights. For example, establish a market for carbon emission rights trading. Second, relevant authorities set a market price that includes the full social value of environmental resources. For example, the collection of pollution taxes and fees. This approach is different from direct regulation and helps government departments use market incentives to control environmental pollution.

Economists typically believe that market incentives are more efficient than direct regulation. However, direct regulation is the main method to control environmental pollution in many countries currently. Direct regulation is better than ignoring environmental pollution, although it raises some concerns. It is expensive for the government to obtain information on pollution control by enterprises, as access to such information is usually asymmetric between the government and firms; firms have more information about the costs of their pollution control technologies than the government does. If the government needs to establish environmental pollution control standards, the prerequisite is that it must obtain relevant information from emission subjects, including technology and pollution control costs that individual enterprises have. However, this method is cost-prohibitive. Concurrently, market incentives offer emitters the flexibility to choose effective methods to reduce emissions and respond to emission standards promptly according to the actual situation of the enterprises. For example, enterprises with low emission costs are increasingly willing to purchase or develop pollution treatment equipment and technologies to reduce their emissions, while enterprises with high emission costs are ready to pay for their emissions. Accordingly, low-cost emitters assume responsibility for pollution treatment, and thus the overall cost of pollution control for society is reduced. Therefore, market incentives are superior to direct regulation in controlling pollution.

Practically, governments usually base their standards for pollution control on what is socially acceptable. This standard is expressed in various political forms and public opinion. Many countries have adopted the "polluter pays principle" as an economic principle for environmental policy. This principle holds that the price of environmental goods or services should reflect the complete cost of production, including pollution emission and resources used to treat it, creating an incentive for emitters to reduce emissions and achieve the optimal level of pollution for society. The general polluter-pays principle requires emitters to pay for emission that exceeds emission standards, with no payment for emitters within the standard. However, the extended polluter-pays principle requires emitters to pay not only for what is outside of the emission standard, but also for what is within it.

2.2.1.2 Direct Control

Two methods of governmental pollution control, direct regulation, and market incentives have been discussed so far. The three main types of policy tools for the government to control pollution are emission standards, emission fees, and emissions trading. Emission standards are direct control methods, while emission charges and emissions trading are market incentives.

Emission standards are the most widely used method of direct pollution control worldwide. It is a legally enforced maximum limit on pollutant emissions set by a regulatory authority. For example, government regulations set emissions limits for pollutants such as sewage and carbon dioxide. Standards are generally set within certain health targets and are often associated with penalties, such as fines. However, such fines are rarely based on economic optimal analysis and therefore usually do not achieve socially-optimal pollution levels.

Direct government control of environmental pollution includes the following elements: legislatures at all levels pass laws on environmental protection and pollution control. Their functions are defined through dedicated administrative agencies for environmental protection. Other government agencies assume part of the environmental protection functions. Legislatures develop, adopt, and promulgate laws and regulations on environmental protection. These laws and regulations set standards for pollution emissions, introduce penalties for violations, approve environmental standards, and implement and monitor pollution control. Emission standards are generally made through environmental regulations of each industry to make specific requirements, and in general, the standards specify the upper limits of pollutants at specific locations for a certain period. Environmental standards in a broad sense also include process standards and product standards. Process standards are primarily limited to pollutant emissions from various aspects of the product manufacturing process. Product standards, meanwhile, specify the conditions that must be met by products that may cause environmental pollution. Currently, environmental protection standards are being implemented globally, such as the European and U.S. automotive emission standards, which are product standards. Most of the emission standards are technology-based and require companies to use technologies that do not increase the cost of environmental pollution. However, the implementation of emission standards will inevitably lead to increased costs for businesses, prompting some businesses to pass on some of these costs to consumers. However, the government is responsible for setting emission standards and directly regulating the discharge of pollutants, which is necessary to control pollution. The government's ability to regulate is the current force in controlling pollution and can provide a relatively effective check on the pollutant emission behavior of emitters.

One of the major reasons why the current environmental situation remains grave worldwide is that certain enterprises, government departments, and local governments are unilaterally pursuing economic goals at the expense of the environment. The government's professional environmental protection departments have shown limited power relative to these enterprises and departments, and have difficulty in navigating environmental issues. From this perspective, the direct control exercised by government departments is weak. Therefore, the rule of law and public opinion monitoring are possible ways to solve this problem. The government should follow the rule of law in controlling pollution and continuously improve environmental protection regulations and emission standards. Simultaneously, the government's pollution control activities should also be combined with the monitoring role of public opinion, which is another aspect that can counteract certain companies and sectors and effectively control their emissions.

However, the government controls environmental quality through direct regulation, which shows slack management. Its problems are two-fold. Firstly, since it is difficult for the government to grasp the information of all enterprises' emission when setting standards, it is challenging to establish a uniform emission standard and promote a socially-optimal emission level. Secondly, different enterprises have distinct pollution control costs, so it is difficult to achieve economic optimization by adopting the same emission standards for diverse enterprises.

2.2.1.3 Market Incentives

Pigovian taxes

Due to the shortcomings of direct control methods. Therefore, many economists prefer to use market incentives to control pollution. One of them is the Pigovian tax or sewage charge.

British economist Arthur C. Pigou (1877–1959) pioneered the idea of taxing pollutant emissions. In his book *The Economics of Welfare* (Pigou & Aslanbeigui, 2017), he proposed taxing emitters so that the social cost would correspond to the private cost, which is called Pigovian taxes. At present, many sewage charges follow the principle of the Pigovian tax and are considered an important tool to control pollution. The principle of the Pigovian tax is shown in Fig. 2.1.

Here, the firm's private marginal net benefit is represented by MNPB and the marginal external cost is represented by MEC. The firm should produce MNPB > 0 if it wants to expand production to Q_2. However, when MEC > MNPC, the firm should stop production and retain it at Q_1 because that production output has reached the social optimum. Tax t represents the tax paid by the firm for each unit of product produced. The amount of tax paid at $t^* >$ MNPB is greater than the firm's marginal net benefit, resulting in a loss of capital. At this point, the firm should also limit its production output to the socially-optimal output level of Q_1. From another perspective, the tax will reduce the firm's marginal net benefit by t^* for each unit of product produced, thus shifting MNPB down to the left to MNPB-t^*. As the firm's

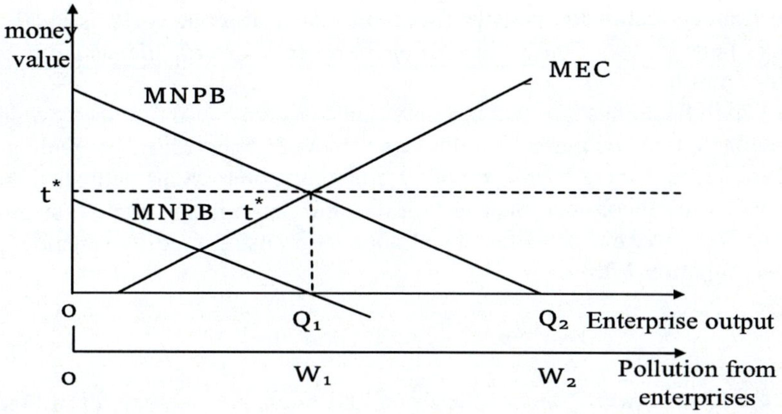

Fig. 2.1 Pigovian taxes

production output decreases, pollution emission also declines from W_2 to W_1. The tax is exactly equal to the marginal external cost corresponding to Q1 (Fig. 2.1), so that the production output reaches the optimal level of marginal damage from pollution to the outside. If the firm produces more output than Q_1, the amount of tax it pays exceeds its private marginal net benefit. Therefore, it is optimal for the firm to produce product Q_1, thus limiting its pollution emissions to the level of W_1. Thereby, t^* in Fig. 2.1 is the optimal amount of tax, at the optimal pollution level such that t* equals MEC. Therefore, optimal Pigovian tax can be defined as the amount of tax paid is the amount when the optimal pollution level equals the marginal external cost.

However, the formulation of optimal Pigovian tax requires the government to have information not only about marginal external costs but also the private marginal net benefits to the firm. However, as mentioned earlier, such information is often difficult for the government to obtain, as is the case with direct regulation. Concurrently, there is no good incentive for firms to provide such information to the government to formulate regulatory measures. This creates information asymmetry, which often hinders the implementation of the Pigovian tax.

Emissions trading

There are two main methods of market incentives to address the problem of negative environmental externalities. One is the Pigovian tax, and the view of welfare school is to use government taxes to influence prices in a way that internalizes negative externalities. The other approach is emissions trading based on the Coase theorem. Coase argues that Pigou's views regarding the externality problem is erroneous because the standard of sewage charges is difficult to determine and requires heterogeneous information about environmental pollution, which is extremely difficult to obtain and often involves high costs. It is also complicated to estimate the cost of pollution control by producers and to monetize pollution losses. Therefore, Coase proposed to internalize negative externalities by making market transactions between polluting producers and polluting losers under the premise of clear property rights and through voluntary negotiations. Coase's idea of environmental externalities is the origin of emissions trading.

The Coase theorem states that if transaction costs are low and property rights are easily defined, then the market equilibrium achieved under private bargaining must be efficient even if there is an externality problem. Common mathematical models illustrate Coase's theorem as the basis for the emission rights market. Typically, the Coase theorem requires a mathematical application using a utility function of the following functional form:

$$V^i(p, r^i, a) = \Phi^i(p, a) + r^i \tag{2.1}$$

where p denotes price, r^i denotes revenue, and a denotes the activity of the firm. The socially-optimal objective can be achieved by solving the externality problem through the clarification of property rights.

Suppose that firm A's discharge activity is represented by a. It then needs to pay firm B an amount of discharge worth v. Assume that the preferences of both firms A and B are quasi-linear and that their respective utility functions are: $V^i(p, r^i, a) = \Phi^i(p, a) + r^i$, when p is fixed, $\Phi^i(p, a)$ can be equal to $\Phi^i(a)$, $i = 1, 2$. Therefore, firm A is willing to pay v while performing a when and only when $\Phi^i(a) - v \geq \Phi^A(0)$. For firm B, it needs to choose two variables: a and v, which are expressed as a function of:

$$\max(\Phi^B(a) + v) \quad a \geq 0, v \tag{2.2}$$

$$s.t \cdots \Phi^A(a) - v \geq \Phi^A(0) \tag{2.3}$$

Transform the constraint into the equation $v = \Phi^A(a) - \Phi^A(0)$ and substitute into the objective function:

$$\max(\Phi^B(a) + \nu) \quad a \geq 0, \nu \tag{2.4}$$

Deriving the optimal a, it follows from the first-order condition that:

$$\frac{\partial \Phi^A(a)}{\partial a} = -\frac{\partial \Phi^B(a)}{\partial a} \tag{2.5}$$

Suppose also that firm A needs to pay a price p and has revenue r^A, and that firm A will choose the optimal quantity a^* such that $\frac{\partial \Phi^A(a)}{\partial a} = 0$ and $a^* > 0$. Since the emission activity of firm A, a has a negative externality effect on firm B, if the socially-optimal emission goal is to be achieved, a^0 must satisfy:

$$\frac{\partial \Phi^A(a^0)}{\partial a^0} + \frac{\partial \Phi^B(a^0)}{\partial a^0} = 0 \quad (a^0 > 0) \tag{2.6}$$

That is:

$$\frac{\partial \Phi^A(a^0)}{\partial a^0} = -\frac{\partial \Phi^B(a^0)}{\partial a^0} \tag{2.7}$$

Therefore, the optimal solution is when $a = a^0$. It shows that when the property rights are clear, the society can reach the optimal state of a equal to a^0.

However, the conditions of the Coase theorem are often difficult to satisfy. For many products with externalities, the transaction costs are high and property rights are often not clearly defined. Therefore, it is difficult for the market to solve all externalities alone.

However, the importance of the Coase theorem is not that it states that markets can solve the externality problem alone, but that it provides a solution—the creation of new markets. If property rights can be clearly defined and transaction costs reduced, then a market can be created to trade externalities. If such a market is created, it is clear from the Coase theorem that the new market will have all the efficiency properties of an ordinary market. Such a market would not only maximize the social surplus, but would also ensure that the product being supplied would be sold by the lowest-cost suppliers and bought by the highest-paying demanders.

Therefore, the government has a role to play in defining property rights and reducing transaction costs, and enterprises can optimize resource allocation by trading emission rights among themselves. Ultimately, regardless of who owns the property rights, the end result will be ever closer to the social optimum. By solving the problem of negative environmental externalities caused by pollution through this trading mechanism, an emission rights market is naturally formed.

Therefore, how do companies choose whether or not to purchase emission rights? Figure 2.2 gives the basic idea of emissions trading. The horizontal axis represents the pollution level and the emission right, and MAC represents the marginal cost of pollution control, that is, the marginal cost of control, which tends to slope down to the right because the lower the emission volume relative to the pollution control volume, the smaller the marginal cost of control. MEC denotes the marginal external cost, when the quantity of emission rights is Q_0, the corresponding price of emission rights is P_0, which is in the optimal state. S denotes the supply curve of emission rights. Since the government controls the issuance of emission rights, the supply of emission rights can be considered relatively stable to some extent, so it is a vertical line. The marginal control cost represents the control cost per pollutant emitted, which is the demand curve of emission rights. When the volume of emissions is Q_1, the discharger purchases it at a price of P_1, because when the volume of emissions is lesser than Q_1, the cost of purchasing the emission rights is lower than the cost of controlling pollution by the discharger itself. When the level of pollution increases from Q_1 to Q_2, the cost of controlling pollution is lower than the cost of buying emission rights, and the firm chooses to control pollution.

How is the price of an emission right determined? As illustrated in Fig. 2.3, the price of an emission right is affected by changes in the demand for the emission right. When a new emitter enters a particular polluting industry, its demand for emission

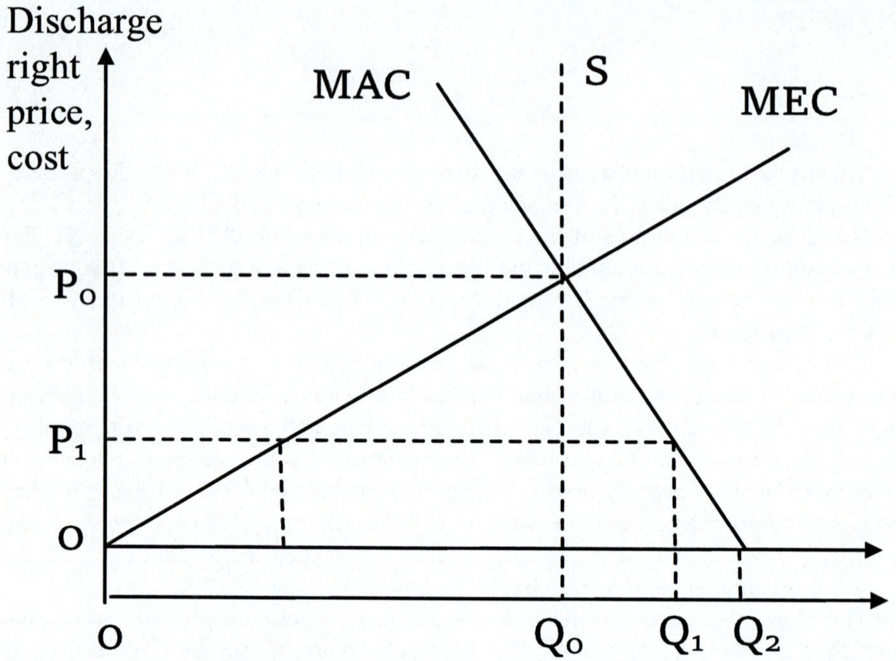

Fig. 2.2 Emissions trading

rights increases, which causes the demand curve for emission rights to shift from D_0 to the right to D_1. If the government regulator does not see the need to increase emissions, then the number of emission rights issued remains the same and the supply curve for emission rights remains S_0. If demand increases while supply remains the same, the price of emission rights increases from P_0 to P_1. If the regulator believes that the entry of new firms requires an increase in emission rights or an increase in the allowable volume of emissions, the regulator can issue more emission rights. For example, by issuing Q_1, the supply curve of emission rights shifts to the right to S_1, which brings the price of emission rights back to P_0. Meanwhile, if the regulator believes that the number of emissions needs to be strictly controlled, the regulator can buy emission rights from the market itself, which shifts the supply curve of emission rights to the left, thus increasing the price of emission rights. In short, the regulator creates a market for emission rights by issuing or buying and selling emission rights, and uses market regulation to control the supply and price of emission rights.

Limits of market incentives

Buchanan (1986) transferred the Homo Economicus Hypothesis in economics to the political field, and extended the concept to the behavior of those who vote or are agents of state affairs to participate in public choice. He argued that the government does not represent the interests of all people but a particular interest, so the

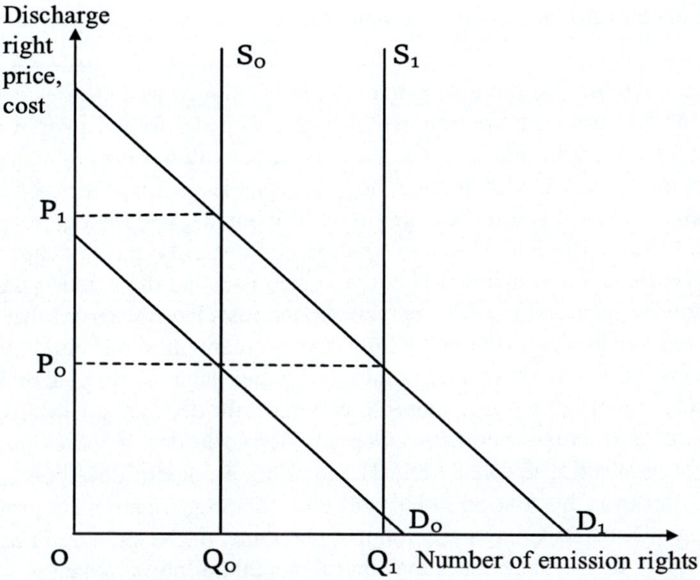

Fig. 2.3 Trading price of emission rights

government also has the same market failure as the market. Similar failures exist in government as in the market. Environmental economists often take public choice theory into account when analyzing environmental problems, arguing that government failure can also worsen the environment, mainly because government policies do not improve the deviation between environmental prices and real prices, and such deviations in economic policy could promote artificial distortions in market prices. The solution to government failure is to require government policy makers to improve their comprehensive understanding and correct judgment of environmental issues to formulate accurate policy measures.

Calabresi (2008) argues that for environmental pollution control, the effectiveness of policy implementation plays a key role in both the government's choice of ex-ante regulation and ex-post punishment. Later, many more scholars began to study information asymmetry. Shavell (2018) pointed out that government policies also exhibit information asymmetry due to, among other things, information asymmetry, which, together with the high judicial costs in a market-based system, aggravates the losses of victims of environmental pollution and makes the system less effective in enforcement.

In terms of economic costs, the free-rider behavior also needs to be considered as society consumes excessive resources without being able to do an accurate economic measurement. The free-rider behavior weakens the individual responsibility for environmental pollution, but the individual overconsumption as an environmental consumer to a large extent is likely to lead to the eventual deterioration of the environment.

2.2.2 *Conservation and Development*

In traditional economics, the environment is not included as a factor in economic development, but growing environmental problems have forced people to reflect on the role of the environment in the process of economic development. Kuznets (2019) first proposed a law on income disparity, that is, in the process of economic development, income disparity reveals a state of widening and then narrowing. This law presents the interrelationship between income disparity and per capita income growth in the shape of an inverted U-curve, which is called the Kuznets Curve.

In the process of economic development, economists have observed that the environment also shows deterioration before improvement in some cities across the globe. For example, environmental pollution in cities such as Mumbai and Brasilia in developing countries have increased compared to a decade ago, while environmental pollution in cities such as London and Tokyo in developed countries have significantly reduced. The urban environmental conditions in developed countries have been improved by first polluting and then treating. Economists proposed an Environmental Kuznets Curve based on this phenomenon, as shown in Fig. 2.4.

In this figure, the vertical axis is the environmental pollution indicator, which can be expressed by indicators such as per capita pollutant emissions. When the economy is at a lower development stage, the severity of environmental pollution is not obvious because fewer products are produced and thus fewer pollutants are emitted. As the economy develops and manufacturing industries expand and multiply, pollution emissions increase, leading to serious environmental problems. As economic development moved to new heights, people's requirements for environmental quality and environmental awareness increased. This triggered changes in the economic structure, thus using resources previously accumulated by economic development to treat the environment, and finally, the environmental pollution situation began to improve, forming an inverted U-shaped Environmental Kuznets Curve.

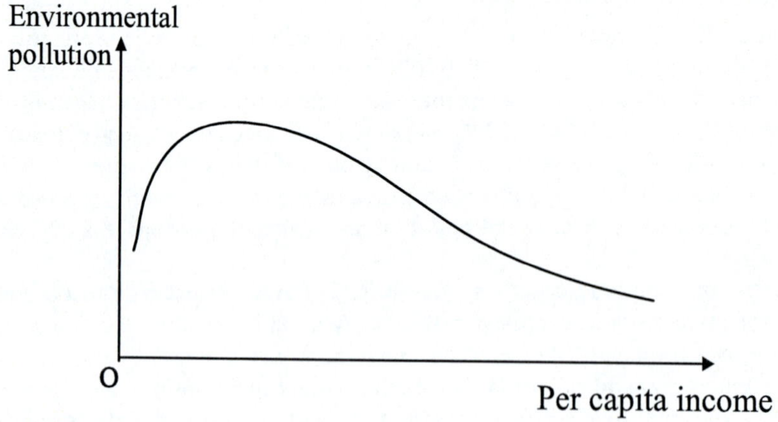

Fig. 2.4 Environmental Kuznets Curve

The natural resources and environmental conditions of a country depend mainly on the following factors. The first concerns the size of the economy. When other conditions are fixed, the size of the economy is proportional to the consumption of natural resources, which leads to increased environmental pollution. The second relates to the structure of the economy, which has a certain degree of influence on the amount of resource consumption and level of environmental pollution. The industrial structure, especially in low-income countries, is mainly the primary industry. In middle-income countries, the proportion of manufacturing industry is expected to increase. In high-income countries, the proportion of tertiary industry, represented by high-tech industry and service industry, is estimated to increase, thus reducing pollution levels. In an agriculture-based economy, there is a high probability of deforestation owing to the expansion of agricultural land, such as the Brazilian rainforest, which if left unchecked could lead to forest decline and severe soil erosion. When a country enters the industrialization stage, it tends to consume a vast amount of resources, thereby depleting the environment. With further economic development, the proportion of tertiary industry represented by high-tech and service industries is increasing, which reduces the level of environmental pollution. The third involves the level of science and technology. Two countries with the same industrial structure but using different technologies have different pollution emissions. Generally, countries with retrograde technology tend to consume more natural resources and emit more pollutants. The fourth applies to government control. In light of the different stages of economic development, government control of environmental pollution will also show different changes, including the importance of environmental protection. Fifth is environmental protection awareness. With the improvement of life quality and income, people have higher requirements for the quality of the environment. People are consciously spending more money for environmental protection through various policy measures and methods. With disposable income increasing, the government and the public manage to save a lot of money, so there is more money available for environmental protection. The first three of the aforementioned five points were recommended by Grossman and Krueger (1995).

The Environmental Kuznets Curve seems to carry an overtone of inevitability. Pollution decreases only after a certain level of economic development is reached, and there is a lack of awareness of environmental protection as well as a weak state of governmental control over the environment. The existence of such a view does not help developing countries to focus on environmental protection while developing their economies. Many developing countries are aware of the environmental consequences of pollution in developed countries and have adopted practical environmental protection strategies. Therefore, developing countries need to consider the overall benefits of economic development while seeking a path to reduce its arc or a horizontal straight line on the original Environmental Kuznets Curve as a way to curb environmental pollution.

2.2.3 *Value Estimation*

2.2.3.1 General Value of Environmental Resources

Whether or not environmental resources have value has been a key issue of discussion in economics for a long time. The understanding of this issue has gone through a long historical process in China. From 1949 to the late 1970s, due to the influence of traditional political economy theories, academics generally believed that environmental and natural resources were worthless and that they had use-value but not the property of value because although they were useful to human beings, they were not generated by labor, such as air, rivers and lakes, natural grasslands, and natural forests (Barnes et al., 2010). Environmental and natural resources in this context refer to the meaning of not being taken as wealth and not being possessed by any person or unit. Environmental resources are mainly generated by nature, and thus have only use-value but no value. This erroneous view led to the deviation of environmental and natural resources in the process of use for decades, resulting in many undesirable consequences, such as high prices for products, low prices for raw materials, and no prices for resources, which have led to a sharp decline in environmental resources, leading to undue environmental damage.

From the late 1970s to the mid-1980s, the environmental consequences of policy oversights and theoretical errors became apparent, and domestic and foreign economists and ecologists, even politicians, and sociologists from various countries jointly placed the issue of resources and environment on the agenda (Maryudi, 2012). Consequently, both developed and developing countries, and even international organizations, considered the importance of resources and the environment when studying international issues. Although countries with mainly planned economies and those with mainly market economies have different interpretations, the view tends to be the same, that is, environmental resources are valuable. Countries with predominantly planned economies believe that only environmental resources such as planted forests with the participation of human labor have value, while primary forests have no value. Countries with predominantly market economies believe that only environmental resources that can enter and be traded in the market have value, otherwise, they have no value.

By the end of the 1980s, Chinese and foreign economists discussing the value of environmental resources fully affirmed that resources and the environment are valuable, especially after the 1990s. People gradually realized that an environmental resource management system coordinated with economic development should be established as soon as possible to fundamentally change the operation mechanism of consuming and polluting inordinately. Changing the possession and use of environmental resources without compensation to possession and use with compensation, and changing high consumption and low output to low consumption and high output can realize the sustainable development of environmental resources. The value of environmental resources is the value expressed by the material production and energy reserves of natural biological groups and their impact on the surrounding environment (Barnes et al., 2010). Its value is mainly expressed in the following three aspects.

First, economic effect, which concerns the value brought by the provision of raw materials, energy, subsistence goods, species genes and other raw materials by environmental resources. Second, ecological effect concerns what is beneficial to the habitat and reproduction of human and biological populations due to the regulating effect of natural ecological environment. It includes climate regulation, soil and water conservation, disaster prevention, and soil improvement. Third, social effect refers to the influence of natural ecological environment on human survival and psycho-emotional, educational, and leisure aspects.

The aforementioned effects of environmental resources are essentially economic benefits, some of which are directly expressed as economic benefits, while some need to pass through certain intermediate transitional links before they can be transformed into economic benefits. Therefore, the economic benefits brought by environmental resources to human beings are actually the total economic benefits that include the superposition of the above three effects. However, for a specific operating entity of environmental resources, due to the wide distribution and complexity of environmental resources and the non-exclusivity of achieving ecological and social effects, the operating entity does not have complete control over the direct and indirect economic benefits that environmental resources can provide, but only has control over some of the possible economic benefits (i.e., the above-mentioned economic effects). Currently, only this aspect of the benefits can bring a realistic inflow of funds to the operating entity through market-based behavior (Goldie, 1997). In the existing legal framework, ecological and social effects are mostly unable to bring financial inflows to the operating entities through market-based behavior. Therefore, if environmental consumption and compensation cannot be measured through means other than market behavior, such as finance and taxation, it will lead to deviations between the economic benefits accounted for by resource and environmental business entities and the actual economic benefits received.

2.2.3.2 Environmental Resource Value Estimation

How does one estimate the economic value of environmental resources? From the three effects contained in environmental resources, it can be concluded that the economic value of environmental resources should include the value of physical goods (economic effect) and the value of environmental services (ecological effect and social effect). However, can these values be used as the value of environmental resources assets of micro-entities? In traditional accounting, an asset is one of the important components of its owner's wealth and an entity with a certain market value (Cairns et al., 1992). Essentially, an asset is a resource that can be converted into money in the future, it has a legal guarantee of the right to benefit a certain micro-subject, and this service can be an asset only if it is useful to a certain micro-subject (Boatsman & Baskin, 1981). Assets are also generally considered internationally as future economic benefits obtained as a result of past transactions concluded (Goldie, 1997). Therefore, it is clear from the definition of asset that the value of an asset that becomes an asset for a micro subject is subject to two conditions. First, it is capable of generating a benefit in the future. Second, there is a corresponding owner for that benefit. For environmental resources, the value of physical goods includes not only the physical products that can be produced by themselves, but also those obtained from environmental resources, such as fruits produced in the forest, while the value of this product can be reflected by the market price. For the value of environmental services of environmental resources, the functions of environmental resources should be analyzed first, because generally only those environmental services that act directly on human beings can be quantified (Elías & Wittman, 2012). In addition, most of the environmental service values present obvious non-exclusivity and non-competitiveness, and thus are considered as public goods and cannot be traded in the market and reflected by market prices. However, as the protection of the environment and the effective use of natural resources have gained increasing attention, people have also explored the realization of environmental services value from the perspective of environmental services. There are currently three ways to realize the value of environmental services of environmental resources. The common method is through taxation or financial subsidies because of the characteristics of public goods, environmental services provided by environmental resources are often excluded from market transactions. Some environmental service values can be realized through ownership and management of environmental resources, such as the landscape value of lake water resources. Some environmental service values can be realized by gradually promoting market transactions, such as the formation of carbon sink trading market, which provides the possibility to realize the carbon absorption function of forest resources.

In that case, what is the value of environmental resource assets in accounting and valuation value? First, we need to understand the meaning of asset valuation, which is the process of estimating and accounting for the value of an asset at a point in time (Cowen & Tabarrok, 2015). In a broad sense, all acts of valuing assets are asset appraisal. In a narrow sense, Article 2 of the Asset Appraisal Law of the People's Republic of China states that "Asset appraisal refers to the appraisal agency and its appraisal professionals who, on commission, assess and estimate real estate, movable assets, intangible assets, enterprise values, asset losses or other economic interests and issue appraisal reports by the professional services." Specifically, asset appraisal refers to the act and process of asset appraisal professionals (or institutions), by the law, analyzing, projecting, and issuing professional opinions on the value of the appraisal object as of the appraisal reference date (Gamarra et al., 2019). Therefore, we believe that the economic value of environmental resource assets should be used as the value in their asset appraisal. It is a tool to help people measure the benefits of environmental resources. The selection of the scope of the value of environmental resource assets assessment should emphasize the inclusion of both the direct and indirect utility of environmental resources to human beings. However, limited by the current level of understanding and technology, only a few indirect utilities of environmental resources have been included in the assessment. In addition, when assessing the value of environmental resource assets for a specific subject, it is important to distinguish the use of different economic interest data to reflect different valuation purposes so that the corresponding transaction value can be clarified.

Types of environmental resource asset valuation values

Since environmental resource assets have both commodity and public product attributes, they can be valued through both micro and macro aspects. From a micro perspective, the value assessment of environmental resource assets is not very different from the general assessment of assets. However, from a macro perspective, environmental resource assets can be valued as a public product. Essentially, when assessing the value of environmental resource assets, it is possible to assess the value of environmental resource assets for micro property rights subjects. Conversely, it is also possible to assess the overall macro value of environmental resource assets, but what needs to be clarified is the defined scope of value, such as the value of a certain kind of environmental resource assets in a certain province or city. Based on the limitations of the current cognition of asset valuation, the environmental resource assets discussed below are mainly valued at the micro-level.

The value of asset appraisal exhibits a variety of value types, which are classified into market and non-market values according to *the classification criteria of the International Valuation Standards* (IVS). Similarly, *the Guidance on Types of Asset Appraisal Value* issued by China in 2017 also classifies them into two types: market value and non-market value. Market value refers to the amount that is voluntarily traded in the market and the value of the appraised object is estimated at the valuation base date. The value other than market value mainly includes liquidation value, investment value, and residual value (Sahide & Giessen, 2015). Since environmental resource assets have both physical commodity value attributes and environmental service value attributes, they exhibit distinct value forms for various situations such as different economic environments, appraisal purposes, and management methods (Maryudi, 2012). The specific value forms are shown in Table 2.1.

As shown in Table 2.1, in the process of valuing environmental resource assets, a variety of value types are involved, most of which carry similar meaning as those in general asset valuation. However, owing to the special nature of environmental resource assets, the following forms of value are emphasized here.

Market value and value other than market value

The economic meanings of market value and value other than market value have been introduced earlier, and for environmental resource assets, their meanings are no different. However, based on the specificity of environmental resource assets and the fact that some market elements are not yet well developed, the use of market value

Table 2.1 Value types of environmental resource asset assessment

Classification basis	Type of value
Valuation approach	Replacement cost, current market value, and so on
Hypothetical perspective	Open market value, liquidation value, and so on
Valuation purpose	Mortgage value, insurance value, investment value, and so on
Usage status	Market value, value other than market value

is largely hindered in the appraisal. However, it cannot be generalized in practice. In some regions of China, certain environmental resource markets are relatively well developed and have largely met the criteria for active markets, such as the forestry market. Simultaneously, for certain environmental resources (e.g., land), transactions for a specific purpose are relatively active. For example, the lease market for forest land is relatively active in the southern region of China. Therefore, due to the specificity of environmental resource assets, the value type of market value used in their assessment process is discussed separately for different regions and elements.

Mortgage value

The International Valuation Standards classify mortgage value as a value other than market value, which is the value when an asset is used as security for a loan. This value is usually considered to be the maximum amount of credit that does not exceed the scope of the asset's security. Creditors usually set the collateral value below the market value of that secured asset for psychological security needs. The use of collateralized loan financing is common in the valuation of certain environmental resource assets, but there is currently no formal regulation of collateralized loan value assessment in China, such as forest rights collateral. Therefore, the type of mortgage value of environmental resource assets should be clarified in the mortgage value assessment.

Value for financial reporting purposes

Currently, the accounting standards in China only regulate the valuation of certain environmental resource assets as biological assets and require that the assets be valued at historical cost, while fair value can be used for measurement purposes provided that reliable fair values are available.

2.3 Environmental Resource Benefit Assessment Methods

2.3.1 Classification of Benefit Assessment Methods

2.3.1.1 Market Price Method

Market price method, also known as market price comparison method, is based on the commodity property of environmental resources. It can use the current market price of environmental resource assets or the current market price of the same and similar environmental resource assets to evaluate the environmental resource benefits being assessed.

The market price method is currently the simplest, most direct, and effective method for assessing the benefits of environmental resources, because all the materials required in its assessment process come from the market, and the economic benefits of environmental resources are estimated directly through market prices.

However, the method is closely related to the degree of marketization of environmental resources. With continuous improvement of the socialist market economy system, it provides an effective space for the application of the market price method in the estimation of environmental resource benefits, and the market price method gradually becomes an important method for the estimation of environmental resource benefits.

When using the market price method to assess the benefits of environmental resources, two points should be noted. Firstly, the premise is that the market for trading environmental resource assets is fully active. In a market economy, the more frequent and active the assets are traded, the easier it is to obtain prices for assets similar to the assessed environmental resource assets. Secondly, reference and comparable indicators and parameters of the evaluated environmental resource assets can be found in the trading market. The basic formula of the market price method is:

$$A = M_R + \sum P_S - \sum P_I \tag{2.8}$$

$$A_V = M_T \times \prod F_C \tag{2.9}$$

where, A is appraised value of an environmental resource; M_R is market reference price; P_S is price difference due to the superiority of the appraisal object over the reference factor; P_I is price difference driven by the inferiority of the appraised object to the reference factor; A_V is an environmental resource assessment value; M_T is market reference transaction price; F_C is factor correction factor.

To apply the market price approach, it is important to determine references in the market that are identical or similar to the environmental resource assets being evaluated. However, it is very difficult to find references that meet these requirements, which forces adjustments to be made to similar asset references in the accounting. Therefore, another key to determining whether the market price approach can be applied is the availability of indicators and parameters for relevant adjustments.

When applying the market price method, different approaches are distinguished depending on different market conditions and differences in the references. If it is possible to find the exact same reference from the market as the environmental resource being evaluated, then the market price of that reference can be obtained and used directly. However, in practice, it is almost impossible to find exactly the same reference. In the case of forest assets of forest resources, for example, trees produced from the same forest land in the same region may also have different levels of depletion due to different natural forces or high or low utilization rates, and thus cannot be exactly the same reference. In most cases, the price obtained is the price of the reference and thus requires adjustment of that price. Price adjustment factors include the following four aspects. First, time factor: since the market price of environmental resources may vary over time, it is necessary to make appropriate adjustments to the market price of the environmental resource reference at different points in time. Second, geographical factor: since similar environmental resources (e.g., land) are sold in the market in different regions, they also show significant price differences.

Therefore, if there are geographical differences between the natural resource assets being evaluated and their market references that result in price deviations, adjustments should be made to the geographical difference for that reference. Third, functional factors: two similar environmental resource assets also have different market prices due to their different functions. For example, although both agricultural and domestic water come from water resources, the price of water is different due to different functions. Generally, the function is high or low in proportion to its market price, but this demonstrates that the function can be fully utilized in the market. Fourth, transaction factor: sometimes the reference of the environmental resource being evaluated is engaged in unfair or abnormal transactions, resulting in a transaction price higher or lower than the normal market value, which requires the price of the reference to be corrected for specific transactions.

In addition, when using the market price method for appraisal, firstly, the environmental resource asset to be appraised should be clearly defined, especially the scope definition. Secondly, an open market survey needs to be conducted to collect market information materials of the same or similar assets and select suitable references. Once again, the collected market information is classified and organized, its authenticity and credibility are verified, its transaction background and process are analyzed, and finally three or more identical or similar assets are judged and selected as references. Thirdly, the differences between the evaluated environmental resource assets and their references are compared and analyzed. The differences are adjusted according to the calculation results of quantitative analysis, and the economic benefits of the evaluated environmental resource assets are finally accounted for and determined.

2.3.1.2 Income Method

The income method, also known as the present value of income method, is a method of estimating the expected future income of environmental resources. It is converted into present value using a discount rate, and the present value of each period is added up to arrive at the economic benefits of the environmental resources being evaluated. The calculation formula is as follows:

$$PV = \sum_{i=1}^{n} \frac{F_i}{(1+r)^i} \quad i = 1, 2, \ldots, n \tag{2.10}$$

where PV denotes the appraised value, F_i denotes the amount of expected return in future period i. When i is infinite, the expected return is also infinite, and when i is finite, F_i also includes its net remaining value at the end of the period, i denotes the return period, and r denotes the discount rate.

In general, environmental resources bring certain benefits to the property owners, and the total amount of money paid by the owner at the time of acquiring the property rights will not be the discounted value of the expected future benefits of that environmental resource. In brief, an investor who wants to invest in an environmental resource generally needs to conduct a feasibility analysis of it, and only if the expected internal rate of return of the environmental resource exceeds the discount rate of the appraisal will the investor be willing to pay for it. The following three points should be noted when using the income approach for appraisal. First, the environmental resources being appraised can be expected to have future returns. Second, the ability to use the money to measure future returns of the environmental resources being appraised. Third, the ability to use the money to measure the risk borne by the property owner.

When the income approach is used for appraisal, it is based on the future benefits of environmental resources, that is, the ability to make a continuous profit after being put into use. The income approach cannot be applied if the investment is not aimed at making a profit or if there are no expected benefits. The income approach is a process of discounting or provincializing the expected future benefits of environmental resources. Meanwhile, in this process, there are various situations such as limited future return, infinite future return, annuitization of future return, and unequal future return. Therefore, different approaches are taken to give estimates, and different situations described in detail in the income valuation analysis of environmental resources.

In addition, when using the income approach for valuation, the various factors affecting the expected future benefits of environmental resources should be clarified first. Second, based on the analysis of relevant indicators, the expected future benefits of the environmental resource's undervaluation should be estimated. The economic parameters related to the benefits, such as discount rate and risk-reward rate, should be predicted and analyzed, and the discount rate should be determined. Finally, the present value of the expected future benefits is estimated, and the assessed benefits of the assessed environmental resources are determined.

2.3.1.3 Cost Method

The cost method, also known as the replacement cost method or replacement value approach, refers to the assessment of the benefits of environmental resources based on their current replacement cost minus the value of various losses. The theoretical basis for using cost method is that the cost of the input determines the benefit of the environmental resource. The higher the cost of the input, the greater the value, and therefore higher the benefit. Based on this principle, when using cost method for appraisal, the replacement cost of the input environmental resources should be determined. The replacement cost is the total amount of money paid to reacquire or create a certain environmental resource or obtain a reference similar to the environmental resource asset being appraised under current market conditions. In terms of content composition, the replacement cost and the original cost are the same, the difference lies in the price level, and the original cost reflects the price level at the time of initial purchase. Other things being equal, the replacement cost of an environmental resource is positively related to its benefits. The following two points should be noted when using cost method.

First, appropriate historical information should be available. The premise of using the cost method is to rely on historical information to obtain additional information, especially the content composition of the cost is key to determining the cost. Secondly, it is necessary for the depletion formed by environmental resources. Since there are effective and ineffective depletion, when using the cost method for assessment, it must be determined that the depletion of an environmental resource is necessary under the average level of the industry or society. According to the definition of the cost method, the basic formula for appraisal using the cost method is:

$$\begin{aligned} &\text{An environmental resource assessment value} = \text{Replacement cost}- \\ &\text{physical devaluation} - \text{functional devaluation} - \text{Economic devaluation} \\ &-\text{Environmental devaluation} \end{aligned} \tag{2.11}$$

Among them, physical devaluation refers to the decline in economic benefits triggered by the continuous decline in the physical properties of an environmental resource after it has been put into use, resulting in a decrease in utility due to the influence of natural forces. For example, a forest tree will gradually exhaust its economic life after its peak production period, leading to the decline of its benefits.

Functional depreciation refers to the progress of science and technology and the application of new technology, which makes the assessed environmental resource assets display obvious technical backwardness compared with other assets in the market, leading to a decline in their value and a reduction in benefits. For example, agricultural land resources, after multiple rounds of doing, will lead to a decline in land fertility. If there is no appropriate technology to maintain the fertility of the land, then that agricultural land resource will face functional depreciation.

Economic devaluation refers to the reduction of environmental resource benefits caused by changes in external environmental factors (e.g., policy factors). For example, the policies implemented for the protection of coal resources restrict the mining and utilization of coal resources, making them less effective.

Environmental devaluation refers to environmental pollution and damage caused by excessive or unreasonable use of environmental resources, which indirectly reduces the utilization of environmental resources. For example, unscientific fishing causes serious damage to marine ecology, resulting in the reduction of marine fish resource species and quantity.

In addition, according to the different replacement methods, replacement cost can be categorized into restoration replacement cost and renewal replacement cost. Rehabilitation replacement cost refers to the cost of reassessing the benefits of the environmental resources being evaluated based on the same cost components and construction processes, among others, at current market prices. Renewal replacement cost refers to the cost of a brand-new reference object with the same or similar function of the environmental resource under the premise of applying new technology, new process, and new design, which is evaluated at the current market price. Based on the above expressions, it can be seen that restoration replacement cost is suitable for use under the premise of no significant technological progress, while renewal replacement cost is applicable in the case of technological progress. The estimation of the replacement cost can be done by the following two methods.

The first is the replacement accounting method, which divides the total cost of the assessed environmental resources into direct and indirect costs to estimate the replacement cost separately. Direct cost refers to the expenses that directly contribute to the production cost of environmental resources, such as the purchase of seedlings in forest resources, direct purchase in land resources, investment in large fixed assets in the energy, and direct labor cost in mining. Indirect costs are the costs that indirectly act on environmental resources, which should be allocated and treated using certain methods, such as transportation costs, management costs, and financial costs.

The second is the price index method. The price index method refers to the method of estimating the replacement cost by adjusting the original cost of environmental resources by the price index. The original cost is expressed as the market value of environmental resources at the time of initial purchase or creation. The calculation formula is as follows:

$$\text{Replacement cost} = \text{original cost} \times (\text{Fixed-base price index for similar environmental resource assets at the valuation base date/Fixed-base price index for similar environmental resource assets at the time of purchase or construction}) \tag{2.12}$$

$$\text{Replacement cost} = \text{original cost} \times (1 + \text{price change index}) \tag{2.13}$$

$$\text{Replacement cost} = \text{original cost} \times \text{Chain price index } a_1 \times a_2 \times a_3 \times \cdots \times a_n \tag{2.14}$$

When using the formula, it should be noted that the original cost of environmental resources is required to be real and reasonable. The price index is the price index on the base date of the appraisal. It is necessary to distinguish different categories or individual price indices for different cost categories for accounting.

A comparison of the price index method and the replacement cost method shows that when using the price index method, only the price change factor is taken into account, and thus the calculated result is the restoration replacement cost. The replacement accounting method not only takes into account the price change factor, but also the technological update and process change, and thus this method can estimate both the restoration replacement and update replacement costs.

It is worth noting that the use of the cost method in the appraisal process of environmental resources is also very different compared to general asset appraisal due to its special characteristics. In addition to the difference in price and technology, the time factor also needs to be taken into account. Generally, environmental resources have a long operating period, almost decades. Therefore, it can cause long-term occupation of funds, and the time value of funds must be considered in the appraisal process.

When using the cost method, the following procedures can be followed. First, estimate the rehabilitation replacement cost or renewal replacement cost of the environmental resources under appraisal. Second, determine the operating and continuing life of the environmental resources. Third, estimate the depletion of the environmental assets under appraisal. Finally, use the replacement cost to deduct the depletion to arrive at the appraisal benefit.

2.3.2 *Benefit Valuation Analysis of Environmental Resources*

The use of the benefit approach for environmental resource benefit assessment has been briefly introduced in the content on the classification of benefit assessment methods. This content will focus on the benefit valuation analysis of environmental resources. In addition to the basic estimation methods of the benefit approach mentioned above, there are usually the following. First, whether the valuation methods limited or unlimited depend on whether the future benefits of the environmental resources being evaluated have a term. Second, the valuation methods of equal benefits and non-equal benefits depend on the specific situation of the environmental resources being evaluated. To facilitate learning, the symbols used in these methods are uniformly defined here.

PV appraised value
i year serial number
PV_n expected realizable value in the *n*th year of the future
F_i expected earnings in the future year *i*
r discount rate or capitalization rate
r_i discount rate or capitalization rate in year *i*
n period of return
A annuity

2.3.2.1 The Case of Constant Returns

In the case of an infinite benefit period, it usually means that an environmental resource can be operated in perpetuity and has infinite benefits. If the expected benefits generated by a certain environmental resource remain constant each year and $r > 0$, with other factors remaining unchanged, these are calculated by the following formula.

$$PV = A/r \tag{2.15}$$

In the case of a limited benefit period, which usually means that the property owner of an environmental resource holds it to maturity or that environmental resource is exhausted, when the benefit period is n and the expected benefit generated by that environmental resource remains constant each year and $r > 0$, all other factors being equal, is calculated by the following equation:

$$PV = (A/r) \times \left[1 - 1/(1 + r)^n\right] \tag{2.16}$$

In the case of a limited benefit period, which usually means that the property owner of an environmental resource holds it to maturity or that environmental resource is exhausted, when the benefit period is n and the expected benefit generated by that environmental resource remains constant each year and r = 0, all other factors being equal, is calculated by the following equation:

$$PV = A \times n \tag{2.17}$$

2.3.2.2 The Case of Change in Benefits

In the case of infinite benefit period, which usually means that the environmental resources can be operated forever and have infinite benefits, if the expected benefits of an environmental resource change up to and including the nth year and remain unchanged after the nth year (excluding the nth year), and $r > 0$, with other factors remaining unchanged, calculated by the following equation:

$$PV = \sum_{i=1}^{n} F_i/(1 + r)^i + A/\left[r(1 + r)^n\right] \tag{2.18}$$

In the case of a limited benefit period, which usually means that the property owner of an environmental resource holds it to maturity or that environmental resource is exhausted, when the benefit period is n and the expected benefit generated by that environmental resource varies each year up to and including year t and remains constant after year t (excluding year t), and $r > 0$, all other factors being equal, is calculated by the following equation:

$$PV = \sum_{i=1}^{n} F_i/(1 + r)^i + A/\left[r(1 + r)^t\right] \times \left[1 - 1/(1 + r)^{n-t}\right] \tag{2.19}$$

If the price of an environmental resource asset can be estimated several years in the future and expected returns remain constant up to and including year n, the asset price is PV_n in year n and $r > 0$, all other factors being equal, calculated by the following equation:

$$PV = (A/r) \times \left[1 - 1/(1 + r)^n\right] + PV_n/(1 + r)^n \tag{2.20}$$

2.3.2.3 Expected Return

When environmental resources are evaluated using the income approach, the income refers to the amount of all economic benefits gained by the environmental resources under normal operating conditions under the principle of return on investment. It is important to note that the amount of this income belongs to its property owner. When evaluating the benefits of environmental resources according to the income approach, it is necessary to first determine whether the environmental resources being evaluated have income and the size of that income. In the process of judging the benefits, it is important to judge its future benefits. Therefore, the expected benefits of the environmental resources being evaluated should be estimated judiciously. Simultaneously, it is necessary to clarify the main factors affecting the return of environmental resources, including subjective and objective factors. The appraiser can estimate the expected return of the environmental resources under appraisal based on this analysis.

Generally, there are two general types of expected earnings to choose from: net profit and net cash flow.

In asset valuation science, both net profit and net cash flow fall under the category of after-tax income for property owners and are commonly used in calculating earnings. The difference between the two is that net profit is determined on an accrual basis, while net cash flow on a cash basis. The correlation between the two can be simply expressed as follows (excluding accounts receivable and payable for the time being):

$$\text{Net cash flow} = \text{net income depreciation/amortization} - \text{additional investment} \tag{2.21}$$

From the perspective of valuing environmental resources, net cash flow is more suitable as an estimate of expected benefits, with the following advantages.

First, net cash flow is important for the survival of environmental resources. The fundamental difference between the two is that net profit is determined on an accrual basis, while net cash flow is determined on a cash basis. Net profit is usually reflected by net cash flow because of the inventory turnover of environmental resources products and accounts receivable and payable in the process of maintaining or operating environmental resources. In the early stage of developing and using or operating environmental resources, capital investment and receivables are often increased, such as the use of land resources and mineral resources, which require a large amount of capital investment in the early stage, in which case, net cash flow is often shown to be significantly lower than net profit. The unrecoverable profits lower net cash flow considerably, resulting in the operation of environmental resources not being perpetuated. Therefore, net cash flow is critical to the survival of environmental resources.

Second, time value can be reflected by net cash flow. Net cash flow is a dynamic indicator that reflects the dynamic benefits generated by environmental resources. It not only represents the amount of benefits, but also reflects the time when the benefits occur. Net profit, however, does not take into account the time difference of net cash flow, and therefore cannot fully represent the disposable earnings of the property owner at a point in time in the future. Since the benefits are estimated using the income approach by converting earnings at a point in time in the future to present value, it is more accurate to use net cash flow.

In addition, it is worth noting that due to the special nature of environmental resources, when calculating net profit, in addition to general expenses (management expenses, operating expenses, etc.), it also includes related pollution expenses caused by improper utilization of environmental resources as well as taxes. Meanwhile, income should also include main business income, other business income, and non-operating income. Taking forest resources as an example, main business income mainly refers to the income from forest products generated by forest resources. Other business income refers to the carbon sink income generated by the carbon absorption function of forest resources. Non-operating income refers to the environmental subsidies received from the positive externalities of environmental protection such as water and soil conservation and biodiversity maintenance of forest resources.

In the actual operation process, the caliber of expected income will be appropriately adjusted with the aim of better reflecting the value connotation of the appraised object. As for cash flow, free cash flow in asset valuation is usually used instead of cash flow in accounting.

There are various methods for estimating expected earnings. However, the root of the estimation is based on the historical earnings of the appraised object or the historical earnings of similar references, combined with future estimation of its utilization and planning and the development prospect of the industry. With the development of social economy and the emergence of different business models, it is increasingly difficult to value expected returns, especially since environmental resources are influenced by various uncertainties, such as natural force majeure, national policies, and cyclical fluctuations. Currently, scholars have researched these partial issues and developed preliminary research results, but how to estimate them specifically needs to be further explored.

2.3.2.4 The Discount Rate and Determination of the Benefit Period

Discount rate

The discount rate is essentially the expected rate of return on investment. The process of discounting is to convert expected future earnings into the value of the current period using a specific rate, which is the discount rate.

The difference between discount rate and capitalization rate

The discount rate and capitalization rate do not differ significantly, but slightly in terms of their application and respective connotations. In terms of application, the discount rate is often used to discount a finite future expected return, while the capitalization rate is applied to the conversion of an infinite future expected return (annuity). In terms of connotation, the discount rate, as the rate of return of an investment, needs to consider the opportunity cost of the investment and the uncertainty. The risk-free rate is generally determined by the interest rate of treasury bonds, while the risk-based rate depends on the risk profile of the asset. The capitalization rate reflects the growth prospect of the asset in addition to the above two rates.

Method of determining the discount rate

I. Cumulative method

The cumulative method is based on the risk-free rate of return and the risk-based rate of return, and the two values are calculated separately and combined to obtain the discount rate. The risk-free rate is usually replaced by the interest rate of treasury bonds. Overall, short-term treasury bonds (e.g., three months) are the best risk-free investment, but in the appraisal, the objects involved are often those with long-term benefits (e.g., environmental resources), making the use of longer-term treasury rates (one year or more) more comparable and alternative.

The risk–reward component must include two types of risk: market risk and the risk associated with the object being assessed. Table 2.2 lists the factors to be considered in determining the risk–reward rate, using forest resources as an example.

Based on the above factors affecting the risk–reward rate, it is difficult to quantify the risk–reward rate. The factors determining the risk–reward rate in valuation practice will vary for different investors and need to be tailored to specific circumstances of the evaluated subject.

II. Capital asset pricing model

The capital asset pricing model (CAPM) is a method of quantifying the risk–reward component of the discount rate by means of a beta coefficient, calculated as follows.

$$R = R_f + (R_m - R_f)\beta \tag{2.22}$$

where R denotes the risk–reward rate of the industry being evaluated; R_m connotes the average market rate of return; R_f represents the risk-free rate of return; and β signifies the systematic risk.

Table 2.2 Factors affecting the risk–reward ratio

Market risks	Risks associated with the evaluated object
General industry conditions	Product or service type
Macroeconomic conditions	Scale of operation
Capital market conditions	Financial status
Regional economic conditions	Management level
Market competition	Assets
Legal and regulatory constraints	Revenue quantity and quality
National industrial policy	Location

In the above formula, the average market return and the risk-free rate of return are relatively easy to obtain, while the process of calculating the β coefficient appears complicated. There are already relevant professional organizations in foreign countries to compile β coefficients for industries and companies. Specialized agencies are also available in China to provide services for beta coefficients of domestic securities market and listed companies. Micro-entities operating environmental resources can determine their own β coefficients by referring to the β coefficients of the industry or similar entities.

In asset valuation, the recognition of assets comes from two sources: long-term liabilities and owner's equity. The interest rate of long-term liabilities and the rate of return on owner's equity affect the calculation of the discount rate. The calculation formula is as follows:

$$\begin{aligned}&\text{Discount rate} = \text{long-term liabilities as a percentage of total assets} \times \text{long-term}\\&\text{liabilities interest rate} \times (1 - \text{income tax rate}) + \text{Owner's equity as a percentage}\\&\text{of total assets} \times \text{Return on investment}\end{aligned} \tag{2.23}$$

$$\text{Return on investment} = \text{risk-free rate of return} + \text{risk-based rate of return} \tag{2.24}$$

III. Market Approach

The market approach is a method of inverting the discount rate by finding the market prices and earnings of comparable similar to the appraised object in the market. The calculation formula is as follows:

$$\begin{aligned}&\text{Discount rate of the appraised object}\\&= \left(\sum_{i=1}^{n} \text{reference object's earnings/reference object's market price}\right) \div \text{n}\end{aligned} \tag{2.25}$$

In this regard, the reference object is similar to the evaluated object in terms of industry, return level, and risk level, among others. The use of the market approach requires finding as many references as possible to reflect the average investment return of the market.

Earnings period

Earnings period refers to the profitability of the appraised object and the duration of its net income generation, and is generally used as a unit of years in appraisal. It is determined by professionals based on the appraised object's own effectiveness, profitability, depreciation and depletion, or by relevant laws, regulations, rules, and contracts, among others. The revenue period can be classified into infinite revenue and limited revenue periods.

If environmental resources are effectively utilized and there is no limitation on the useful life, it can be considered an infinite benefit period. If the resources are subject to laws, property rights, and contracts, then the years specified above should be used as the benefit period. If no benefit period is specified, it can be calculated according to the normal life of the resource, which is usually the maximum number of years that the resource can generate benefits (e.g., mineral resources). When the utility of the resource is exhausted, its life ends from an economic point of view.

2.3.3 *Cost–Benefit Analysis of Environmental Resources*

2.3.3.1 Static Cost–Benefit Analysis

In this section, economic theories on individual choice are first introduced, followed by a discussion of the social costs of environmental resources.

Individual choice

Economics is the study of rational people making decisions, which are arrived at by choosing among different options. When there are multiple options, a rational person usually chooses the option that maximizes net benefit (the difference between benefit and cost), which results in the simplest cost–benefit analysis. Suppose rational person *i* chooses between option A and option D if:

$$(B_A - C_A) - (B_D - C_D) > 0 \tag{2.26}$$

where *B* is the benefit and C is the cost. Rational person i will choose *A* only if the net benefit of option *A* exceeds that of option *D*.

Mainstream economics infers that human survival and development cannot be achieved without production. In the case of limited technology, the expansion of production will inevitably increase pollutant emissions and environmental pollution. Therefore, the trade-off between environmental pollution and economic benefits is a major concern for human beings, and cost–benefit analysis is one of the bases for making trade-off choices.

It is necessary to analyze cost–benefit analysis of environmental resources based on the theory of supply and demand in the general commodity market. Assuming the existence of a market, benefits can be derived from the demand curve for a given environmental resource. The quantity of goods or services that consumers are willing to buy at different prices is represented by the demand curve. Price is generally inversely proportional to the quantity of goods that consumers are willing to buy. Therefore, the demand curve is a line that slopes down to the right (Fig. 2.5).

A consumer's willingness to pay for the last unit of goods at a given production output can be represented by the corresponding point on the demand curve. The amount of money paid by the consumer for n units of the goods is equal to the sum of the 1st, 2nd, …, the sum of the amount of money paid for n units of the good. In Fig. 2.5, the area below the demand curve between 0 and n is the total amount of money that consumers are willing to pay, which can be used to represent total earnings.

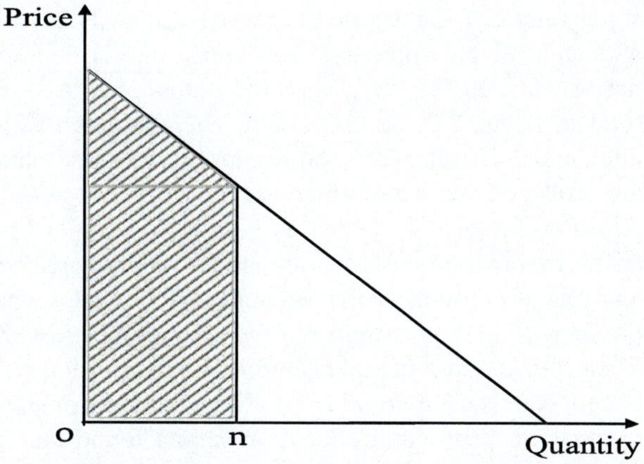

Fig. 2.5 Demand curve (Muth, 1964)

For the marginal cost of a product, it can be expressed as the cost consumed per additional unit of product produced. In the case of goods and environmental services provided by environmental resources, costs can be understood as opportunity costs, which are obvious in the case of environmental services. The opportunity cost of an environmental service is the welfare that is sacrificed by providing this service and not being able to provide other services. Environmental resources can often be used for different purposes, but in most cases, they are incompatible. For example, when land is used as a building site, it loses its purpose as agricultural land. The marginal opportunity cost curve is used here to represent the opportunity cost consumed per additional unit of product produced. In a perfectly competitive market, the marginal opportunity cost curve is equivalent to the supply curve. Total cost is the sum of the costs of producing all products. If the supply curve is continuous, then the area between the origin and n in Fig. 2.6, below the supply curve, is the total cost.

Net benefit is that portion of total benefit that exceeds the total cost. In Fig. 2.7, net benefit is expressed as the portion of the area below the demand curve and above the

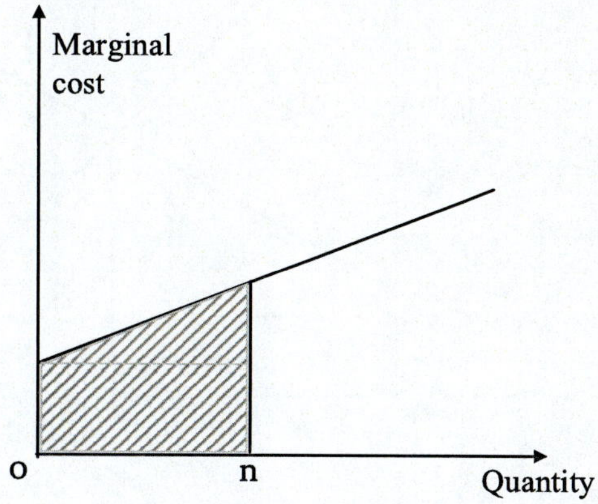

Fig. 2.6 Supply curve (Muth, 1964)

supply curve. The purpose of cost–benefit analysis is to determine the most efficient allocation of environmental resources, that is, the level of output that maximizes net benefits. In Fig. 2.7, when the output is 8, A + B is the total benefit, B is the total cost, and A is the net benefit. Therefore, can yield 8 be considered the optimal allocation? The answer is no because it does not maximize net benefits. If 10 units are produced, net benefit increases by C. When production continues to increase to 12 units, net benefit decreases because, for units 11 and 12, the final increase in cost is higher than the benefit received, reducing the benefit by D. After comparative analysis, it is found that net benefit is maximized when 10 units of the product are produced, and thus 10 units of production is the most efficient allocation of resources. Therefore, environmental resources are most efficiently allocated when the marginal benefit equals the marginal cost. This allocation of environmental resources is Pareto optimal when no other allocation can be found that improves the welfare of some people without compromising the welfare of others.

Social cost of resources

Externalities are one of the manifestations of market failures and are the main focus of economics discussions, and have always been a concern of environmental economics. In the framework of traditional economics, market economy is considered to be perfect, as the "invisible hand" works to maximize the interests of the majority, thus achieving optimal allocation of social resources. However, in his 1920 book "Welfare Economics," Pigou pointed out that it is almost impossible to achieve optimal allocation of social resources. Pigou noted that optimality can be achieved if one relies exclusively on market economy, adding that private marginal costs and private

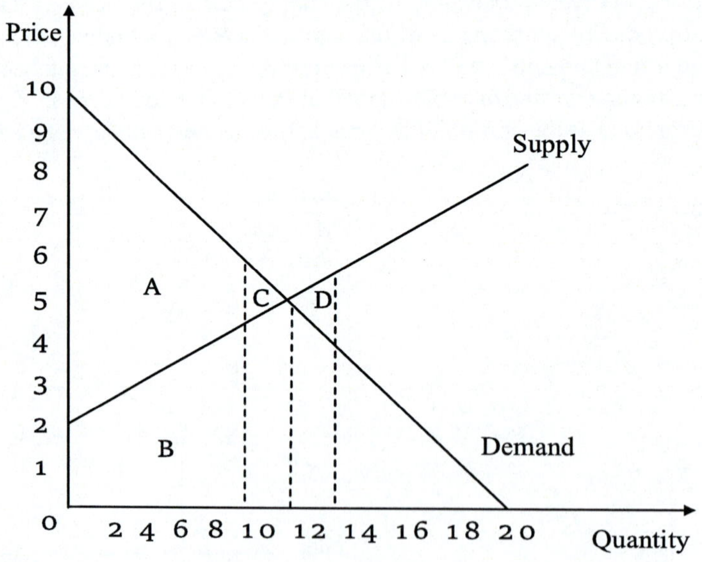

Fig. 2.7 Net income

marginal benefits are not always equal to social marginal costs and social marginal benefits, and the difference between them constitutes an externality. The difference between private marginal cost and social marginal cost is called negative externality, while the difference between private marginal revenue and social marginal revenue is positive externality.

The external effects of negative or positive externalities are not corrected by market mechanism, thus preventing efficient allocation of market resources, and therefore the Pareto optimum. In the case of negative externalities (Fig. 2.8), the agent conducting economic activity does not bear all its costs, as economic activity leads to higher social marginal cost (MSC) than private marginal cost (MPC), forming the difference between the two—marginal external cost (MEC). Under the production optimization principle that marginal revenue equals marginal cost, the socially-optimal output QS is lower than the output QP produced by private individuals, that is, private output exceeds social demand. In the case of positive externality effects (Fig. 2.9), the economic agent performing economic activity does not enjoy all its benefits, as social marginal revenue (MSR) created by its economic activity is greater than private marginal revenue (MPR), creating the difference between the two—marginal external revenue (MER). Under the production optimization principle that marginal revenue equals marginal cost, socially optimal output QS is greater than the output produced by private individuals QP; that is, private output cannot satisfy social demand. MSC and MSR are expressed as follows.

$$MSC = MPC + MEC \tag{2.27}$$

$$MSR = MPR + MER \tag{2.28}$$

From the above, MSC of environmental resources can be understood as marginal social opportunity cost (MSOC), so the following equation is given:

$$\mathrm{MSOC} = \mathrm{MPC} + \mathrm{MEC} = \mathrm{SS} \tag{2.29}$$

The social marginal opportunity cost is the cost to society of producing one more unit of a product, which is the social supply curve SS. In addition, marginal external cost can be categorized into marginal user cost and marginal environmental cost. Therefore, MSOC can be expressed as the sum of private marginal cost, marginal user cost, and marginal environmental cost:

$$MSOC = MPC + MUC + MEC' \tag{2.30}$$

where *MUC* represents marginal user cost, which is the cost incurred by a natural resource that is not left to future generations but utilized now, and *MEC* represents marginal environmental cost. Similarly for social marginal revenue, which also represents social demand function and denotes the price that consumers are willing to pay for another product, the following equation is therefore available:

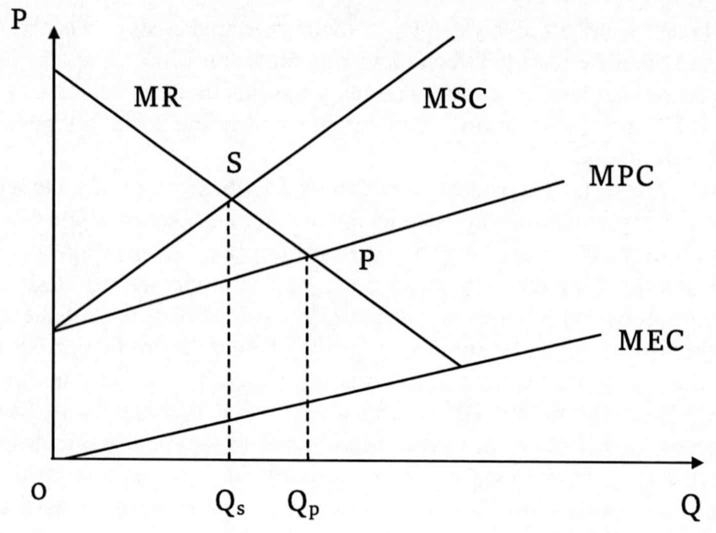

Fig. 2.8 Negative externality impact

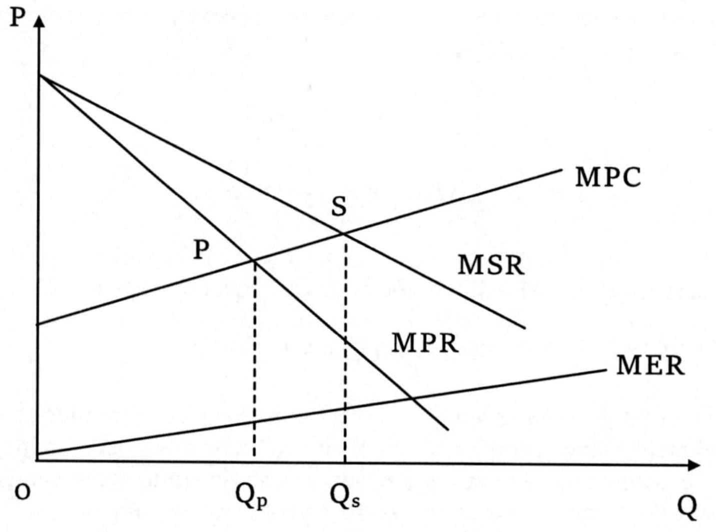

Fig. 2.9 Positive externality impact

$$MSB = MWTP = DS \tag{2.31}$$

where *MWTP* is the marginal willingness to pay and *DS* is the social demand curve. Social marginal revenue is the marginal willingness to pay of society, which is the social demand curve. From this, when *SS* and *DS* are determined, the equilibrium price found by the intersection of the two is the cost price of the whole society.

Therefore, the cost price of society differs from private cost, which is often lower than social cost. Since micro-entities in the market usually consider only private costs, it leads to an unreasonable allocation of resources, resulting in wastage of resources and pollution of the environment.

2.3.3.2 Dynamic Cost–Benefit Analysis

The above-discussed static cost–benefit analysis of environmental resources is the basic criterion for measuring the benefits of environmental resources. When time factor is not considered, a static cost–benefit analysis is effective for comparing different resource allocation options. However, in reality, most environmental resource allocation problems involve different times and even different eras. To address this, economists have expanded the static concept into a dynamic one, and dynamic cost–benefit analysis considers not only the problem of benefits and costs but also time value. Methods to assess the benefits of environmental resources considering the value of time have been described in detail in the Classification of Benefit Assessment Methods and Benefit Valuation Analysis of Environmental Resources. Here, the concept of dynamics is illustrated using a case of two different periods and non-renewable resources. Assuming that a resource can only be used in two periods, one needs to decide how to allocate the resource (e.g., mineral resources) between the two periods so that the present value of the net benefits of resource use is maximized.

First, assume that: The marginal cost of resource development is fixed at ¥4 per unit. Essentially, no matter how many units of resources are developed, the marginal cost of each unit of resources developed is the same. In reality, the situation may be more complicated, for example, the deeper a mineral resource, the higher the cost of exploitation. The total supply of resources is fixed in two periods, which can also be expressed as the marginal cost of developing resources. The supply curve of the resource is the marginal cost curve of resource development. To simplify the calculation, a horizontal resource supply curve is assumed in this case, which is different from the upward-sloping supply curve when discussing statics. The demand function (marginal willingness to pay) is fixed in both periods and is expressed by the equation $p = 16–0.4$ g. The demand function says that when the demand is 0, the price is ¥16; for every 1 unit increase in demand, the price decreases by ¥0.4. That is, the demand curve is downward sloping.

If the total supply of resources is 60 units or more, we simply use the static formula so that marginal cost equals marginal revenue to obtain the value of efficient allocation of resources. According to the intersection of the supply curve and demand curve, the optimal number of resources allocated in the first and second periods is 30 units (Fig. 2.10). In the case where supply meets demand and the number of resources allocated in the first and second periods is irrelevant, static cost–benefit analysis is sufficient to solve the problem because the resources are sufficient and the time factor does not play a key role.

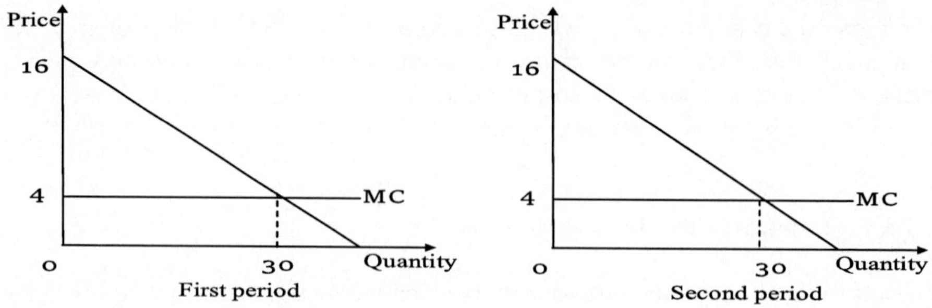

Fig. 2.10 Static resource allocation for two phases

However, if the total supply of resources is only 40 units, then resources become scarce between the two periods. The static concept is not sufficient for the analysis and thus the dynamic concept needs to be introduced to maximize the present value of the net benefits between the two periods. Therefore, it is necessary to first discuss the calculation of the present value of net benefits in both periods, followed by maximization of the present value of net benefits.

Assuming that the first period is arbitrarily chosen to be 30 units and the second 10 units, the present value of the net benefits of the first period is the area of the triangle below the demand curve and above the supply curve and between 0 and 30: $(1/2) \times (16–4) \times (30) = 180$. The present value of the second period is the area below the demand curve and above the supply curve and between 0–10 and divided by $(1 + r)$. Assuming $r = 10\%$, the present value of net income in the second period is $[(1/2) \times (16–4 + 12–4) \times (10)]/(1 + 0.1) = 100/1.10 = 90.91$. The sum of the present value of income in both periods is $180 + 90.91 = 270.91$.

The method of calculating the present value of net income in the two periods is known from the above, but is the quantity 30 units in the first period and quantity 10 units in the second period the optimal allocation of resources? According to the principle of economics, a simple method can be found. The dynamic and efficient allocation of resources must satisfy that the present value of the marginal net return of the last unit of resources developed in the first period is equal to the present value of the marginal net return of the last unit of resources developed in the second period. Therefore, the sum of net returns of the two periods can be maximized. How does one compare the present value of net benefit in the two periods? Economists introduced the benefit periods of the two periods in a graph (Fig. 2.11).

The two diagonal lines in Fig. 2.11 represent the present value of marginal net benefits in each of the two periods. The first-period marginal net return curve is from left to right, and the second from right to left. Correspondingly, the horizontal axis indicates the mining volume of the first period from left to right and the mining volume of the second period from right to left. The total scale of the horizontal axis is 40, which means that the total amount mined in the two periods cannot exceed 40 units. If the amount mined in the first period is q, then the amount mined in the second period must be 40-q. Note that the marginal net revenue curve is used here,

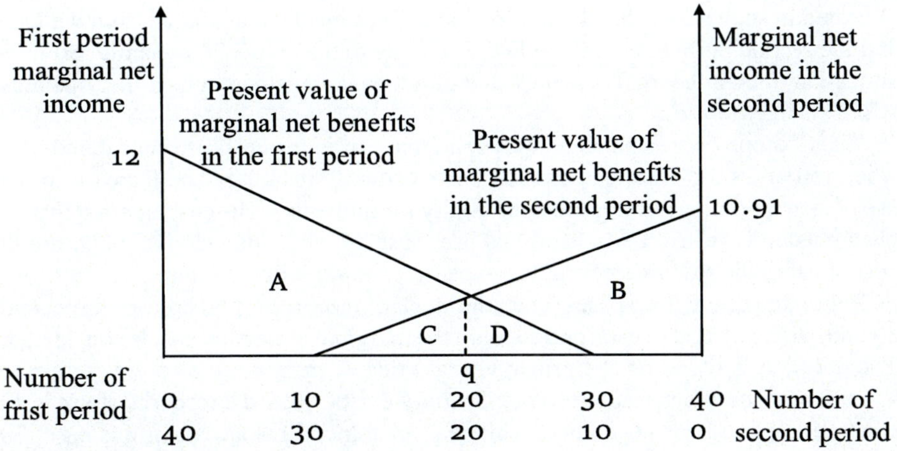

Fig. 2.11 Dynamic cost–benefit analysis of resources

not the demand curve. The net marginal revenue curve is the part of the demand curve above the marginal cost (supply) curve in Fig. 2.10; that is, the triangle where the demand curve is above the horizontal supply curve is cut off in Fig. 2.11. The intersection of the present value line of the first-period net marginal revenue with the left vertical axis of the figure is 12. When the extraction volume is 30, the first-period marginal willingness to pay is equal to the marginal cost and the net marginal revenue is 0. Therefore, the intersection of the first period's net marginal revenue curve with the horizontal axis is 30. Since the current value of the net marginal revenue in the first period is equal to the present value of the net marginal revenue in the first period, no discounting is required.

However, the present value line of marginal net benefits in the first period is asymmetric to the present value line of marginal net benefits in the second period. The reason is that the second-period marginal net income must be discounted. Therefore, the intersection of the second-period net present value of benefits line with the right vertical axis is lower than the intersection of the first-period net present value of benefits line with the left vertical axis. The present value of marginal net benefit is 10.91 when the number of resources extracted is 0 and the discount rate is 10%, and the present value of marginal net benefit is 0 when the number of resources extracted in the second period is 30 units. The higher discount rate causes the present value of the marginal net benefit line in the second period to twist around the point ($q_2 = 30$, $B = 0$), thus making it smoother. Here q_2 denotes the quantity of the second period and B denotes marginal net benefit. In this way, the dynamic resource allocation point q is determined based on the intersection of the lines of the present value of marginal net benefits in both periods. The total present value of net benefits is equal to the area below the line of the present value of marginal net benefits for the first period to the left of point q plus the area below the line of marginal net benefits for the second period to the right of that point. That is, 40 units of resources should be allocated in such a way that the first period is developed q (calculated from the left) and the second period is developed 40-q. Owing to the choice of dynamic resource allocation points, the total present value of net benefits is maximized (i.e., the area ABCD is maximized).

When resources are abundant, only marginal extraction costs are considered, and when resources are scarce, extraction costs certainly increase. The increase in the use of resources now reduces the opportunity for future use. The cost incurred by this phenomenon is referred to as marginal user cost, which is the marginal opportunity cost of using the resource now at the expense of using it in the future.

Return to the previous case. When the stock of resources is 60 units or more, and 30 units are extracted in each period, only the marginal extraction cost is considered. The decision is based on the principle that price = marginal cost of extraction = 4, and there is no marginal user cost. Scarcity exists when the resource stock is 40 units. Two periods or generations compete for mineral resources. In this case, 30 units cannot be mined in each period, and the dynamically-effective extraction will be less than 30 units. The dynamic effective mining quantity q determined from Fig. 2.11 is 20.476 for the first period and 19.524 for the second period, and the prices determined by these two mining quantities are 7.81 and 8.19, respectively. The excess above the marginal mining cost is the current value of the marginal user cost. After discounting the excess cost of the second period, the present value of the marginal user cost for both periods are 3.81. Figure 2.12 gives the marginal user cost for the two periods. Note that the difference between P and MC is the current value of marginal user cost.

The discount rate directly affects marginal user cost and the allocation of resources over different time periods. Since the discount rate is greater than 0, the second period is smaller than the first. If the discount rate is greater than 10%, the line of present value of marginal net benefits for the second period in Fig. 2.11 continues to reverse in the right direction. A change in the discount rate (increase or decrease) causes future weights (decrease or increase) to increase or decrease the number of current resources extracted.

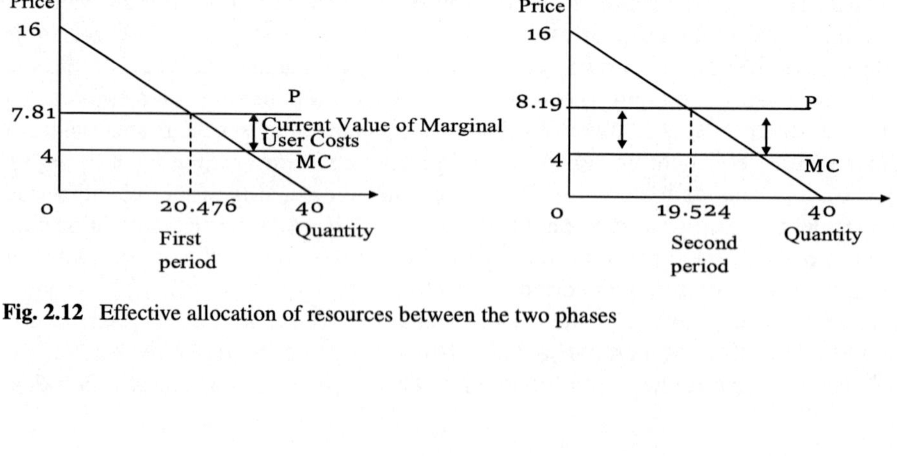

Fig. 2.12 Effective allocation of resources between the two phases

2.4 Case and Practice

With the development of social economy, tourism has initially become a new hot spot for consumption. Earlier, tourists mainly visited well-known tourist attractions and scenic spots, although similar man-made facilities and humanistic tourism content also influenced people's enthusiasm for tourism to a certain extent. Meanwhile, ecological and forest tourism, as well as a large number of less artificially-sculpted tourism projects such as farm tours, which are close to nature and help people understand nature in tourism, are becoming increasingly popular.

New tourism projects, including ecotourism, have attracted a large number of tourists with their unique, natural, and simple characteristics. However, the development of this type of tourism projects has yet to appeal to tourists. Tourism facilities and equipment are not yet sound, services are still far from meeting the needs, and management is not yet in place. However, tourism services, like other products or services, tend to increasingly get personalized and differentiated. In this process, having a wealth of natural resources and beautiful natural environment gradually become a valuable asset for economic and social development. Based on effective utilization, scientific management, and administration, it can bring considerable economic and social benefits for the socio-economic development of relevant regions, while also promoting the cause of natural resources and environmental protection.

2.5 Summary

In the wake of a deteriorating global environment, it is imperative to focus on pollution control. In the process of controlling environmental pollution, both direct control and market incentives should be used to ensure that environmental pollution is controlled at the socially-optimal level. Concurrently, the relationship between environmental protection and economic development should be properly understood. The strategy of polluting first and treating later is obviously no longer feasible, making it challenging to balance environmental protection and governance with economic development. In this case, it is necessary to assess the benefits of environmental resources. The environmental resource benefit assessment methods, benefit valuation analysis, and cost–benefit analysis introduced in this chapter provide some effective tools for measuring the benefits of environmental resources and offer reliable methods for maximizing the benefits of environmental resources.

References

Barnes, J. I., MacGregor, J. J., Nhuleipo, O., & Muteyauli, P. I. (2010). The value of Namibia's forest resources: Preliminary economic asset and flow accounts. *Development Southern Africa, 27*(2), 159–176.

Boatsman, J. R., & Baskin, E. F. (1981). Asset valuation with incomplete markets. *Accounting Review*, 38–53.

Buchanan, J. M. (1986). *Liberty, Market and State: Political Economy in the 1980s*. New York University Press.

Cairns, J., Elliot, R., & McAvinchey, I. (1992). *Dictionary of modern economics*. Macmillan International Higher Education.

Calabresi, G. (2008). *The cost of accidents: A legal and economic analysis*. Yale University Press.

Cowen, T., & Tabarrok, A. (2015). *Modern principles of microeconomics*. Macmillan International Higher Education.

Elías, S., & Wittman, H. (2012). State, forest and community: Decentralization of forest administration in Guatemala. In *The politics of decentralization* (pp. 296–309). Routledge.

Gamarra, N. C., Correia, R. A., Bragagnolo, C., Campos-Silva, J. V., Jepson, P. R., Ladle, R. J., & Malhado, A. C. M. (2019). Are protected areas undervalued? An asset-based analysis of Brazilian protected area management plans. *Journal of Environmental Management, 249*, 109347.

Goldie, F. J. (1997). *The Fitzroy Dearborn encyclopedia of banking and finance* (10th ed.). Reference Reviews.

Grossman, G. M., & Krueger, A. B. (1995). Pollution and growth: What do we know. *The Economics of Sustainable Development, 19*, 41.

Kuznets, S. (2019). Economic growth and income inequality. In *The gap between rich and poor* (pp. 25–37). Routledge.

Maryudi, A. (2012). Restoring state control over forest resources through administrative procedures: Evidence from a community forestry programme in Central Java, Indonesia. *ASEAS—Austrian Journal of South-East Asian Studies, 5*(2), 229–242.

Muth, R. F. (1964). The derived demand curve for a productive factor and the industry supply curve. *Oxford Economic Papers, 16*(2), 221–234.

Pigou, A. C., & Aslanbeigui, N. (2017). *The economics of welfare*. Routledge.

Sahide, M. A. K., & Giessen, L. (2015). The fragmented land use administration in Indonesia—Analysing bureaucratic responsibilities influencing tropical rainforest transformation systems. *Land Use Policy, 43*, 96–110.

Shavell, S. (2018). A model of the optimal use of liability and safety regulation. In *Economics and liability for environmental problems* (pp. 77–86). Routledge.

Chapter 3
Environmental Resource Allocation Efficiency and Sustainable Development

3.1 Introduction

By introducing the overview of environmental and natural resource economics in Chap. 1 and the analysis of environmental resource benefit assessment in Chap. 2, we have mastered the relevant concepts and two practical evaluation methods to quantitatively study environmental resources.

The two methods are static and dynamic. When time is not a core factor affecting the allocation of environmental resources, the first factor can be used to measure the efficiency of environmental resource allocation. Representative examples include water and solar-energy resources. A common feature of these resources is that there is no obvious correlation between the flow of the next year and the usage of the previous year. Dynamic efficiency is suitable when time is a key factor affecting environmental resource allocation. Typical examples are depletable resources, such as oil and coal, which are characterized by the inability of future generations to use the reserves that have been used.

These standard assessment methods are very effective for studying environmental resource problems. They can help us explore the source of environmental and resource problems and provide us with the correct methods. In public sector decision-making, these standardized evaluation methods can help government leaders design optimal environmental resource policies to balance the inherent contradiction between economic and environmental resources.

However, these standardized assessment methods are not the only tools used to find a balance between economic and environmental resources. Generally, efficiency standards can provide a reliable benchmark for developing the economy and preventing environmental pollution and resource waste, but efficiency indicators also have obvious defects. For example, we may be concerned not only with the value of environmental resources but also with how these values are distributed and the size of each person's share. In other words, fairness is an important criterion for balancing economic and environmental resources.

X. Deng et al., *Environmental and Natural Resources Economics*,
https://doi.org/10.1007/978-981-99-9923-1_3

In this section, we focus on discussing and researching specific fairness issues. We will start with a specific and ethically representative situation regarding our descendants' survival and reproduction. We first consider the intertemporal distribution of exhaustible resources. Specifically, we use a dynamic efficiency approach to evaluate the allocation of exhaustible resources at different points in time and examine how changes in the discount rate affect these allocations. We define the conceptual connotations of environmental intergenerational fairness to create a theoretical basis for fairness standards. Finally, we demonstrate how environmental intergenerational fairness is actionable on practical measures and investigate its relationship with dynamic efficiency.

3.2 Efficiency and Equity

3.2.1 A Two-Period Model

Dynamic efficiency balances the present and future use of an exhaustible resource by maximizing the present value of the net benefit of resource use, implying that the use of a resource is scaled across periods but not a special configuration. We used a simple numerical case—a dynamic intertemporal resource allocation over two epochs—to illustrate the characteristics of this intertemporal configuration. Later chapters show how these conclusions can be generalized to more complex situations.

Suppose we want to allocate the supply of a fixed exhaustible class resource efficiently to the two epochs. Assuming that the demand functions for these two periods are the same, the function $P = 8 - 0.4q$ expresses the marginal willingness to pay, and the supply cost of the resource supply is a constant value, assumed to be \$2 per unit. A schematic of this is shown in Fig. 3.1.

Note that if the total supply of resources is 30 units or more, we consider only an efficient allocation for two periods. Regardless of the discount rate, an efficient configuration produced 15 units per period. Thus, under the condition of a total supply of 30 units, the consumption demand in each period is satisfied, and consumption in period 1 does not affect consumption in period 2. In the above analysis, because we assume that the configurations of the two periods are independent, the static efficiency criterion can be used to allocate resources efficiently.

However, if the total supply of resources is less than 30 units (e.g., 20 units), how can resources be effectively allocated over time? According to the dynamic efficiency criterion, the objective of efficient allocation is to maximize the present value of the net benefit. The present value of the net benefit for both periods is the sum of the present values for each period. As a simple numerical example, we consider the present value of a specific configuration with 15 units in period 1 and 5 units in period 2. How can we calculate the present values for this configuration?

The present value in Period 1 is the area of the triangle below the demand curve and above the supply curve. It is not difficult to see that the height of this triangle

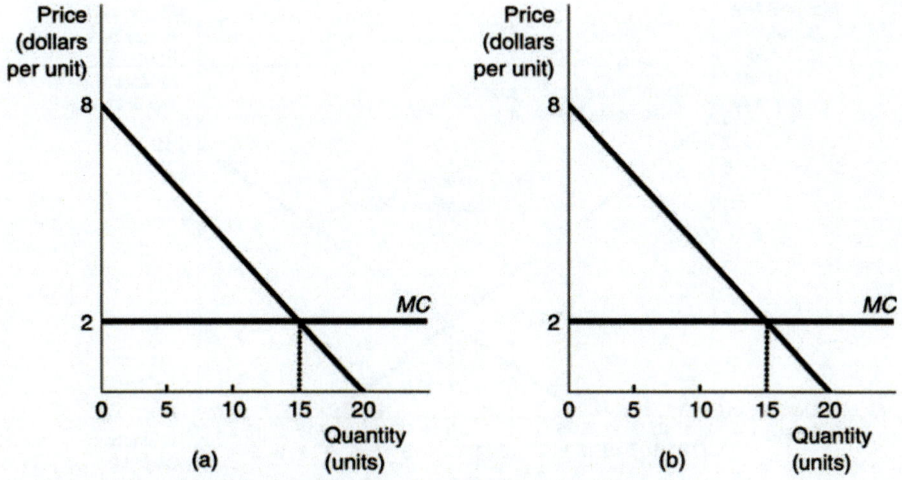

Fig. 3.1 Configuration of exhaustible resources in **a** period 1 and **b** period 2

is 6 US dollars (8 US dollars − 2 US dollars), and the base is 15 units; therefore, the area can be calculated as $(1/2) \times 6 \times 15 = 45$ US dollars. The present value in period 3 is the geometric area of the portion below the demand curve and above the supply curve from the origin to the five units, multiplied by $1/(1 + r)$. If r = 0.1, the present value of the net benefit obtained in period 2 can be calculated as $25/(1 + 0.1) = 22.73$. After summing, the present value of the total net benefit for the two periods was \$67.73.

Through the specific introduction of the above case, we have mastered how to solve for the present value of the net income of any intertemporal allocation. How do you discover the configuration that maximizes the present value of net benefits? One method is to use a computer to exhaust all the configuration combinations of q_1 and q_2 whose sum is 20 and select the one with the largest present value of the net benefit. This method is simple and mechanical, lacks skill, and may be time-consuming.

Another simpler and more direct calculation method uses the relationship between the intertemporal marginal net returns. We know that efficient dynamic resource allocation should satisfy the following conditions. The present value of the net marginal benefit of the last unit in the previous period is equal to the present value of the marginal net benefit in the next period. We can intuitively understand this principle without resorting to a complex mathematical derivation, which we illustrate here using a simple illustration of a two-phase configuration problem (see Fig. 5.2).

Figure 3.2 shows the changes in the present value of the marginal net benefit over the two periods. Looking at the present value curve of the marginal net benefit in period 1 from left to right, we find that the present value curve of the marginal net benefit intersects the vertical axis at \$6 per unit; The cost is \$2, so the present value of the difference, the net marginal benefit, is \$6. The present value of the marginal net

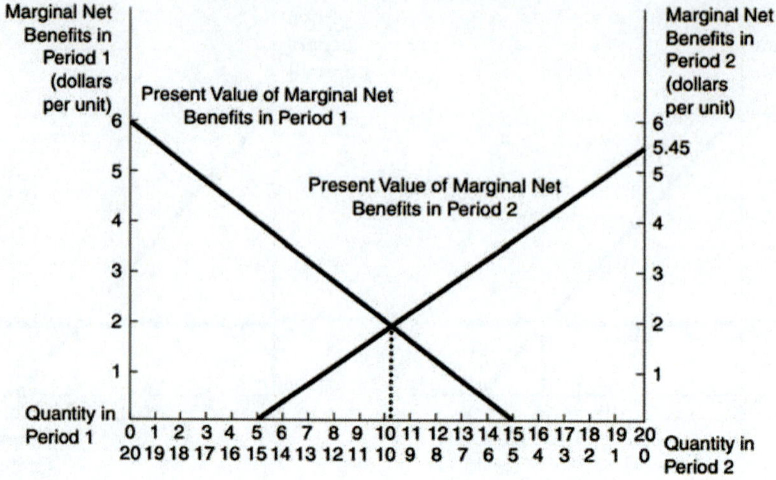

Fig. 3.2 Dynamic efficiency configuration (Tietenberg & Lynne, 2012)

benefit at 15 units in period 1 is zero because, at that amount, the willingness-to-pay just covers the cost.

A challenge with the graphical approach is constructing a present-value curve of the marginal net benefit for period 2. We need to consider two issues. First, the starting point of period 2 net returns is on the right, not the left. Therefore, the increase in the scale of period 2 is marked from right to left so that all the points on the horizontal axis can be obtained. Similar to period 1, 20 units in total can be allocated to two periods, and each point on the axis corresponds to a configurational combination of one characteristic of the two periods. Second, the intersection of the present value of the marginal net benefit curve corresponding to period 2 with the y-axis does not coincide with the intersection of the present value of the marginal net benefit curve corresponding to period 1 with the y-axis. Because the marginal benefit of period 2 occurs one year later, it must be discounted, that is, multiplied by $1/(1+r)$ and converted into the form of the present value. Thus, when we apply a 10% discount rate, the marginal net benefit is \$6, which is 6/1.1 = \$5.45 after discounting. Note that a larger discount rate causes the present value curve of the marginal net benefit for period 2 to rotate on the right axis around the zero point of net benefit ($q_1 = 5$, $q_2 = 15$), as we will use this proof shortly.

We can easily observe that the intersection of the two present-value curves of marginal net returns is an efficient allocation. At this point, the present values of the marginal net benefit in both periods are equal. The present value of the total net benefit is the area below the present value of the marginal net benefit curve in period 1 up to the point of effective allocation, plus the area below the present value of the marginal net benefit curve in period 2 from the right axis to the point of effective allocation. As we achieved an efficient configuration, the sum of these two areas was the largest.

Because the efficiency standard is strictly based on economic principles and has nothing to do with the institutional background, we can use this standard to evaluate the effectiveness of various resource allocation schemes proposed by the market, government designs, or researchers. Although any efficient allocation method must consider the principle of resource scarcity, using this principle in practice requires specific treatments.

Opportunity cost is caused by the scarcity of intertemporal resources; therefore, we call opportunity cost the marginal user cost. In reality, when resources are scarce, people typically use them more in the present and less in the future. The marginal user cost is essentially the marginal opportunity cost discounted to its present value in the current period. More specifically, when there is no scarcity of resources, the use of resources does not need to consider inter-temporal allocation; however, once scarcity occurs, the original resource allocation is no longer used. For example, in areas with abundant water resources, green spaces with a large amount of water may be appropriate. Failure to include water scarcity in the present value of the resource results in inefficiencies in allocating use because of the additional cost of increasing its scarcity value in the future. Therefore, the additional marginal value brought about by scarcity is the marginal user cost.

Returning to the numerical example given earlier, we illustrate the concept of marginal user cost. If the total supply of resources is 30 or more units, 15 units can be allocated in each period, the resources are not scarce, and the marginal user cost is zero. However, when the total supply of resources is only 20 units, scarcity occurs, and the resources allocated to each period are less than the amount in the absence of scarcity. Due to scarcity, the marginal user cost will no longer be zero. As shown in Fig. 5.2, the present value of the marginal user cost (i.e., the additional value created by scarcity) is represented on the graph as the vertical distance from the intersection of the two present value curves to the horizontal axis, which is equal to the marginal net benefit in each period. Present value. It is important to point out that the present value of the marginal net benefit in period 1 is equal to the present value of the marginal net benefit in period 2, which we can accurately calculate to be \$1.905.

When we apply the concept of marginal user cost to a market economy environment, we can intuitively deepen our understanding of its connotations. An efficient market not only needs to consider the marginal cost of resource extraction but also needs to weigh the level of marginal user cost. When there is no scarcity of resources, the market price equals the marginal cost of resource extraction. However, when scarcity occurs, the market price equals the sum of the marginal cost of extraction and marginal user cost.

When faced with the problem of intertemporal resource scarcity, an efficient market solution is to use prices to solve the problem. We substitute the effective quantities of the two periods, corresponding to the above numerical examples as $q_1 = 10.237$ and $q_2 = 9.762$, into the willingness to pay function, namely $P = 8 - 0.4q$, to obtain $P_1 = 3.905$ and $P_2 = 4.905$. The corresponding supply and demand curves are shown in Fig. 5.3. By comparing Figs. 5.1 and 5.3, we observe the impact of resource scarcity on the final market price formation.

In an efficient market, the marginal user cost in each period is the difference between the market price and the marginal cost of extraction. Corresponding to the above numerical example, the marginal user cost of period 1 is \$1.905, and the marginal user cost of period 2 is \$2.095. In both periods, the present value of the marginal user cost was \$1.905. From this, we calculate that the actual marginal user cost is \$1.905 $\times (1 + r)$. Initially, we take $r = 0.1$, and we can calculate the marginal user cost for period 2 to be \$2.095. We can infer that when the present values of the marginal user costs over time are equal, the actual marginal user costs increase.

The magnitude of the marginal user cost and the efficient allocation of resources in two different periods are affected by the scarcity of resources and the discount rate. In Fig. 5.2, due to the discount rate, efficient allocation always results in more resource use in period 1 than in period 2. Discount rates greater than 0.1 are shown in the illustration as rotating the period 2 curve to the right by a specified amount along its intersection with the horizontal axis. The higher the discount rate, the greater is the rotation.

We can see that, with an increase in the discount rate, the number of resources allocated to Period 2 inevitably decreases. In all the intertemporal models considered, a general conclusion is that a higher discount rate makes the resource more inclined to be used in the current period. Result of less weight. Therefore, when making intertemporal resource allocation, an appropriate discount rate is important.

3.2.2 *Environmental Equity and Environmental Intergenerational Equity*

The concept of environmental equity originated in the United States and is a mature product of developing environmental movements to a certain stage. First, they mainly focused on protecting wild animals, calling for environmental protection, saving resources, and reducing the harm caused by environmental pollution. Issues such as environmental fairness have not received the attention of relevant environmental protection organizations in the United States. During this period, the white upper class in the United States had high social status, and their influence gave them the absolute right to speak about the style of the environmental movement. By the 1980s, this "one-word" phenomenon began to improve. During this period, black people at the bottom of society began actively participating in environmental protection and brand attention. Their extensive participation affected the direction of future environmental movements to a certain extent because the problem they urgently wanted to solve was the life of their harsh environment. Owing to their social status and economic income, their living environments are consistently exposed to serious environmental hazards. When they realized the seriousness of the survival problem, they began to express their dissatisfaction and anger and were asked to safeguard their legal rights. As an increasing number of people stand up bravely, the issue of maintaining environmental justice has gradually evolved into an important social issue.

Academic discussions on environmental fairness focus on three aspects: the regulatory system, geographic location, and social rights. The first is fairness at the rule and regulation levels, emphasizing the equal applicability of laws and regulations to legal residents. For example, residents should have equal voting rights when participating in the election of cadres within their jurisdiction; the second is fairness at the geographical level, which emphasizes that everyone who contributes to the community should receive equal returns rather than being unfairly treated. For example, in real production, some people will build factories or companies in processing parks or free trade zones to enjoy low tax rates. Still, the solid waste and sulfur dioxide emitted during the production process will be passed on to a third party through a third party, other people, or in other regions. However, the regions that receive the transfer of pollutants share little of their production profits. Finally, fairness exists at the social rights level, emphasizing the legitimate rights of individuals or groups in society. Inequity in environmental protection policies is seen as a reflection of the division of social rights, which in American society is reflected in the color discrimination between blacks and whites. It is precisely because of the institutional racism inherent in American society that many black communities' social rights and interests are violated unjustly.

In addition, with the wider application of sustainability criteria in the field of environmental equity, discussions on intergenerational equity in the environment have attracted increasing public attention. American scholar, Weiss, E. B. (1988) was the first to propose a theory of intergenerational equity. She advocated that the primary principle of intergenerational equity should be set as the preservation choice and believed that human beings should protect and maintain biodiversity for future generations and avoid excessive consumption by future generations. Environmental resources that people should enjoy. On this basis, some scholars have further summarized and expanded the three key principles of intergenerational equity: "preservation choice," "preservation quality," and "preservation access and use."

3.2.3 How to Realize the Static and Intergenerational Allocation of Natural Resources

Through the two-phase model example in Fig. 5.1, we can see that, for the static allocation of natural resources, we can determine the market price based on the intersection of the marginal revenue and marginal cost curves. However, for the allocation of exhaustible resources, owing to scarcity, we must consider intergenerational allocation, where the dynamic efficiency criterion is a key concept when considering the intertemporal allocation of resources. A central assumption of the dynamic efficiency criterion is that society aims to maximize the present value of the net benefits generated by these resources. To achieve this, it is necessary to balance current and future resource usage. To illustrate this balance in defining the dynamic efficiency criteria, we further analyze the example in Fig. 5.1 and extend our findings to more complex multiperiod models.

3.2.3.1 Further Discussion of the Two-Period Resource Allocation Model

In the previous section, the two-period allocation of limited resources under constant marginal mining costs were analyzed. Under constant resource demand curve conditions, efficient allocation means that more resources are used in period 1, and fewer resources are used in period 3. The two key factors affecting the outcome of this configuration are marginal mining costs and marginal user costs.

The reserves of depletable resources are limited, and their supply tends to be fixed. Using one more unit of oil resources in the current period often implies that the consumption of one unit of oil resources must be reduced. Therefore, decision-makers must consider the net benefits of future losses when considering how to use current resources. The marginal user cost is the opportunity cost of considering this intergenerational balance. In the two-period resource allocation model, we assume that the marginal mining cost is a fixed value and that the marginal user cost in period 2 is greater than that in period 1 to a certain extent. If the transition of the demand curve from period 1 to period 2 does not occur abruptly, the intertemporal growth rate of the marginal user cost is equal to the discount rate, and the marginal user cost of period 2 is $1 + r$ times that of period 1. Therefore, if the inter-period resource allocation of periods 1 and 2 is to be kept balanced, then the marginal user cost of period 1 needs to increase at rate r.

In general, the model deduction results above show that when the marginal mining cost is fixed, the efficient allocation of exhaustible resources increases the marginal user cost and decreases the number of consumers. We extend this conclusion to more complex multiperiod models.

3.2.3.2 Resource Allocation Model with Constant Marginal Mining Cost in N Periods

As in the previous article, we still assume that the marginal mining cost is a fixed value but consider a longer period, that is, a multi-period resource allocation problem involving n periods. Figure 3.3a plots the trend of the demand curve, and Fig. 3.3b plots the trend of the marginal cost over time. It should be noted that the X-axis represents the period, we can see that the total number of periods has been extended from two to many at this time, and the number of mineable resources has increased from 20 to 40 units.

Figure 3.3a plots the trend of resource extraction over time, such as Fig. 3.3b plots the trend of marginal user cost and marginal extraction cost over time. By definition, the total marginal cost can be obtained by adding the marginal user cost

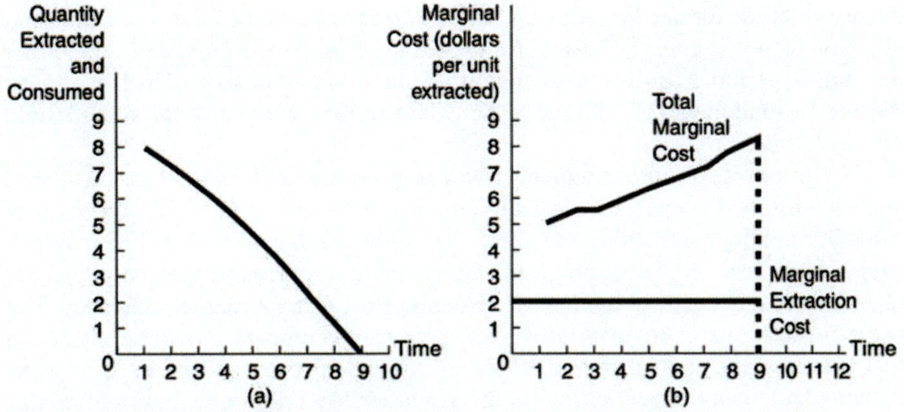

Fig. 3.3 Resource allocation of multi-period marginal mining costs (Tietenberg, 1984)

and the marginal extraction cost. It should be noted that the line parallel to the X-axis represents a fixed marginal cost of extraction. It can be found that although the marginal mining cost is fixed, the marginal user cost still maintains an upward trend. This means that as the reserves of available resources become less and less, the opportunity cost of resource consumption will continue to increase.

Over time, the total marginal cost rises, whereas the extraction volume decreases until it eventually drops to 0. From Fig. 3.3a, the mining volume dropped to zero during Period 9. Figure 3.3b shows the total marginal cost increases from $5 to $8. At this point, the market reaches equilibrium; that is, the resource supply and demand are equal, and the ideal price that consumers are willing to pay is also equal to the total marginal cost. Even with a fixed marginal mining cost, efficient allocation continues to consume resources until they are exhausted. It is worth pointing out that, while in this example, we assume that resources eventually tend to run out, this process does not happen suddenly.

3.2.4 Efficiency Analysis of Natural Resources Allocation and Sustainability Criteria

There are no cases in which efficient allocations meet sustainability criteria in the cases we have constructed. In the two-period resource allocation example, more resources are allocated to period 1 than period 2. Thus, the net benefit in period 2 is lower than in period 1. Sustainability standards do not allow previous generations to seek immediate benefits at the expense of future generations. Therefore, one question that we cannot help but ask is whether an effective allocation is fair.

However, the appearance of things may be deceptive, and choosing a specific resource extraction path did not prevent people in Period 1 from storing a portion of the net proceeds for use by people in Period 2. If resource allocation is dynamic and efficient, then it is always possible for people to set aside a sufficient portion of the net gain in period 1 for use by people in period 2, who choose period 2 in at least the same situation as they did for resource extraction, and one of the times would improve.

To illustrate this more intuitively, we use several examples to show the above analysis results. Compare dynamically efficient allocation and equal allocation of resources among generations and allow sharing between generations. For example, suppose you believe that setting aside all available resources per period in general (10 units) is a better way to allocate resources than using dynamic efficiency. The net gain from this alternative is $40 per period. We compare this to an allocation that takes advantage of dynamic efficiency allocation to achieve a net gain. If the dynamic efficiency allocation meets the sustainability criteria, we must show that each generation is at least as good as the equivalent allocation.

In the dynamic efficiency allocation with no sharing, period 1's net income is $40.466, while period 2 net income is $39.512. By comparison, it is not difficult to see that if part of the net benefit is not shared between the two periods, this number violates the sustainability criterion; the living conditions of the second generation deteriorate compared to the allocation of equal shares. We can also calculate that in an allocation that shares resources equally across two periods, it will receive $40, and in a dynamic efficiency allocation without any benefit sharing, it will receive $39.512.

If the first generation was willing to share some of their net gains from resource extraction with the second generation, the outcome would change. Assume that the first generation retains a net gain of $40 (the same amount as when allocated in equal amounts per period) while storing an additional $0.466 for the second generation at a 10% interest rate (obtained by period 1 in the dynamic efficiency allocation $40.466 minus $40 reserved for itself). These savings increase to $0.513 in period 2, adding this amount to the net gain from dynamic efficiency allocation ($39.512), and the second generation receives $40.025. It can be seen that under the dynamic efficiency allocation of sharing part of the net income, those in period 2 are better off than when the resources are allocated equally.

This simple example shows that although dynamic efficiency allocation does not automatically achieve sustainability criteria, it is consistent with sustainability criteria even in economic situations that rely heavily on lucrative resources. Under the sustainability criteria, the situation in which people in period 2 can improve does not necessarily happen, and it requires people in period 1 to share a certain percentage of the net benefit. As the above examples demonstrate, although net benefit-sharing sometimes occurs, it is more likely to be accidental and abnormal. One of the difficulties of using these sustainability criteria to assess the fairness of intertemporal resource allocation is that they are difficult to apply in practice. If we want to know exactly whether the welfare level of future generations is lower than that of contemporary people, we must not only choose how people allocate resources across time but also understand the preferences of future generations. Fortunately, the well-known "Hartwick Rule" makes it possible to defuse this difficulty. In an earlier article, Hartwick (1977) pointed out that resource consumption can be maintained at a fixed level if all the gains from extracting the resource and utilizing environmental endowments are invested in capital superiors. In other words, the investment level should be sufficient to ensure that the value of the total capital stock does not decrease.

Sustainability standards can be understood from two important perspectives. The first view claims that based on sustainability criteria, we judge the sustainability of resource allocation over time by examining whether the value of the total capital stock has declined. Capital stock can be checked annually without knowing any information about the configuration or preferences of the offspring. Second, this analysis implies a specific degree of sharing required for sustainable outcomes; that is, all scarce benefits must be fully invested.

Consider the following simple example: Suppose your grandparents left you an estate worth $10,000, and you deposit it in a bank to earn $1,000 per year of interest. If you spend exactly this interest every year, the principal in the bank will remain at $10,000, and this income will last forever; if you spend more than $1,000 a year, the principal will inevitably grow over time. It will continue to decrease over time, and your account balance will eventually become zero. In this scenario, an annual spending of $1,000 or less would meet the sustainability standard, but spending over $1,000 would violate it.

In this context, the "Hartwick rule" represents a method for identifying whether a spending pattern is sustainable by examining the changes in the value of the principal over time (Hartwick, 1977). If the principal is reduced, the payout becomes unsustainable. Conversely, the payout is sustainable if the principal increases or remains constant. Generally speaking, the Hartwick rule holds that contemporary people are endowed with natural endowments consisting of the environment, resources, and man-made capital (such as buildings, equipment, schools, roads, etc.). In other words, we should not cut down all trees, use all the oil, or put future generations at risk of extinction. Instead, we need to ensure that the value of the total capital stock is maintained, increased, and not decreased.

The desirability of these sustainability criteria must be strongly based on sustainable approaches to both types of capital. If man-made capital is an easy substitute for natural capital, then it is sufficient to maintain the sum of the values of both types of capital. However, if man-made capital becomes a good substitute for natural capital, then investing in man-made capital alone may not meet sustainability requirements. Manmade capital cannot completely replace certain environmental resources. Although natural air can be replaced by a massive air-conditioning system in a city with a dome roof, this method is a promising remedy, and we should carefully look for effective methods of intergenerational compensation.

As the possibility of substitution is ultimately limited, people realize the shortcomings of the definition of constant total capital, and some economists have proposed new and supplementary definitions. According to the new definition, if the value of natural capital stock is perpetuated during the allocation of resources, the allocation can be defined as sustainable. This definition assumes that natural capital determines the welfare of future generations and that the substitution between man-made capital and natural capital is small or non-existent. To effectively distinguish these two standards, the sustainability that maintains the total capital value unchanged is called "weak sustainability" (less restrictions), and the sustainability that maintains the value of natural capital unchanged is called "strong sustainability."

"Environmental Sustainability" is the last supplementary definition, which requires keeping the specific material flow of some important individual resource constant. According to this definition, maintaining a constant overall value is insufficient for sustainability. For example, for a fishery, the definition requires that the catch should not exceed the growth of the fishery's fish stock. For wetlands, this definition requires the maintenance of the original ecological function during this period.

To use sustainability and efficiency criteria in policy design, certain compatibility requirements must be met. As we will see in the subsequent chapters of this book, not all dynamically efficient configurations are sustainable, and not all sustainable configurations are dynamically efficient. But at the same time, some sustainable configurations are dynamic and effective, and some effective dynamic configurations are sustainable. Market allocation may involve a combination of efficiency, inefficiency, and sustainability. These differences have important implications for formulating environmental policies under specific scenarios.

Among all possible ways of using resources that meet the sustainability criteria, we choose the one that maximizes the most appropriate dynamic or static efficiency. In this process, sustainability criteria are an overwhelming constraint in social decision-making. However, the sustainability criterion is insufficient because it does not tell us which one to choose when faced with an infinite number of sustainable configurations. Thus, we turn to the Efficiency Criterion. Efficiency, as a means, can help us find the allocation of wealth maximization from all possible sustainable allocation methods. A combination of efficiency and sustainability criteria can be beneficial for guiding policy development. Many unsustainable configurations originate from behavioral inefficiencies. Correcting these inefficiencies can restore sustainability or force an economy to move in a more sustainable direction. Importantly, correcting inefficiencies often produces win–win results. A win–win situation improves the situation of all affected parties after the change. This starkly contrasts other changes that some policymakers gain while others lose. A win–win situation is possible because the process of moving from a low-efficiency to a high-efficiency configuration increases the net benefit. The increase in net benefit offers the possibility of compensating those hurt by the change. Compensating losers reduces resistance to change and increases the likelihood of change. Are our economic and political systems capable of producing efficient and sustainable results? In the subsequent sections, we provide clear answers to this important question.

3.3 Sustainable Development

In the first two weeks of June 1992, delegations from 178 countries and regions gathered in Rio de Janeiro to draft a plan for sustainable economic development in the future. The United Nations Conference on Environment and Development, also known as the Earth Summit, is listed by organizers as the world's largest summit in history with the aim of reaching a consensus on solving global environmental resource problems. Sustainability was the focus of discussions at this conference.

The academic community has yet to reach a consensus on the definition of sustainable development. According to *the Brundtland Report*, sustainable development must meet the basic needs of contemporary people for resource extraction and utilization of environmental endowments while not jeopardizing the sustainable survival of future generations. Some dissenting critics argue that the definition of sustainable development must be redefined, partly because the original definition is not sufficiently precise. Although there are many supporters, they also have fundamental defects. After careful deliberation, we found that this definition is seriously divorced from reality and that things are not always like the appearance that people see.

In this section, we examine the definition of sustainable development and determine whether it has guiding significance for the future. What are the basic principles of sustainable development? What is the significance of sustainable development in changing the operations of existing systems? How can the transition to sustainable development be realized? Can the global economic system automatically evolve into sustainable development, or need to change the relevant policies? What policies should be implemented?

3.3.1 Market Allocation

Many market failures, such as intergenerational externalities, public goods, and market forces, can have negative incentive effects that seriously hinder people's visions of sustainable development.

Allowing free access to resources or the use of public goods often promotes the unsustainable allocation of resources. With free access to resources, no renewable energy can guarantee sustainable use. Uncontrolled use will decrease the stock of resources left for future generations. Under extreme conditions, some species may have been overharvested or hunted for extinction.

Generational externalities also severely undermine markets' ability to produce sustainable development outcomes. Greenhouse gas emissions pass the cost to future generations, and their impact on the current generation is external. Current mitigation actions are being implemented by the present generation; however, the benefits of the observation period will be far into the future. Environmental science and technology accurately predict that future greenhouse gas emissions will far exceed the amount required by sustainable standards.

While market failures are often detrimental to sustainable development, we cannot draw the general conclusion that market failures invariably contribute to unsustainability. When renewable resources face a supply shortage, the market can act as a safety valve to protect sustainability. In fisheries, a decline in the supply of one renewable resource triggers the availability of another alternative renewable resource. Even if the government sacrifices the interests of future generations to meet the needs of the present and conducts unfavorable interventions, such as the government with natural gas in the 1970s, the market will mitigate the damage of this external intervention to the interests of future generations by substituting products. Although government intervention makes the transition to alternative energy sources less desirable, it does not completely prevent it.

The flexibility and responsiveness of market prices to resource scarcity signals are important benchmarks for transitioning to sustainability. However, although the market has performed quite successfully in the past, it also has many of the above-mentioned failures and linearity; therefore, sustainable development can be automatically achieved only by relying on the market mechanism, which is impossible.

3.3.2 Efficiency and Sustainability

It is assumed that, in the future, government interventions can eliminate all market failures and make the global economic system work efficiently. In this idealized world, intergenerational and intragenerational externalities are reduced to efficient levels. By restricting the consumption of general resources to an efficient level that does not exceed the regeneration capacity of the species, the previously formed cartel natural resource market will resume competition. One way to answer this question is to examine the different models that embody the essence of international resource allocation. For each model, the question becomes "whether efficient markets are automatically sustainable." The conclusion is clear: Restoring market efficiency is insufficient for guaranteeing sustainable development.

For example, we considered a multiphase configuration that can exhaust resources. Suppose there is a simple economic operation model in which people's production and consumption revolve around the same exhaustible resources. Even if the assumptions of a constant population size and no sudden change in the demand curve are satisfied, the dynamic efficiency criterion suggests that consumption continues to shrink over time. In this hypothetical environment, the survival of descendants will continue to deteriorate unless current generations distribute part of their net benefits to future generations. Otherwise, efficient market allocation cannot guarantee sustainability.

Although the reserves of alternative renewable resources are sufficient, they cannot fundamentally solve this problem. Even in an ideal living environment, the storage capacity of exhaustible resources continues to decline until they are exhausted. Under the pressure of transfer without compensation, even an efficient market will use exhaustible resources that are far greater than its sustainable supply level to maintain a higher standard of living. An important study by Dasgupta and Heal (1979) obtained similar results using a more realistic model. They assume that, in an economy, a single consumer product is produced by capital and an exhaustible resource. A limited supply of exhaustible resources can be used to generate capital and produce consumer goods together with capital. The more capital generated, the higher the marginal product of the remaining exhaustible resources to produce consumer goods. They demonstrated the existence of a sustainable and constant level of consumption in this model. Increased capital accumulation (obtained through increasing marginal product of exhaustible resources) can compensate for the decreasing availability of exhaustible resources. However, their model also shows that any positive discount rate inevitably leads to a decline in consumption levels, which goes against the sustainability criterion. Discounting is an endogenous factor in dynamically efficient allocation.

In this model, sustainable development is possible but not the result of a spontaneous allocation of markets (or even efficient markets). Why not? What guarantees sustainable resource allocation? Hartwick (1977) argues that sustainable per capita rents can only be achieved if all scarce rents earned are used for capital investment, and people are not using them now in the consumption trajectory (meeting our definition of sustainability).

At a positive discount rate, it is critical that we consume scarcity rent, violating the Hartwick rule. Restoring efficiency often shows an approach toward sustainability, but it is not sufficient on its own, and further policies must be developed to guarantee sustainable results.

However, not all the economic models were satisfactory. In particular, a class of endogenous technological progress models assumes that efficient markets exist for sustainable outputs. Endogenous technological progress refers to the fact that internal economic incentives in the growth process can bring about a rate of technological progress that can benefit future generations (technological progress can move production possibilities outward). As long as the rate of technological progress is sufficient to offset the damage caused to future generations by the current generation, efficient markets can guarantee sustainability.

However, endogenous technological progress rates do not necessarily maintain sustainable output. In our analysis, we pointed out that maintaining a non-declining value of capital stock (both physical and natural) provides an observable method for judging the sustainability of current activities. If the value of capital stock falls, the activity is not sustainable. According to Pezzey (1992), the answer is no. When capital stock is misestimated, an increase in net income may coexist with a discontinuity. If we consume nonrenewable resources too quickly, this will cause their prices to continue to decline, which will further lead to the nominal price illusion that the value of the consumed resources is less than the value of the newly added capital, thus increasing the value of the capital stock. However, with an accurate price assessment, the net gain from new capital may be lower. Howarth and Norgaard (1990) arrived at similar conclusions from different perspectives. They assumed that each generation was allocated a specific share of exhaustible resources, changed the resource share, calculated the resource allocation of each generation, and obtained a competitive resource allocation plan across generations, thus reflecting the intergenerational distribution of resource property rights. Two of their conclusions are relevant to the purpose of our study: allocation outcomes are sensitive to the initial distribution of property rights across generations, and allocating property rights to the first generation does not produce sustainable outcomes. This study supports our conclusion from another perspective: efficient allocation of exhaustible resources does not necessarily lead to sustainable results. What about renewable resources? In principle, the flow of renewable resources is continuous.

Is efficient market allocation of renewable resources consistent with sustainable development? Pezzey (1992) studied the long-term sustainability of renewable resource allocation. In this model, welfare can only continue to increase if both of the following conditions are true: (i) the growth rate of resources exceeds the sum of the discount rate and the population growth rate, and (ii) the initial food supply is sufficient to satisfy existing population needs. In reality, the first condition is sometimes difficult to meet, especially in the case of rapid population growth and the slow growth of ecological resources. When the population is growing rapidly, owing to population pressure, it is difficult for people to control the harvest below the sustainable limit; therefore, sustainable development of renewable resources is difficult to achieve.

The second condition is attracting increasing attention. This implies a clear possibility that if the initial conditions are far from sustainable paths, it will be difficult to achieve the development goals without any external intervention. Let us consider the simplest example to understand this problem. A certain country is very poor; people must sacrifice future development and eat all the corn seeds to survive. Two lessons can be drawn from this. It is important to act as soon as possible to ensure that the situation does not deteriorate to the point where an existential crisis affects investment, and international aid may be a sustainable policy for the poorest countries.

We must carefully distinguish between the aspects we have discussed. Efficiency may contribute significantly to sustainable development but is not necessarily a sufficient or necessary condition for sustainability. Three different cases are used to illustrate this. In the first case, where both private inefficient and efficient outcomes are sustainable, restoring efficiency can improve welfare but is not necessary to achieve sustainability. This conclusion is even more pronounced when the stock of resources is more abundant relative to the amount of use. In the second case, the private inefficiency equilibrium is not sustainable, and the efficient outcome is sustainable; thus, restoring efficiency would increase the current welfare level and ensure sustainability. In the last case, neither private inefficiency nor an efficient outcome is sustainable, at which point, restoring efficiency is insufficient to produce a sustainable outcome. Therefore, the present generation must make sacrifices to protect the interests of future generations.

Efficient markets do not always lead to sustainable development paths, which does not imply that unsustainability is the norm. Historical developmental experience shows that the contradiction between efficiency and sustainability norms is an exception rather than a norm. Capital accumulation and technological progress are the two core paths to increase the total amount of resources and use efficiency and still improve the welfare level of future generations, even when resources are reduced. However, these two standards are incompatible. With a reduction in resources and an increase in global externalities, it is foreseeable that the conflict between the two standards will be further highlighted and valued.

3.3.3 Trade and Environment

Openness to trade is a traditional development approach. Benefiting from the convenience of international trade and the resulting market competitiveness and inclusiveness, a free and open international market provides consumers with more affordable choices and opportunities for domestic producers to explore foreign markets. The theory of comparative advantage assumes that trade can benefit both parties and that things might become more complicated.

3.3.3.1 The Role of Property Rights

Numerous studies have shown that when countries (presumably their underdeveloped southern counterparts) have poorly defined property rights or do not internalize external influences, international trade undoubtedly creates adverse and inefficient impacts. In this case, intertrade can significantly exacerbate the tragedy of the commons. The property rights of exporting countries are not clearly defined, and the consumption of these low-priced resources is greatly expanded by artificially driving down their prices. In this scenario, trade increases the pressure on public resources and accelerates their depletion, exacerbating environmental resource problems.

3.3.3.2 Pollution Haven Hypothesis and Race to the Bottom

The failure to control environmental externalities provides another explanation for environmental degradation caused by international trade, which is the so-called "pollution haven" hypothesis. According to this hypothesis, producers transfer their production bases to countries with relatively loose environmental regulations (usually low-income developing countries) to transfer pollution sources after being subjected to strict environmental regulations in their home countries. Attracted by price concessions, consumers in countries with strict environmental regulations favor products produced in pollution havens.

Structural, technological, and scale effects are the three key factors that affect the pollution levels of safe havens. Regarding structural effects, pollutant emissions change with different combinations of polluting and clean industries. When the proportion of polluting industries relative to clean industries increases, the level of pollution increases even if the total output remains the same. The technology effect includes the pollutant emission rate of each industrial unit output. If pollution havens increase their pollution owing to the opening of trade, then pollution in the country will increase. Finally, the scale effect examines the impact of output levels on pollution emissions. Even without structural and technological effects, pollution emissions from pollution havens increase with the total output.

The pollution haven hypothesis explains environmental degradation and provides a reason for developing countries to adopt lower environmental standards that help address unemployment and stagnant economic growth. In other words, it proposes a "race to the bottom" feedback mechanism: Country competition stimulates developing peers to maintain low environmental standards to attract employment, and employment shifts to low environmental standards in exchange for regions with low production costs.

Is there a realistic basis for the pollution haven hypothesis and its drive-to-bottom implications? Early empirical studies suggest that there is no evidence that environmental regulations have impacted trade or financial flows. However, Copeland and Taylor (2004) show that after controlling for other factors, the impact of environmental regulations on trade flow and enterprise location is small and almost negligible. Studies that have attempted to separate the structure, technology, and scale effects have generally found that structural effects (the most important effect in confirming the pollution haven hypothesis) are relatively small relative to scale effects and that technology effects generally reduce rather than increase pollution. Although trade can increase pollution through scale effects, these findings are far from what would be expected from a bottom-out effect.

These results are unsurprising because pollution control costs constitute a relatively small fraction of production costs. It would be strange if low environmental standards became the decisive factor for enterprises in choosing a site or trade direction. Meeting these environmental standards has become a significant part of production costs.

3.3.3.3 Porter's Induced Innovation Hypothesis

Michael Porter, a professor at Harvard Business School, proposed that a certain degree of environmental regulation will not hinder production and employment but can also promote enterprises to carry out green technology innovation, that is, the "Porter-induced innovation" hypothesis. This non-traditional view holds that when a country's environmental regulation intensity is appropriate, enterprises will actively seek production technology innovation to reduce the cost of environmental governance, which will further play a role in shaping their competitive advantage. Because strict environmental regulations can force enterprises to carry out technological innovation, enterprises with innovative abilities will become more competitive. This advantage is particularly significant for companies that manufacture pollution control equipment. Other companies will also find this advantage: meeting environmental regulations will reduce production costs.

3.3.3.4 The Environmental Kuznets Curve

Although free trade advocates are aware of its possible environmental impacts of free trade, especially in the face of externalities or property rights issues in exporting countries, they believe that these issues can be self-corrected. They believe that free trade can increase income, which can promote environmental protection.

The special functional relationship that illustrates this point comes from the early work of Harvard Professor Simon Kuznets, known as the environmental Kuznets curve. According to this relationship, the relationship between per capita income and the degree of environmental degradation follows an inverted U-shape. Before the turning point, the degree of environmental degradation increased with an increase in per capita wealth; however, after the turning point, the degree of environmental degradation increased with an increase in per capita wealth. Decrease as wealth increases. Some earlier studies used the country as the unit of observation (data point) to map the relationship between per capita income level and indicators such as sulfur dioxide concentration, which supports this view.

This theory argues that increasing income through trade provides a self-correction mechanism. However, if remediation involves the outward transfer of polluting industries, its implications are quite different. If the environmental Kuznets curve represents pollution transfer rather than pollution reduction, it would change its meaning significantly. For many developing countries in the low-income stage, production pollution cannot be transferred; therefore, they may never be able to cross the turning point of the environmental Kuznets curve.

How does trade affect Kuznets' model? Cole (2004) studied this problem and found that considering the impact of trade in detail when establishing the Environmental Kuznets model will not lead to the disappearance of the turning point for most pollutants but will affect the time of the turning point. In particular, controlling the transfer of pollution-intensive industries causes the turning point to occur later than when these factors are not considered.

Is there a general conclusion that pollution problems self-correct during development? Earlier studies used different countries as study subjects but concluded that individuals in individual countries could ultimately enhance environmental protection as income increases. Subsequent studies have focused on whether environmental protection has improved over time and on per capita wealth in the same country (List & Galle, 1999). However, most empirical studies fail to support this conjecture (Vincent, 1997; Deacon & Norman, 2006). Other studies have found that environmental Kuznets curve relationships exist for some pollutants (e.g., sulfur dioxide) but not for others (such as carbon dioxide).

Environmental regulations are not determinants of a business's location or trade direction. This implies that sound environmental regulations do not threaten polluting production companies and thereby reduce local employment. Except for a few special cases, the relocation of enterprises is not related to local environmental regulations. If environmental degradation is caused by a country's inappropriate property rights system or the internalization of externalities, it is urgent to correct the root causes of these market failures and not prevent trade. These trade-related inefficiencies can be addressed using appropriate property rights systems or pollution control mechanisms. However, if it is not politically feasible to establish an adequate property rights system or pollution control mechanism, other avenues must be found, including restricting unfavorable trade. However, these trade restrictions must be used carefully, as they are a suboptimal policy tool in this context and have the potential to backfire.

3.4 Sustainable Environmental Resource Policy

3.4.1 Intertemporal Effective Allocation of Environmental Resources

In policy design, we must consider sustainability and efficiency criteria to achieve an intertemporal effective allocation of environmental resources. As seen in the earlier chapters of this book, not all efficient allocations are sustainable or efficient. At the same time, some sustainable configurations are effective, and some are sustainable. Market allocation may involve a combination of efficiency, inefficiency, and sustainability.

These differences have important policy implications for future research. Specifically, they suggested specific policy strategies. Among all possible ways of using resources that meet the sustainability criteria, we choose the one that maximizes the most appropriate dynamic or static efficiency. In this process, sustainability criteria are an overwhelming constraint in social decision-making. However, sustainability criteria are insufficient. Therefore, we must rely on an efficiency criterion. Thus, the efficiency criterion can help determine the allocation with the maximum benefit from all possible sustainable allocation methods.

A combination of efficiency and sustainability criteria is beneficial. Many unsustainable configurations originate from behavioral inefficiencies. Correcting this can restore sustainability and force an economy to move in a sustainable direction. Importantly, correcting inefficiencies can often lead to win–win outcomes. The win–win situation improves the situation of all affected parties after the change because the process of moving from inefficient to efficient allocation increases the net benefit. The increase in net gain offers the possibility of compensating those hurt by the change. Compensating losers reduces resistance to change and increases the likelihood of change. Are our economic and political systems capable of producing efficient and sustainable results? In the subsequent sections, we provide clear answers to this important question.

3.4.2 Market Allocation of Renewable Resources

Human energy demand depends entirely on renewable energy sources because of the high environmental costs of using exhaustible resources. Several renewable energy sources are used to generate electricity, including hydropower, wind, solar, and ocean tidal power. These energy sources not only make electricity production more sustainable but also drive the widespread use of certain electric drive technologies in home heating and electric vehicles, such as ground source heat pumps, air source heat pumps, and plug-in car battery chargers, and reduce the need for fossil fuels. Degree of dependence.

Renewable energy can be generated using several methods. The time-matching (peak supply and peak demand) and energy forms (gas, liquid, or electricity) of some renewable energy sources make it impossible for humans to rely on only one renewable energy source as a long-term solution to energy problems. Different types of renewable energy sources offer various advantages. Therefore, different renewable energy sources must cooperate to meet demand through combined applications.

Producers and consumers in the early stages of a new technology or product are pioneers and face the issues of low reliability and high cost. From a personal perspective, the optimal strategy is to delay consumption until all technical deficiencies have been addressed, costs have been reduced, and investment decisions have less risk. However, from a societal point of view, if every producer and consumer delays replacement, this new industry will not be able to operate at an efficient scale, learn enough lessons to improve reliability and reduce costs, and will not be able to guarantee a large and stable market. What measures should the government introduce to solve these problems?

The first policy instrument subsidizes early-stage pioneers by publishing tax deductions and codes. In practice, this is usually achieved through production tax relief or other tax incentives (tax incentives). Once the market is sufficiently large to take advantage of economies of scale and eliminate unreliability, governments can eliminate these tax policies. Large-scale tax incentives implemented at both federal and state levels in the United States have played a significant role in guiding producers to accept the financial and technical risks of wind power development. However, this approach was not always effective. Tax breaks expired in 1985 and have since been reinstated several times. In contrast to the "intermittent and repeated" tax subsidy policy of the United States, European countries have continued to increase their economic incentives to encourage wind energy development. Wind energy is currently the dominant energy source worldwide.

The second is the Renewable Energy Quota System (RPS) and Renewable Energy Letter of Credit (REC). The renewable energy quota system stipulates the minimum proportion of renewable energy power generation (such as wind and solar) in the total power generation. Power generation entities can meet institutional requirements in two ways: one is to directly generate electricity from renewable energy in accordance with regulations, and the other is to purchase renewable energy electricity from independent power generators.

The third policy tool is the feed-in tariff subsidy. This policy instrument originated and is widely used in European countries, especially Germany, and is now being gradually adopted by the US government. For example, in April 2012, the Los Angeles City Council passed CLEANL, a feed-in tariff program designed to require the Los Angeles Department of Water and Power to purchase electricity generated by solar rooftop batteries for residential, commercial, and public organizations. The feed-in tariff subsidy policy sets the price of electricity for electricity generation users who supply electricity to the grid using compliant renewable energy sources. The subsidy standard (determined in advance) is sufficient to cover the cost of electricity supply and ensure a reasonable return on investment. Each power generation facility in Germany enjoys a 20-year incentive subsidy. Simultaneously, to reflect the technological update and economic scale effect, the subsidy scale for newly operating facilities is reduced annually.

3.4.3 Market Allocation of Exhaustible Resources

In the previous section, we analyzed in detail how alternative, exhaustible, and renewable resources can be effectively allocated over time under various assumptions. Can an efficient configuration be achieved in the real-world market environment? This question is the focus of analysis in this section. With millions of consumers and producers acting according to their preferences, what kinds of institutional guarantees are needed to achieve dynamic and efficient allocation?

3.4.3.1 Reasonable Property Rights Structure

Most people believe that even a perfectly competitive market cannot achieve an efficient allocation of environmental resources because producers will exploit as many resources as possible as quickly as possible to maximize profits. This kind of thinking creates a false perception that the market is shortsighted and does not care about future conditions.

According to the learning content in Chap. 2, as long as the property rights system for environmental resources is clearly defined in terms of exclusivity, transferability, and coercion, such transactions in the open market will not necessarily lead to short-sighted behavior. Producers can efficiently maximize profits by incorporating marginal user costs into the production function.

Environmental resources have two potential values for owners: the first is the profit value obtained when selling resources, which is usually regarded as the only source of value for resources, and the second is the storage value of resources. The value of a resource fluctuates around its price as long as its price continues to rise, as does its store value. However, the owner must store resources to increase the value of assets. Producers lose the opportunity to sell all of their resources at an early stage.

Profit-maximizing producers seek to maximize the resource value by balancing current and future production volumes. It is expected that rising future resource prices will stimulate resource conservation behaviors. If producers ignore this stimulus, they will be unable to maximize the value of their resources. Producers seeking to maximize resource value recycle resources from short-sighted producers. As long as the market satisfies the social discount rate and the private discount rate are equivalent—a clearly defined property rights structure; open and transparent futures price information, and other conditions—producers can maximize their profits while maximizing the present value of net social benefits. Therefore, a reasonable property rights structure is a key institutional guarantee to ensure the effective allocation of environmental resources over time.

3.4.3.2 Environmental Costs

However, not all realities have the necessary conditions to achieve harmonious effects. One such situation is the failure to establish a clear property rights structure, such as a cost function for producers that incorporates the environmental costs of natural resource extraction. Environmental costs include aesthetic damage from mining, health risks from uranium remnants, and acid leaching into rivers from mining operations. Environmental costs not only have important practical significance but also very important theoretical significance. This is an important bridge linking environmental economics and natural resource economics, which traditionally belong to different disciplines.

For example, suppose that an exhaustible resource causes a certain degree of environmental pollution during the extraction process and that these environmental costs are internalized in the producer's production function. According to the discussion in Sect. 3.2, this is the external cost. Resource owners only enjoy the rights to exploit, process, and use resources but do not assume corresponding environmental protection responsibilities. What is the difference between resource allocation based on resource owner costs alone and efficient resource allocation based on full costs?

Figure 3.4a, b plot the dynamic intertemporal allocations considering the environmental cost constraints. The environmental cost affects the equilibrium price of dynamic allocation from the demand and supply sides, and the two effects are exactly opposite. On the demand side, environmental costs increase the equilibrium price of resources, thereby reducing demand. Then, under the condition of keeping other factors unchanged, the speed of resource consumption will slow down, and its use period will be extended accordingly. However, other factors are also affected by environmental costs. For example, under efficient resource allocation, an increase in the marginal cost of extraction accelerates the decline in the cumulative extraction of depletable resources. In the examples shown in Figure 3.4a, b, we can observe that after considering the environmental cost, the effective cumulative extraction volume decreased from 40 to 30 units. Therefore, from the supply side perspective, under the premise of controlling for other factors, environmental costs will accelerate the transition from exhaustible resources to renewable resources.

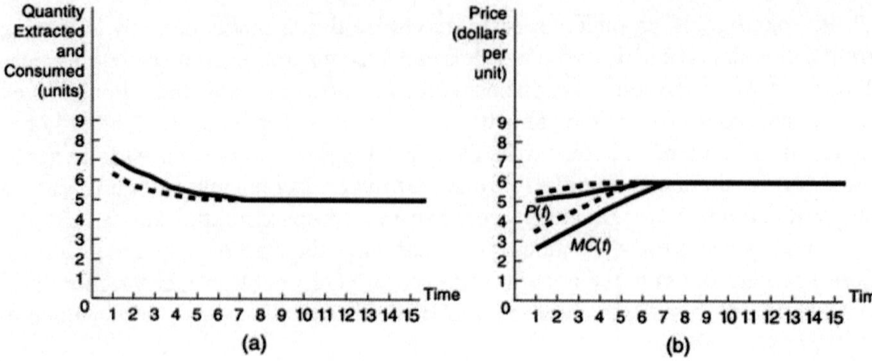

Fig. 3.4 Dynamic intertemporal configuration considering environmental costs (Tietenberg & Lynne, 2012)

Through mathematical derivation, we know that the impact of environmental costs on the supply side is greater than that on the demand side, which leads to the emergence of an alternative time point for effective allocation earlier than market allocation. Generally, the impact of environmental costs on the supply and demand sides of resources is closely related to the marginal mining cost function. If the marginal cost is constant, it cannot affect the supply side of market allocation, and the timing of resource substitution is delayed accordingly. However, if the environmental cost comes from renewable resources rather than exhaustible resources, the replacement timing of efficient allocation will be later than that of market allocation. Policies that internalize external costs can affect extraction volumes and price trends, sometimes with unexpected results. Sinn published an intriguing article in 2008 arguing that such demand-reducing policies can have a price effect that reduces real welfare (under certain conditions) in response to global warming. As this analysis argues that policies aimed at internalizing externalities actually reduce economic welfare, this result is known as the "green paradox." This confirms the interdependence of the decisions that people make regarding the future.

3.4.4 The Impact of Environmental and Resource Policies on High-Quality Economic Development

Chinese state leaders pointed out at the 19th Congress of the Communist Party of China held in 2017 that China's economy has moved from a stage of high-speed growth to a stage of high-quality development. High-quality development refers to the quality and efficiency of economic development, and not just the speed and scale of economic growth. Many studies have pointed out that in contrast to the previous evaluation goals of officials based solely on total GDP, the evaluation criteria for the high-quality economic development stage consider all aspects and multiple dimensions, such as resource utilization, environmental protection, people's life satisfaction, and opening to the outside world. Undoubtedly, this change will inevitably have a wide-ranging and far-reaching impact on policy, social, and economic systems and is a key change related to China's future development. In this section, we briefly describe the latest processes by which China's environmental and resource policies affect its high-quality economic development from five perspectives: finance, price, equity, ecological compensation, and value accounting.

3.4.4.1 Fiscal Policy on Environmental Resources

In recent years, China's financial expenditure on environmental pollution control has increased continuously; however, it still faces the serious problem of insufficient investment in ecological and environmental protection. A sustainable and stable supply of ecological and environmental protection funds is a prerequisite for promoting various ecological and environmental protection efforts. Although China's financial sector has continuously increased investment in ecological and environmental protection in recent years, there is still a big gap compared with actual governance needs. According to data from the National Bureau of Statistics of China, the total investment in environmental pollution control in China increased only six times from 2001 to 2019. Still, the intensity of environmental pollution control showed the opposite trend, decreasing from 1.2% in 2001 to 0.7% in 2019. In addition, China's financial sector is actively promoting resource tax reforms. *The resource tax law of the people's Republic of China* was successfully promulgated and implemented on September 1, 2020. Over the past 27 years, China has lacked a formal legal system for resource tax and has followed *the Provisional Regulations of the people's Republic of China on resource tax.*

3.4.4.2 The Price Policy of Environmental Resources

In recent years, China has actively promoted market-oriented reforms in the pricing policies of environmental resources. In the first half of 2018, China's National Development and Reform Commission issued *the Opinions on Innovating and Improving the Price Mechanism for Promoting Green Development*. The document issued a sewage treatment charging policy, a solid waste treatment charging mechanism, and several new measures to improve the environmental resource price policy. Additionally, many regions in China are optimizing and improving their price policy systems, including sewage treatment fees, solid waste treatment fees, water prices, electricity prices, and natural gas prices. For example, China's National Development and Reform Commission and other important government departments jointly issued a *Notice on Continuing the Comprehensive Reform of Agricultural Water Prices* in 2020. Many regions have carried out pilot work on agricultural water price reform and have formed many valuable policy experiences and practical plans. At the same time, China is also actively promoting the market-oriented reform of the electricity price policy and has introduced a series of important measures to improve the tiered electricity price system for residents and the peak-valley time-of-use electricity price policy. For example, China's National Development and Reform Commission and other departments took the lead in issuing the *Notice on Doing a Good Job in Signing Mid- and Long-Term Electricity Contracts in 2021* in November 2020. The document specifically states that peaks and valleys must be included in the agreed-upon terms of electricity purchase and sale contracts. Difference. Guangdong, Fujian, Zhejiang, and other provinces successively adjusted their peak-to-valley electricity price rules.

3.4.4.3 Rights and Interests Policies on Environmental Resources

The Chinese government has actively improved its property rights system for natural-resource assets. In April 2019, important government agencies, such as the General Office of the State Council, issued *Guiding Opinions on Coordinating and Promoting the Reform of the Property Rights System of Natural Resources Assets*, which requires all departments and units to deepen the reform of the property rights system for natural resource assets. Simultaneously, the original *Land Administration Law of the People's Republic of China* was comprehensively revised, and the unified confirmation and registration of natural resources was accelerated. Local governments at all levels are also working hard to promote the reform of property rights systems for natural-resource assets. According to the author's incomplete statistics, by the end of 2020, opinions on implementing the reform of the natural resource property rights system had been successively promoted in more than ten provinces, including Hebei, Shanxi, and Shandong. Among them, Qingdao, Shenyang, Dalian, and other cities have issued supporting pilot programs and policy initiatives. This model, which combines top-down top-level design and bottom-up local advancement, fully considers objective conditions such as the level of local economic development and resource endowment, which are conducive to achieving targeted and cause-based reform in the process of property rights system reform. Land reform.

Many regions in China are also making efforts to promote the trading of environmental rights in the secondary market. For example, Shaanxi Province officially opened its emission rights market in September 2020 and improved the original trading rules and accounting system for emission rights trading. Shaanxi Province has always been a pioneer in this regard. In July 2020, it took the lead in realizing full coverage of fixed pollution source pollution discharge permits, laying a solid foundation for establishing a pollution discharge rights market. In addition, referring to the mature experiences of developed countries in Europe and the United States, China is addressing the problem of climate change by building a carbon emission-trading platform. In 2013, the Chinese government launched a pilot study on carbon emissions trading in eight regions, including Beijing, Shanghai, and Guangdong. At present, the carbon trading markets in these eight pilot regions are running steadily, and they have accumulated mature experience that can be replicated and promoted for implementing carbon emissions trading at the national level.

3.4.4.4 Ecological Compensation Policy in Environmental Resources

In China, rules and regulations in the field of ecological compensation are constantly improving. In November 2020, China's National Development and Reform Commission took the lead in issuing a draft of *the Regulations on Compensation for Ecological Protection.* In 2020, the General Office of the Central Committee of the Communist Party of China and other departments issued *the Notice on Printing and Distributing the Reform Plan for the Division of Central and Local Fiscal Powers and Expenditure Responsibilities in the Field of Natural Resources.* The central and local governments share compensation and spending responsibilities. On December 18, 2020, China's taxation department issued the *Notice of the Ministry of Finance on Issuing the Budget for Transfer Payments from the Central Government to Local Key Ecological Function Areas in 2020.* The regulations and specific allocation of ecological transfer payments to local governments at all levels totaled approximately 79.5 billion yuan. Among them, Gansu Province received the largest number of transfer payments for ecological function zones, at approximately 6.7 billion yuan. Regarding different uses, the transfer payment for the key subsidy areas was the largest, reaching 62.6 billion yuan. In addition, the Chinese government introduced a series of measures to encourage local governments to actively carry out pilot projects in key national ecological function zones. The regions selected as key ecological function zones can enjoy financial subsidies, tax incentives, and other policy benefits.

3.4.4.5 Value Accounting Policies for Environmental Resources

In recent years, China has been working hard to explore pilot work on ecological product value and has achieved positive results. Lishui City, a representative case, formulated China's first local standard of *Guidelines for Ecological Product Value Accounting* in April 2020, which provided a good template for establishing and improving an ecological product value accounting index system. In addition, Lishui City has been actively improving its policy framework for the value conversion of ecological products. Currently, it has designed several green financial products closely related to the value accounting and evaluation of ecological products, and the balance of green loans has reached 19 billion yuan. China is conducting a grassroots exploration of national ownership of natural resource balance sheets. To effectively improve the preparation system of natural resource balance sheets, the Ministry of Natural Resources of China has successively conducted a trial filling of balance sheets of natural resources owned by the entire population in Jiangsu, Qinghai, Hubei, and other provinces and cities. Applicability of regulations, scientific methods, and the effectiveness of results.

Overall, China is constantly improving and perfecting its environmental resource policy system in the stage of high-quality development. It has achieved a series of positive results in many fields, which play a crucial role in promoting the construction of an ecological civilization and practicing the concept of green and low-carbon development. Important role. However, it should be noted that it is very important to formulate appropriate high-quality environmental policies as China's economy moves towards a high-quality development stage. On the one hand, China's previous high-speed economic growth model has been difficult to maintain. Therefore, technological innovation and industrial transformation is a key problem that needs to be solved urgently, as it will have a broad and far-reaching impact on China's future economic growth prospects. However, China's extensional development model has led to serious resource waste and environmental pollution problems. In the high-quality development stage, solving other problems is particularly difficult. Achieving economic development in the context of tightening environmental resource constraints maximizes output.

3.5 Summary

Focusing on efficiency and equity can guide us in making appropriate private and social choices to protect the environment and conserve resources. According to Hartwick's rule, if the current generation shares all scarce rents obtained from the use of scarce resources with the next generation, the result of the allocation will satisfy the sustainability criterion. However, this is not always feasible in practice. Likewise, configurations that meet sustainability criteria are not always efficient. In summary, the market allocation results may have the following situations: (i) efficient but not meeting the sustainability criteria; (ii) sustainable but not efficient; (iii) neither sustainable nor sustainable; (iv) both sustainable and effective. A win–win situation provides an opportunity to simultaneously increase the welfare of present and future generations at the same time. Specifically, we determine how to induce market allocations to produce allocations that satisfy sustainability criteria. We also examine how government departments can formulate policies to effectively promote sustainable development transformation.

Market failure often diminishes sustainable development prospects. Intergenerational externalities such as climate change impose enormous costs on future generations. The free use of public resources can lead to the overexploitation of resources and even the extinction of species. However, efficient markets do not necessarily lead to sustainable development. Efficiency-enhancing practices are desirable and useful for sustainable development but are not sufficient to constitute a necessary condition for achieving sustainable development. In principle, dynamic allocation can produce a rational allocation framework; however, this is not necessarily the case. Thus, we must be careful when trade is part of a development strategy. Trade's impact on the environment is neither entirely beneficial nor harmful.

The policy instruments to support the market allocation of renewable resources mainly include subsidies for early-stage pioneers by publishing tax reduction and tax exemption classifications and codes, renewable energy quota system (RPS), renewable energy Letter of Credit (REC), and feed-in tariff subsidy.

The efficient allocation characteristics of exhaustible and renewable resources depend on the pre-conditioning settings. If the marginal cost of extraction remains constant, the effective amount of exhaustible resource extraction will continue to decline over time. For a clear property rights structure, the market allocation of exhaustible resources is efficient. However, resource extraction generates external environmental costs, leading to inefficient market allocation.

China is actively improving its environmental and resource policies. Recent environmental and resource policies have played pivotal roles in promoting an ecological civilization and high-quality economic development. There are two key issues in the high-quality development stage: whether the environmental resource policy is conducive to promoting economic transformation and upgrading and whether the environmental resource policy can minimize the cost of transformation and upgrading.

References

Cole, M. A. (2004). Trade, the pollution haven hypothesis and the environmental Kuznets curve: examining the linkages. *Ecological Economics, 48*(1), 71–81.

Copeland, B., & Taylor, M. S. (2004). Trade, Growth, and the Environment. *Journal of Economic Literature, 42*(1), 7–71. https://doi.org/10.1257/.42.1.7.

Dasgupta, P. S., & Heal, G. M. (1979). *Economic theory and exhaustible resources.* Cambridge University Press.

Deacon, R. & Norman, C. (2006). Does the environmental Kuznets curve describe how individual countries behave? *Land Economics, 82*(2), 291–315.

Hartwick, J. M. (1977). Intergenerational Equity and the Investing of Rents from Exhaustible Resources. *The American Economic Review, 67*(5), 972–974.

Howarth, R. B., & Norgaard, R. B. (1990). Intergenerational Resource Rights, Efficiency, and Social Optimality. *Land Economics, 66*(1), 1–11. https://doi.org/10.2307/3146678.

List, J. A., & Gallet, C. A. (1999). The Kuznets Curve: What Happens After the Inverted-U? *Review of Development Economics, 3*(2), 200–206. https://doi.org/10.1111/1467-9361.00061.

Pezzey, J. (1992). Sustainability: An interdisciplinary guide. *Environmental Values.*

Tietenberg, T. (1984). *Environmental & natural resource economics* (Vol. 27, No. 1, pp. 31–52).

Tietenberg, T., & Lynne, L. (2012). *Environment and natural resources economy.* China Renmin University Press.

Torell, L. A., Libbin, J. D., & Miller, M. D. (1990). The market value of water in the Ogallala aquifer. *Land Economics, 66*, 163–175.

Vincent, J. R. (1997). Testing for environmental Kuznets curves within a developing country. *Environment and Development Economics, 2*(4), 417–431. https://doi.org/10.1017/S1355770X97000223.

Weiss, E. B. (1988). In fairness to future generations: International law, common patrimony, and intergenerational equity.

Chapter 4
Carbon Neutrality and Environmental Governance

4.1 Introduction

In 2015, the 21st United Nations Climate Change Conference adopted the Paris Agreement (which came into force in October 2016). The Paris Agreement proposes that, for sustainable global development and in response to the impending global climate catastrophe, global warming should be kept at pre-industrial levels of 2 °C and work toward limiting global temperature rise to 1.5 °C. The primary approach of the Paris Agreement to achieve global climate mitigation is to reduce greenhouse gas emissions (Eggleston et al., 2006). Greenhouse gases refer to all gas components that can cause global temperature rises, such as carbon dioxide (CO_2), methane (CH_4), and hydrofluorocarbons (HFCs). In 2018, the Intergovernmental Panel on Climate Change (IPCC) proposed that achieving the Paris Agreement's greenhouse gas emission targets would require net-zero emissions or carbon neutrality by 2050. Since 2016, many economies worldwide have issued carbon neutrality targets and action plans for the period up to 2050.

In 2020, the global economy was significantly affected by Covid-19. On September 22, 2020, Chinese President Xi announced that China would increase its national contribution, adopt stronger policies and measures, and strive to achieve peak carbon emissions by 2030 and carbon neutrality by 2060. China explicitly aims to achieve peak carbon and carbon neutrality in current global climate governance and economic revitalization under the influence of Covid-19. The peak carbon and carbon neutrality goals provide directions for China to address global climate change and promote green and sustainable development. This lays out an ambitious roadmap and presents powerful strategies. China has contributed to the global response to climate change, global challenges, and sustainable development through wisdom and solutions. China has also taken the initiative to assume international responsibility for climate change.

China takes global climate change and environmental protection seriously. At the Copenhagen Climate Conference, China declared its position on greenhouse gas

X. Deng et al., *Environmental and Natural Resources Economics*,
https://doi.org/10.1007/978-981-99-9923-1_4

emissions. China announced that carbon emissions per unit of GDP would be reduced by 40–45% by 2020 compared to 2005. As a Party to the UNFCCC in 2012, China pledged to reduce its total carbon emissions to less than 10 billion tons between 2016 and 2020. In addition to the peak carbon and carbon neutrality target for 2020, China clarified its carbon emission targets and initiatives at the "Climate Ambition Summit" to commemorate the fifth anniversary of the Paris Agreement. China announced its National Determined Contributions (NDC) Target for 2020. The target plan includes a reduction in carbon emissions per unit of GDP by more than 65% by 2030 relative to 2005. The share of non-fossil energy sources in primary energy consumption is expected to increase by approximately 25%. At the same time, the plan provides specific provisions for forest stock, wind power, and solar energy toward the project goals. Peak carbon refers to the point at which carbon emissions reach their peak, with no further growth and a gradual decline thereafter. Figure 4.1 shows that the platform period is also known as peak carbon. China is currently in the "rising" stage of carbon emissions. According to this plan, China will reach the "platform period" by 2030. Then, China will gradually reduce its carbon emissions and finally reach carbon neutrality.

Carbon neutrality is the second goal of emissions reduction after peak carbon, which means that carbon emissions are equal to carbon absorption. As Fig. 4.2 demonstrates, the carbon emitted by industries, residents, and energy consumption would be equal to the carbon dioxide absorbed by forests and carbon capture. In this context, carbon refers to carbon dioxide in a narrow sense and greenhouse gases in a broader sense. Carbon neutrality requires a combination of carbon emissions reduction, natural means such as forests to absorb carbon emissions, and carbon management technologies such as carbon capture.

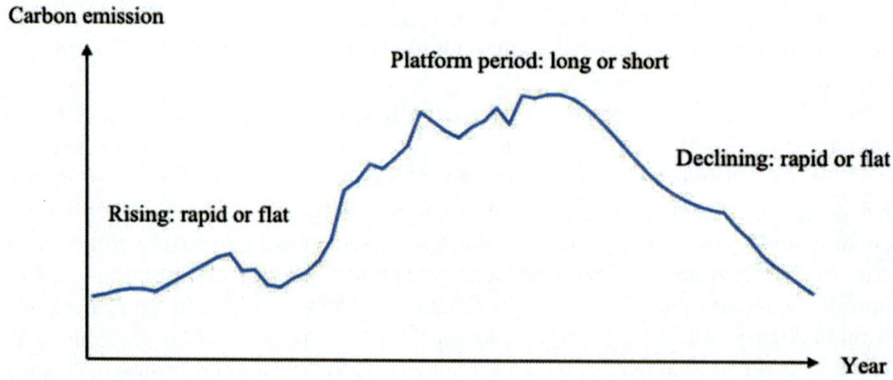

Fig. 4.1 Peak carbon illustration

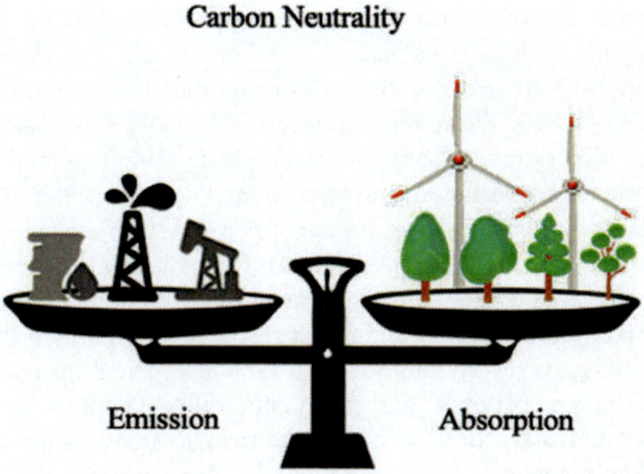

Fig. 4.2 Schematic diagram of carbon neutralization

4.2 Research Progress on Carbon Emissions

4.2.1 Global Carbon Neutrality Trend

4.2.1.1 Carbon Neutrality Has Become a Global Trend

Some economies have already taken the lead in achieving carbon neutrality. Some countries have made carbon neutrality a critical goal for future economic development. They also established an implementation plan for carbon neutrality in the form of legislation. Simultaneously, with the global recognition of green development and sustainable development, an increasing number of countries proposed timetables and roadmaps for achieving carbon neutrality following their economic development. According to the "2020 Status Report" released by the Task Force on Climate-related Financial Disclosures (TCFD) in October 2020, more than 120 countries and regions worldwide are striving to achieve zero greenhouse gas emissions by 2050. Carbon neutrality is therefore a global trend.

According to the final global net-zero emissions data released by the Energy and Climate Intelligence Unit (ECIU), Bhutan and Suriname were the first countries to achieve carbon neutrality. The UK, France, New Zealand, Sweden, and France set carbon neutrality targets by law, while the European Union, South Korea, and Spain set carbon neutrality targets for 2050. They began the legislative process, and Sweden set a carbon–neutral target of 2045, five years ahead of the carbon–neutral targets of other significant countries. Finland, Austria, Italy, and other countries also announced their plans to gradually achieve carbon neutrality, though without providing specific implementation plans as of yet.

To achieve the goal of carbon neutrality, several countries focused on the central theme of green development. These countries implemented policies and measures to reduce carbon emissions and achieve green development. According to a study published in August 2020 by the Organisation for Economic Co-operation and Development (OECD), at least 30 member countries' economic partners in the base period adopted green development as a significant economic transformation or response to COVID-19. These measures can be broadly classified into (i) measures in the form of direct grants, loans, or tax breaks to stimulate a circular economy, clean energy, green transportation, and other areas; (ii) fiscal policies such as financial subsidies for households or enterprises to support renewable energy utilization and fossil energy transformation; (iii) special funds for ecological restoration with the aim of providing more employment opportunities through ecological restoration, and (iv) protection of ecological diversity and forest resources to maintain the overall ecological balance. Energy, ecology, and transportation are critical for countries to achieve the goals of a "green economy" and carbon neutrality. There are still a few green fiscal measures for industry, forestry, and waste management. In terms of policy type, the policies implemented at this stage focus on loans or financial subsidies.

This study analyzes the specific measures implemented by each country to achieve a green economy and carbon neutrality.

4.2.1.2 The Carbon Neutrality Plan in the US

In 1993, the US Climate Change Action Plan set the goal of reducing total greenhouse gas emissions from 1990 to 2000 (Houghton et al., 1990). However, in 2001, the US unilaterally announced its withdrawal from the Kyoto Protocol. Moreover, the US relaxed the timeframe for greenhouse gas growth to 2025. Under the Obama administration, the US set targets and timetables to reduce total greenhouse gas emissions. The 1993 Climate Change Action Program was reintroduced and committed to 2020 greenhouse gas (GHG) emissions parity with the 1990 levels. The US announced its withdrawal from the Paris Agreement in 2017 and return in 2021. The action plan to promote "carbon reduction" and neutrality has been continuously adjusted. The US designated a "3550" plan to promote carbon neutrality. The US "3550" plan divides carbon neutrality into three phases: the first phase, in 2030, will ensure that new sales of light- and medium-duty vehicles reach zero emissions and all commercial buildings customize zero-emission standards; the second step, by 2035, is to achieve carbon neutrality in the power sector and a 50% reduction in the carbon footprint of the building stock; and the third step, by 2050, is to ensure 100% clean energy use and achieve carbon neutrality.

Although the US lacks international responsibility, it has been promoting the reduction of GHG emissions and carbon neutrality since the implementation of the US Clean Air Act in 1963. When the US withdrew from the Kyoto Protocol and the Paris Agreement, its efforts to reduce emissions continued. During this period, the US government promulgated a series of laws, regulations, and policy measures, including the Clean Energy Revolution and Environmental Justice Plan; the Building a Modern, Sustainable Infrastructure and an Equitable Clean Energy Future Plan; and the Executive Order on Responding to the Climate Crisis at Home and Abroad. The Biden administration plans to spend an additional $2 trillion on transportation, buildings, and clean energy. In addition, US state governments have taken more comprehensive actions than federal governments in carbon neutrality. For example, California passed the Global Warming Solutions Act in 2006 to reduce its carbon emissions. In 2006, California passed the Global Warming Solutions Act to reduce carbon emissions by 20% of the 1990 levels by 2050. In 2007, California, Arizona, and Washington State jointly launched the Western Climate Initiative (WCI). The WCI has developed a more mature carbon trading mechanism that includes GHG emissions reduction, allowance trading and auctions, and emission offsets.

Under the influence of COVID-19, the US did not propose an apparent "green recovery" plan. However, the Biden administration identified four pillars of economic recovery. One policy emphasizes responses to the global climate crisis. They are investing in clean energy and sustainable infrastructure, among other things. The Biden administration also promulgated a "Green New Deal," proposing five planned measures for energy transition and environmental protection. The first is the formation of 100% clean energy and zero carbon emissions by 2050, the carbon–neutral goal. The second is to promote smart infrastructure to ensure that the building, water, transportation, and energy sectors can withstand the impacts of the climate crisis. Third, it will return to the Paris Agreement to unite other countries in the fight against climate threats. Fourth, it aims to avoid environmental harm to low-income communities. The fifth goal was to assist residents affected by the energy transition. These five plans revolve around climate change, green development, and carbon neutrality.

4.2.1.3 Carbon Neutrality Action in Europe

In 2021, the European Union (EU) proposed a carbon neutrality program to combat global climate change. This is currently the most ambitious plan among the world's major economies for reducing GHG emissions. The proposal plans to reach 40% of renewable energy consumption in the EU by 2030 and reduce the proportion of new fuel cars registered by 2030 by 55% compared to 2021. By 2035, no new fuel cars will be registered, and by 2026, a carbon border tariff will be established, among other goals. The EU is targeting carbon neutrality in six primary areas. The first is tightening the carbon emissions trading system, proposing to gradually abolish the carbon emissions offset system in 2021–2030 and start reducing carbon allowances. The second is the clean energy development plan, which focuses on hydrogen energy for technological development while simultaneously proposing an offshore renewable energy development plan, including offshore wind power projects. The third is creating green buildings and releasing the "Energy Alliance and Innovation Wave" initiative. One plan proposes to achieve zero energy consumption in buildings by 2030. The fourth is the layout of new energy transportation for smart cars, new energy vehicles, digital transportation, and other critical layout areas. Fifth, the bloc aims to develop a circular economy and tackle industrial pollution. The sixth step involves strengthening carbon reduction in the field of waste disposal.

In 2005, the EU introduced a series of policy measures for climate change and GHG emission reduction. In 2005, the EU established the European Union Emission

Trading Scheme (EU ETS), which sets the total amount of emissions and allocates quotas to 45% of the EU member states' gas emission areas. In 2011, the EU published the "2050 Energy Roadmap," which set the GHG emission target for 2050. The 2019 "European Green New Deal" set active targets for a climate-neutral circular economy. In 2020, in response to the impact of COVID-19 on the EU economy, the European Commission issued a €750 billion Next-Generation EU recovery plan. These 750 billion euros are in the form of "new spending" in the long-term budget of the EU. Together with the 1.074 trillion euro budget previously earmarked by the European Commission, the total EU medium-term budget for 2021–2027 was 1.8 trillion euros. It will be used to help "green talks" and "digital" recovery in epidemic areas. According to a plan announced by the EU, 25% of this funding will be allocated to climate-friendly development. In the latest version of the EU Next-Generation Recovery Plan, this percentage increased again to 30%. This was the highest percentage of the EU budget allocated to the environment and climate. The EU Next-Generation Recovery Plan also includes biodiversity conservation, with a focus on climate change. A total amount of 1.8 trillion euros is spent on climate and digital transformation, as well as supporting research and innovation activities, in addition to the EU's daily business.

In 2020, the German government proposed a 130 billion USD economic recovery plan for 2020–2021 in response to the impact of the "new epidemic." The plan focuses on supporting and subsidizing key areas, such as 5G, taxation, and industry. Fifty billion euros of this plan is classified as a separate "future package" aimed at promoting the development of electric mobility, quantum programs, and artificial intelligence. The focus of support is on the "digital transformation" on the one hand and "climate transformation" on the other. The climate sector includes electric vehicles, hydrogen energy, and buildings. For example, subsidies for electric cars and trains, the construction of charging facilities, and subsidies for new energy vehicles. Low- and zero-emission vehicles are subsidized and supported by vehicle taxes. The German government focused on providing strategic support for hydrogen energy. They are expanding the market for "green hydrogen." Nearly 10 billion euros has been invested in promoting hydrogen energy and international cooperation.

France enacted a 100-billion-euro stimulus plan in 2020 for a "green recovery" in response to the COVID-19 pandemic. Of this, 30 billion euros will be used to support the development of environmentally friendly industrial sectors in France. The first area of focus is the development of hydrogen energy. France invested 2 billion euros to support the hydrogen energy industry and achieve its carbon neutrality goal. Second, in the energy-efficient building sector, 6 billion euros will be invested in upgrading the energy systems of buildings to reduce GHG emissions from the building sector. Third, a "decarbonization" plan for the industrial sector is proposed in the industrial development sector. Two billion euros will be invested in upgrading energy-efficient equipment for industrial enterprises. Fourth, in the area of green infrastructure, 1.2 billion euros will be provided to support renewable energy transportation projects and green infrastructure. The aim is to reduce the carbon emissions of transportation projects and public services.

4.2.1.4 Carbon Neutrality Plans in Other Countries

The UK enacted the Climate Change Act in 2008, making it the first country to legislate a medium-to long-term carbon reduction target. It was also the first country in the world to embark on the quest for a carbon neutrality target. The British Standards Institution (BSI) issued a carbon neutrality specification in 2010. In 2019, the UK enacted the Climate Change Act, formally setting a net-zero carbon–neutral emission target for 2050. Green Resilience" plan. The plan focuses on the following investment areas. The first is the energy efficiency retrofitting of housing. The UK invested more than £2 billion in this project. This project is estimated to provide more than 100,000 green jobs. Second, the public sector "decarbonization plan–clean growth strategy." Investments in the project exceeded £1 billion. The money invested is used to upgrade energy efficiency and low-carbon heat in the public sector. Third, the Green Job Fund was developed, with over £40 million invested in funding environmental charities. The fourth is the development of air capture technology. This new clean technology is designed to capture carbon dioxide in the air directly. Fifth, it supports the development of new energy vehicle technologies. The project focuses on funding batteries, engines, and fuel cells to promote the development of new energy vehicles.

In 2020, the South Korean government enacted a new policy, totaling 160 trillion KRW (approximately 1,300 billion USD). This new policy focuses on supporting the "Green New Deal." It is expected to invest 42.7 trillion KRW (approximately $35 billion USD) in renewable energy and low-carbon infrastructure. Specifically, it focuses on new renewable energy sources such as solar, wind, and hydrogen; the construction of green infrastructure, including charging stations and hydrogen recharge tanks, and so on. Russia officially joined the Paris Agreement in 2019. It plans to establish a carbon trading system and commits to achieve its carbon neutrality goal by 2025. In 2019, the Japanese government proposed to reduce GHG emissions by 80% by 2050. In 2020, the Kan administration announced that Japan would achieve carbon neutrality by 2050. This is the first time that Japan has explicitly stated its goal of carbon neutrality. The Japanese government has pledged to designate corresponding policy measures as soon as possible.

4.2.2 Carbon Emissions and the Significance of Achieving Carbon Neutrality in China

4.2.2.1 Carbon Emissions

Figure 4.3 shows the total carbon emissions of major countries from 2015 to 2019. China's total carbon emissions ranked first in the world and showed a continuing increasing trend. According to BP (2020), China's total carbon emissions reached 9,186 Mt in 2015. In 2019, the total carbon emissions grew to 9,826 Mt. In the US,

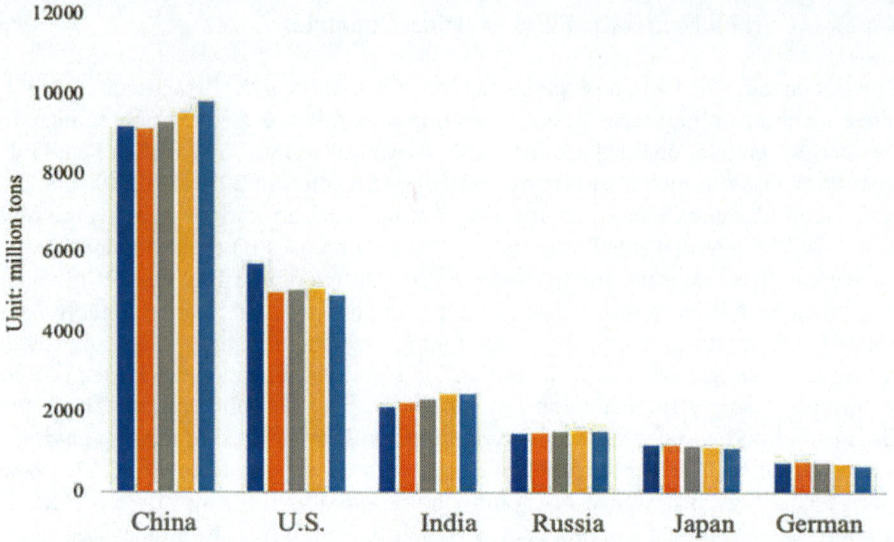

Fig. 4.3 Total carbon emissions of major countries, 2015–2019. *Note* China's carbon data are based on raw coal, BP Statistical Review of World Energy, June 2020

carbon emissions reached 5,741 Mt in 2015. However, since then, carbon emissions have decreased. India, a developing country with a large population, emitted 2,149 Mt of carbon in 2015. While showing a continued increase, India's total carbon emissions reached 2,480 megatons in 2019. Total carbon emissions are third in the world, much lower than those in China. This puts more significant pressure on China's commitment to reach its carbon peak by 2030 and carbon neutrality by 2060 to achieve this goal.

The reason for high energy consumption in China, especially coal resources, is the demand for energy for social and economic development. Another significant reason is that thermal power constitutes a large proportion of the power supply system. Figure 4.4 shows the carbon emissions of each province, region, and city in China for 2000, 2010, and 2019. According to the development trend of carbon emissions in each province, region, and city in these three years, Shandong has had the highest level of carbon emissions in China since 2010. The eastern region includes three of the five provinces with the highest total carbon emissions in China: Shandong, Hebei, and Jiangsu. The central region comprises two provinces: Inner Mongolia and Shanxi. The five provinces with the lowest total carbon emissions in 2019 were Hainan, Qinghai, Beijing, Gansu, and Chongqing. Therefore, spatially, carbon emissions show a pattern of being high in the east and low in the west.

From the carbon emissions trend in Fig. 4.4, most provinces showed a clear upward trend from 2000 to 2019 to. Even Hainan and Qinghai provinces, which have minor total carbon emissions, show an increasing linear trend. Shandong Province, Xinjiang, and Hebei exhibit the most apparent growth trends in carbon emissions. Beijing and Shanghai show more pronounced increases in total carbon emissions

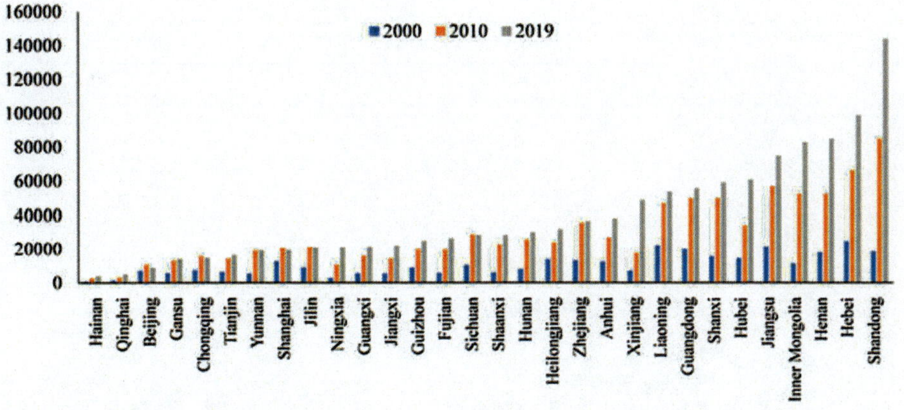

Fig. 4.4 Carbon emissions in China

from 2000 to 2010. In 2019, there was a considerable decrease in the total carbon emissions in these two regions compared to 2010. Jilin exhibits a similar trend. However, the total carbon emissions from Jilin in 2019 are the same as in 2020. There is a significant difference in the volume of total carbon emissions between provinces and municipalities. Moreover, with the development of the national economy and society in each province and city, the gap in the total carbon emissions continues to increase. The carbon emissions of all provinces show the development trend of "high emission and high growth, low emission and low growth."

According to Fig. 4.5, Shandong Province had the highest total carbon emissions in 2019, which were much higher than those of other provinces and cities. Furthermore, the per-capita carbon emission level of Shandong Province was higher than that of most provinces and cities. Hence, the cost of reaching the carbon neutrality goal in Shandong Province is much higher than that in other provinces. Hebei Province is second only to Shandong Province in terms of total carbon emissions, and its per-capita carbon emissions are the same as those of Shandong Province. Therefore, Hebei Province also has significant carbon emissions. According to the calculations, the carbon emissions of Shandong Province and Hebei Province are 143,586,400 tons and 985,374,500 tons, respectively, in 2019. The total carbon emissions of the two provinces accounted for 21.80% of total carbon emissions.

Figure 4.5 shows that Inner Mongolia and Jiangsu are the second-largest carbon emission provinces after Shandong and Hebei. However, Jiangsu Province has lower per-capita and total carbon emissions than Hebei and Shandong. Moreover, Inner Mongolia has the c highest carbon emissions per capita, at 32.55 tons per person. The next highest is in Ningxia, with per-capita carbon emission of 30.05 tons per person. The total carbon emissions of the Xinjiang region were lower than those of Hedong, Hebei, Inner Mongolia and Jiangxi. However, Xinjiang's per-capita carbon emissions ranked third in China, indicating that this region is also under more significant pressure to reduce carbon emissions. In terms of per-capita carbon emissions, Inner Mongolia, Ningxia, and Xinjiang are all located in Northwest China with a

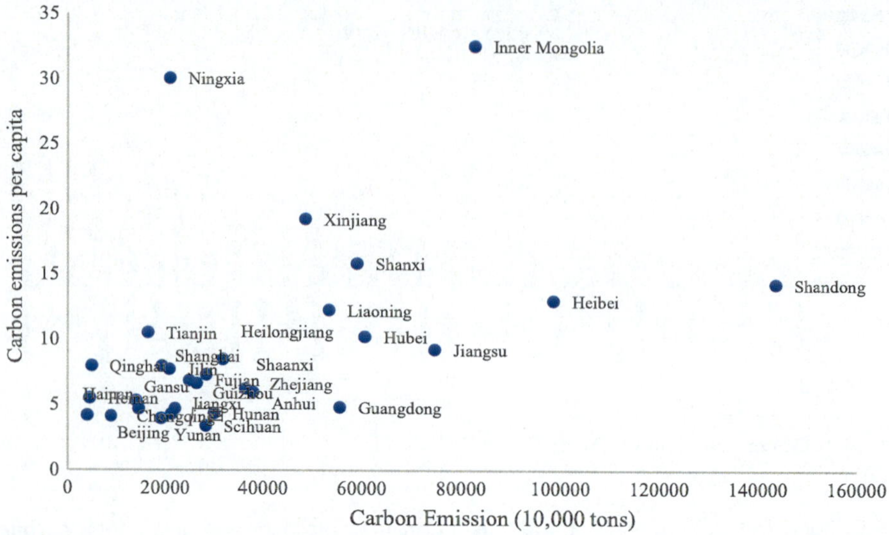

Fig. 4.5 Total carbon emissions and per-capita carbon emissions in China, 2019

relatively sparse population. However, per-capita carbon emissions are among the highest in China, indicating that industrial development in the western region of China, especially in the Northwest region, caused serious carbon emission problems in recent years. The ecological environment in the Northwest is fragile, and restoring ecological damage is becoming more expensive and complicated. Therefore, there is a need to strengthen ecological governance in the western regions. The total carbon emissions of Shanxi, Liaoning, and Hubei are at the same level, and the difference in per-capita carbon emissions is not significant. Although Guangdong Province has the same total carbon emissions as these three provinces, its per-capita carbon emissions are much lower.

Figure 4.6 shows the energy and emissions intensities for each province, region, and city in China in 2019. The emissions and energy intensities of Ningxia, Xinjiang, Shaanxi, and Inner Mongolia show a clear trend of "double high." The economic development levels of these provinces does not have a significant advantage in China. However, their total carbon emissions and energy consumption have been increasing. Higher carbon emissions and energy intensities are observed in these regions. The energy intensity of the Qinghai Province is also higher than that of most regions in China. This is because the total economic volume of Qinghai Province is similar to those of Ningxia and Xinjiang. Qinghai is located in inland Northwest China and has a low level of economic development. However, the rapid growth of the economy and industrialization level in recent years has increased total energy consumption. Therefore, the energy intensity is high. However, Qinghai belongs to China's important protected area of the "Three Rivers Source." Therefore, the ecological environment is protected to a high degree and total carbon emissions are controlled such that the emission intensity is not too high in relative terms.

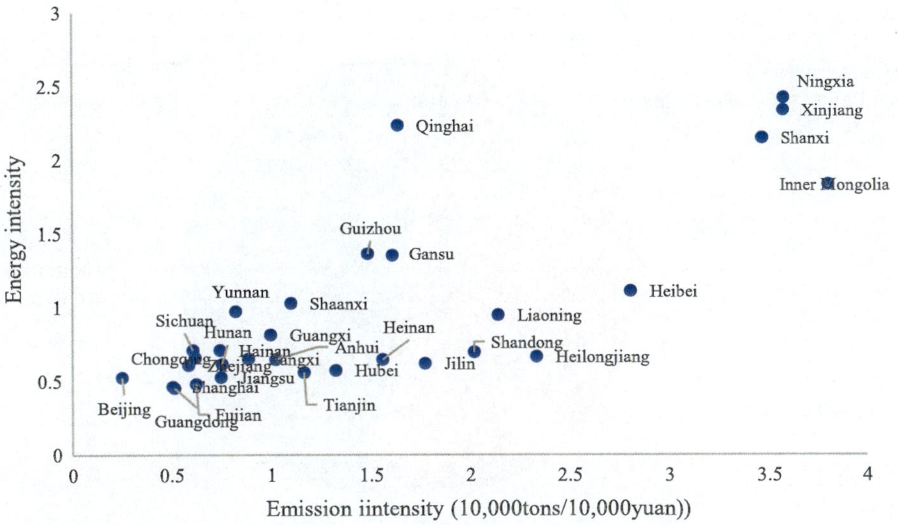

Fig. 4.6 Energy and emissions intensity in China, 2019

Reviewing Fig. 4.6 also reveals that the emission intensity is higher in Hebei Province, given its total carbon emissions Shandong Province, which has higher total carbon emissions than Hebei Province, has a much lower emission intensity. This means that, for the same carbon emissions, Hebei Province has a weaker ability to convert its economy than Shandong Province. The energy and emissions intensities of Liaoning, Heilongjiang, and Jilin provinces are similar. Liaoning Province has a higher energy intensity than Heilongjiang and Jilin provinces. Jilin province has a lower emission intensity, which means that Northeast China has similar characteristics in terms of economic development level and carbon emissions. These can be cross-referenced when formulating carbon emission reduction policies.

Figure 4.7 shows China's primary carbon emission sources in 2018. The most important source of carbon emissions in China is electric heat production, which accounts for 46.61% of the total carbon emissions. Electricity and heat are primarily used in power generation and heating. Thermal power accounts for most of China's current power generation. Industry and transportation are second only to electric heat production, accounting for 37.29% and 7.68% of all carbon emissions, respectively. In contrast, the residential, construction, agricultural, and service sectors accounted for a smaller share of total carbon emissions. Therefore, strengthening carbon emissions from electric heat production and industrial sources is the best way to achieve peak carbon and carbon neutrality in China.

According to the statistics released by the National Bureau of Statistics of China (Table 4.1), the total power generation in China in 2019 was 7,503.43 billion kWh, of which thermal power accounted for 69.6% of the total power supply. Compared to the growth rate of 4.7% in 2018, thermal power was still the primary source of electricity. Although the total amount and proportion of hydropower, nuclear power,

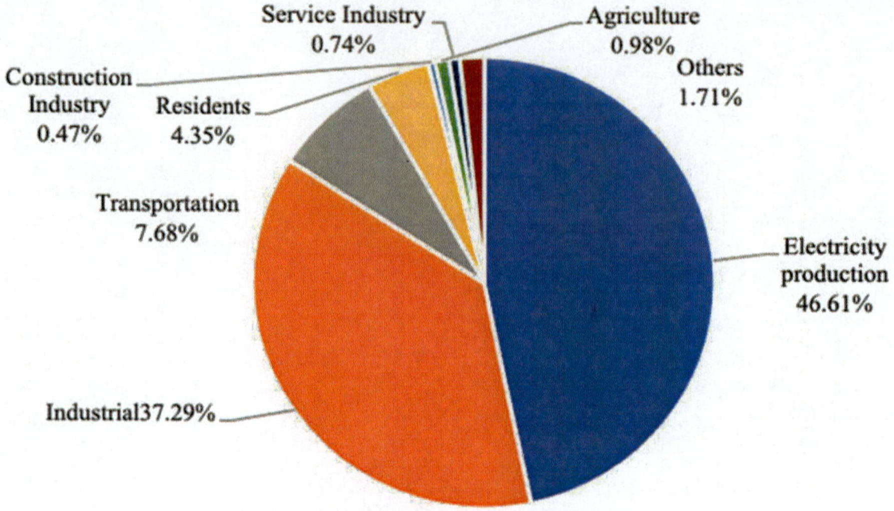

Fig. 4.7 Main sources of carbon emissions in China, 2018

wind power, and solar power generation are increasing, they still cannot mitigate the environmental pollution and energy consumption problems caused by thermal power. In China, photovoltaic and biomass power generation were 224.3 billion kWh and 111.1 billion kWh, respectively, in 2019, and the proportion of total power generation is still low. Fossil energy, especially coal and oil, is the primary source of thermal power in China, making it difficult to replace the consumption of coal and oil and solve environmental problems in the short term when meeting the total demand for electricity in society.

4.2.2.2 The Necessity of Carbon Neutrality

First, global climate change has become severe. Currently, the mainstream academic community generally believes that GHG emissions cause significant climate change. Global climate change has serious consequences. For example, rising sea levels, threatened biodiversity, and frequent extreme weather have had more adverse effects on global human society. With the rise in global temperatures, extreme climate phenomena have seriously threatened human survival and development. Forest fires, droughts, heavy rain, and typhoons are associated with global climate change. According to a study by the United Nations Intergovernmental Panel on Climate Change (Intergovernmental Panel on Climate Change, IPCC), if the goal of carbon neutrality is not achieved by the end of the twenty-first century, the global temperature will escalate beyond the ecological red line of 1.5–2 °C, leading to a human survival crisis. This implies that carbon neutrality is an action that humans must take for sustainable development. Currently, the international community is actively

Table 4.1 National electricity production, China, 2009–2019

Year	Thermal energy		Hydroelectricity		Nuclear energy		Wind energy		Solar energy	
	Electricity generation	Percentage	Electricity generation	Percentage	Electricity generation	Percentage	Electricity generation	Percentage	Electricity generation	Percentage
2010	33,319	79.2	7222	17.2	739	1.8	446	1.1	1	0.0
2011	38,337	81.3	6990	14.8	864	1.8	703	1.5	6	0.0
2012	38,928	78.1	8721	17.5	974	2.0	960	1.9	36	0.1
2013	42,470	78.2	9203	16.9	1116	2.1	1412	2.6	84	0.2
2014	42,687	75.6	10,643	18.8	1325	2.3	1561	2.8	235	0.4
2015	42,842	73.3	11,303	19.4	1708	2.9	1858	3.2	385	0.7
2016	44,371	72.2	11,934	19.4	2133	3.5	2371	3.9	662	1.1
2017	46,627	71.8	11,898	18.3	2481	3.8	2950	4.5	967	1.5
2018	50,769	71.0	12,342	17	2944	4.0	3660	5.0	1775	3.0
2019	52,201.5	69.6	13,044.4	17.4	3483.5	4.6	4057	5.4	2243	3.0

cooperating to launch global climate governance. Examples include the international promulgation of the United Nations Framework Convention on Climate Change, the Kyoto Protocol, the Paris Agreement, and other relevant international guidelines. However, it still has not reversed the trend of global climate deterioration and ecological crisis. As a responsible power, China takes responsibility for global climate change, with the goal of achieving carbon neutrality. This is important for China's development and the fate of humanity.

Second, carbon neutrality has become a tool in international political competition. Unilateralism and trade protectionism are on the rise, and some countries use climate issues and carbon neutrality targets to establish carbon trade barriers. Some countries hope to implement exclusionary policies in the field of carbon emission reduction and undermine global trade and integration processes. Developed countries have set carbon thresholds for developing countries, which can be used to suppress the growth of developing countries and lead to a loss of their financing. As the world's largest developing country, China is actively pursuing carbon neutrality. China is undergoing economic transformation, which could effectively promote global equity. Moreover, it plays an excellent leading role in the international community and promotes effective and orderly international governance.

Finally, the goal of carbon neutrality reflects China's wisdom and its role as a great power. Climate change affects the fate of humanity and the development of future generations. Addressing climate change is the joint responsibility of all humanities. Western countries had a head-start in development and emitted significant GHG since the Industrial Revolution. Current GHG emissions in some countries are low. However, historically, accumulated GHG and per-capita emissions are much higher than those of developing countries. Carbon emissions are the right of all the countries. National actions to reduce carbon emissions should be considered in the context of history, the present, and the future. However, some developed countries use this technology to suppress the growth of developing countries. Thus, carbon neutrality is a consequence of global political games. China's economy is undergoing a transition from "high-speed growth" to "high-quality development." China is facing the dual pressure of economic development and resource protection. However, China is still implementing its global carbon neutrality target. This demonstrates China's ability and determination to assume global responsibility. China does not look at the problem from a historical perspective but from a developmental perspective. This is an excellent example for both developed and developing countries.

4.2.2.3 The Strategic Significance of Carbon Neutrality

First, carbon neutrality aligns with current national security and sustainable development strategic goals. From the perspective of the international environment, China's status continuously improved after the 2008 financial crisis. However, developed capital-dominated countries continue to advocate the China threat theory, which is increasing geopolitical risks. China is highly dependent on oil and other energy sources. According to the 2019 Domestic and Foreign Oil and Gas Industry Development Report released by the China National Petroleum Corporation, China's current dependence on foreign oil exceeds 70%. External dependence on energy has led to instability in the energy supply, endangering energy and national security (Zhao et al., 2022). The goal of carbon neutrality changed many countries' energy utilization structure. Fossil energy consumption, which can effectively promote renewable energy development, is the primary source of GHG emissions. Second, from a domestic perspective, the goal of carbon neutrality is to optimize the industrial structure, force its transformation of the industrial structure, and prevent economic growth from being locked in an extensive economic growth path (Liu et al., 2022). It effectively improves the efficiency of energy utilization and enhances its position in the international division of labor and the global value chain, thereby promoting the transformation of economic growth into a better direction, by overtaking the curve and heading toward high-quality economic growth to improve people's lives. The contradiction between the needs of life and unbalanced and inadequate development.

Second, carbon neutrality aims to disrupt international unilateralism. Economic development is currently facing a superposition of internal contradictions, especially in the current trade war between China and the US. Some countries led by the US suppressed foreign trade and launched trade wars against China due to climate change issues. Carbon neutrality is a solution for US-led countries pursuing unilateralism in the field of climate change. Through international cooperation in the field of climate change, an effective international communication mechanism can be established to prevent trade of carbon emissions. At the same time, the carbon trading mechanism is also an essential part of the national trading mechanism. The impact of unilateralism, such as carbon tariffs, on imports and exports can be effectively avoided through the carbon market.

Finally, carbon neutrality could effectively promote RMB internationalization, which is indispensable is the process of China's rise. It can effectively improve the ability to resist global financial risks and is a manifestation of comprehensive national strength. To achieve carbon neutrality, it is necessary to rely on the carbon market to establish a sound domestic market and actively participate in the international market. In the carbon market construction and trading process, relying on the RMB for settlement can bypass US hegemony, facilitate international opening, and promote the international development of the RMB.

4.3 Peak Carbon and Carbon Neutrality

4.3.1 Drivers of Carbon Emissions

4.3.1.1 Driving Factor Measurement Model

To analyze the carbon–neutral path, a regression model of the factors influencing carbon emissions was established. Owing to the disparity in the total populations of different regions under the same emission level, the total carbon emissions in regions with concentrated populations may be higher. Per-capita carbon emissions were used as the explained variable (Pco2) to eliminate this factor. To eliminate the possible endogenous row problem caused by missing variables, panel data are used for the analysis, and the basic model is

$$Pco2_{it} = \beta_0 + \beta_j Contral_{it} + \mu_{it} + \varepsilon_{it}, \tag{4.1}$$

where *i* indicates one of the 30 provinces, and *t* is the time dimension. To describe the time variable of carbon emissions more accurately, the sample interval was 2000–2019. β_0 is the intercept term, and β_j is the coefficient to be estimated to determine the direction and magnitude of the impact of the corresponding variables on carbon emissions. μ_{it} represents the fixed effect, which actually means that we control for such unobservable, time-invariant factors, such as the unique cultural characteristics of a region, geographic location, and other time-invariant individual heterogeneity. ε_{it} is a random disturbance term. Because of the requirement for model simplification, some factors are not added to the model, but will have an impact on the model. Such variables are left out here due to model limitations or because the factors cannot be quantitatively measured currently.

Because carbon emissions are path-dependent, the regions where the total carbon emissions of the previous year were paid may also have more significant total carbon emissions this year because carbon emissions reduction or increase is not achieved overnight. Therefore, the model was extended to the dynamics by adding a lag period to the explained variables. Simultaneously, to overcome the model's possible endogeneity, the generalized method of moments (GMM) is used for estimation. The basic form of the model is

$$Pco2_{it} = \beta_0 + \alpha L.Pco2_{it} + \beta_j Contral_{it} + \mu_{it} + \varepsilon_{it}, \tag{4.2}$$

where $L.Pco2_{it}$ represents the variable whose $Pco2_{it}$ lags by one period, α is its coefficient to be estimated, and the meanings of other variables remain unchanged. The following control variables in this study were mainly considered from among the potential factors affecting carbon emissions.

Total Population (number of individuals). Areas with concentrated populations require greater production and better living facilities. Population agglomeration brings about the growth of economic aggregates and negative impacts such as carbon emissions. This variable is expressed as the total resident population of each region. Foreign Direct Investment (FDI). Foreign direct investment plays an essential role in economic growth in the early stages of reform, opening-up, and solving the problem of lack of capital. However, despite continuous economic development, the impact of foreign investment on carbon emissions has always been controversial. However, foreign direct investment promotes industrial agglomeration in China, encourages the development of traditional industrial sectors to industrialized sectors, and brings in many industries with high pollution and high energy consumption. It is dominated by labor- and resource-intensive industries, which have higher environmental pollution and carbon emissions. On the other hand, foreign direct investment has a technological spillover effect that can bring about advanced production technology and management levels, and objectively promote the reduction of total carbon emissions. Therefore, discussing the impact of foreign investment on carbon emissions is important when introducing foreign investment. This variable is expressed as the proportion of total foreign direct investment in GDP, in which the foreign direct investment denominated in US dollars is converted according to the average price of RMB against USD in the current year.

Industrial structure (IS). According to the carbon emission sources in Fig. 4.7, the industry contributed 37.29% of the total carbon emissions in 2018; therefore, the industrial structure occupies an important position in the carbon emission process. Unreasonably high-polluting industries in some areas of China and the agglomeration of high-pollution and high-energy-consuming industries are fundamental reasons for the high total carbon emissions. The development level of the industrial structure is represented by the proportion of the secondary industry's added value in the GDP to measure the secondary industry's contribution to carbon emissions.

Economic development level (GDP). Economic development is inseparable from the input of various resources and factors; in particular, the current economic structure, with fossil energy as the primary energy source, determines the close relationship between the level of regional economic development and carbon emissions. Higher levels of economic development imply greater energy consumption. Industrial agglomerations, population agglomerations, and energy consumption. The regional economic development level was expressed as per-capita GDP.

Urbanization level (URBAN). Urbanization, industrialization, and population agglomeration often occur simultaneously. On the one hand, a high level of urbanization means the agglomeration of a large number of people and industrial sectors, which increases total carbon emissions. Maintaining the operation of the city requires a large amount of electricity and heat, and thermal power production accounts for 46.61% of the total carbon emissions (Fig. 4.7). However, urban construction produces high carbon emissions. According to Fig. 4.7, the construction industry contributes 0.47% of carbon emissions, and cities also contain transportation and service industries, which also produce a large amount of carbon emissions. This variable is expressed as the ratio of the urban resident population to the total regional population.

Energy Intensity (EI). Energy intensity determines energy efficiency. Generally, a higher energy intensity implies a higher energy conversion rate, which reduces energy consumption. However, in some cases, higher energy intensity may also lead to an increase in total energy consumption, and carbon emissions from reduced energy intensity cannot offset carbon emissions from increased energy consumption, thereby increasing carbon emissions. This variable is expressed as the ratio of total energy use to GDP.

Technical level (RD). While the technical level can improve energy intensity and efficiency and ensure the reduction of energy input under a certain economic level, technical efficiency can affect carbon emissions in terms of pollution reduction. The development of polluting equipment and carbon capture technologies has resulted in reduced carbon emissions. However, technical level and energy intensity face the same problem. The performance is the "Jevons paradox" or the "rebound effect" of energy, meaning that increasing the technical level improves the efficiency of energy utilization and the extraction of energy. Moreover, the efficiency of transportation means that more fossil energy can be exploited and utilized, so the technological level not only reduces carbon emissions but also increases carbon emissions. This variable is expressed as the proportion of R&D expenditure in each region as a proportion of the GDP.

The sample period of this study is 2000–2019, and the data sources were the 2001–2020 Annual Review of China Statistics, EPS database (https://www.epsnet.com.cn/), and annual statistical inspections and bulletins of various regions. Non-percentage data were logarithmized to eliminate heteroscedasticity. The value index deflates variables involving prices, with 2000 as the base period. GDP and per-capita GDP are real GDP after the price index is equalized. The descriptive statistics of the variables are presented in Table 4.2.

Table 4.2 Descriptive statistics

Variable	Symbol	Observation	Average	Standard	Min	Max
Carbon emissions per capita	People	600	6.5705	5.0016	0.2678	33.5376
Population	FDI	600	8.1624	0.7548	6.248	9.3519
Foreign direct investment	IS	600	0.1109	0.2999	0.0000	4.3422
Industry structure	PGDP	600	45.1085	8.4094	15.134	61.5000
Economic development level	RD	600	10.2343	0.8364	7.9707	12.013
Technology level	URBAN	600	1.1148	1.0295	0.0162	6.2799
Urbanization	RI	600	1.1528	1.6852	0.0233	13.3035
Energy intensity	People	600	1.5692	0.9878	0.4540	5.7508

4.3.1.2 Analysis of Driving Factors

Table 4.3 provides the results of the fixed effect (FE), random effect (RE), and GMM regressions. Static panels can be analyzed using mixed, fixed, and random effects, and the Hausman test is used to select the estimation method. The Hausman test value in Table 4.3 is 21.41, and the corresponding *p*-value is 0.003, which is less than the critical value of 0.01; that is, the Hausman test rejects the null hypothesis at the 1% level, so the fixed effects model is better than the random- and mixed-effects models.

Although fixed effects estimation is better than random and mixed-effects estimations, it cannot solve the endogeneity problem and cannot effectively estimate Formula (4.2), which has lag terms. Therefore, based on FE, GMM estimation is introduced to estimate the dynamic panel model. According to the GMM estimation test, AR (1) is significant at the 1% level, but AR (2) is not significant, indicating that the model has first-order serial autocorrelation, but no second-order serial autocorrelation, and the lag term of the explained variable added at this time is reasonable. Static panels without lags and dynamic panels with second-order and higher lags were incorrect. The Hansen test value was 25.51, and the corresponding *p*-value was 0.379, indicating that there was no weak instrumental variable or over-identification problem with the lag term of the variable as the instrumental variable; that is, the GMM estimation result was robust and credible. Therefore, the GMM is the benchmark model analyzed in this study.

According to the GMM estimation results in Table 4.3, the coefficient value of $L.PCO_2$ is 0.5184, which is significant at the level of 1%, which means that carbon emissions are "path-dependent," and the total amount of carbon emissions in the previous period was relatively large. This is consistent with the growing trend in overall carbon emissions in China. The population coefficient is 1.4019 and is significant at the 1% level, indicating that total population growth has led to an increase in total carbon emissions. Every one percentage-point increase in population increases or decreases carbon emissions by 140.19 points. Population thus has a more significant impact on carbon emissions.

The coefficient of foreign direct investment is −2.0316, which passes the 1% significance test, indicating that every percentage point increase in foreign direct investment can reduce carbon emissions by 203.16 percentage points. This shows that current foreign direct investment positively affects China's carbon emissions reduction. Foreign direct investment has brought about advanced production technology and management experience and produces a technology spillover effect. Economic development is also significantly positive at the 1% level, indicating a synergistic effect between economic growth and carbon emissions, and that economic growth is at the expense of carbon emissions. Therefore, dealing with the relationship between carbon emissions and economic growth is an urgent issue for China. This result also shows that China's current economic growth overemphasizes the growth rate while ignoring the quality of economic growth. With the transformation of China's economy, achieving high-quality growth and synergy between economic growth

Table 4.3 Factors influencing carbon emissions

	(1)	(2)	(3)
Variable	FE	RE	GMM
L.PCO_2			0.5184*** (9.84)
People	1.0467 (0.48)	−0.2752 (−0.61)	1.4019*** (4.92)
FDI	−0.3108 (−0.64)	−0.4342 (−0.86)	−2.0316*** (−7.88)
IS	−0.0221 (−0.82)	−0.0272 (−1.27)	0.0358** (2.51)
PGDP	2.9018*** (10.40)	3.8153*** (18.22)	1.6451*** (18.02)
RD	0.1851 (0.98)	−0.2205 (−1.29)	0.4983*** (8.14)
URBAN	0.3501*** (3.21)	0.3207*** (2.95)	1.1319*** (11.12)
EI	0.1301 (0.27)	2.1174*** (6.64)	1.3900*** (4.76)
Constant	−31.4533* (−1.85)	−32.4009*** (−6.69)	−27.2648*** (−13.58)
Observations	600	600	570
R-squared	0.431		
Number of ID	30	30	30
Hausman test		21.41***	
P-value		(0.003)	
AR(1)			−2.81***
P-value			(0.005)
AR(2)			1.11
P-value			(0.268)
Hansen test			25.51
P-value			(0.379)

Note ***, **, and * indicate significance at the 1%, 5%, and 10% levels, respectively

and carbon emission reduction is necessary to achieve the peak carbon and carbon neutrality goals.

The coefficients of technology level (RD) and energy intensity (EI) are significantly positive, indicating that improvements in technology level and energy intensity led to the Jevons paradox. There has been a large-scale increase in the amount of energy consumption, and the carbon emissions brought about by the improvement in the technological level, and the reduction in carbon emissions cannot offset the carbon emissions from the increase in energy consumption. The coefficient of the urbanization level (URBAN) is 1.1319, which is significant at the 1% level, indicating that urbanization is also an essential factor in the growth of carbon emissions. Urban development led to population growth and industrial agglomeration. Population and industrial agglomerations require significant energy consumption. Maintaining urban production and living requires considerable electricity and heat, thereby increasing the total carbon emissions.

4.3.1.3 Contribution of Driving Factors

According to the results in Table 4.3, in addition to foreign direct investment (FDI), which can significantly reduce carbon emissions, population (People), industrial structure (IS), economic level (PGDP), technology level (RD), urbanization (URBAN), and energy intensity (EI) can also significantly increase carbon emissions. However, the regression model cannot determine which factor is the leading cause of carbon emission growth, because the coefficients of the regression model are not comparable. To analyze this problem and reduce carbon emissions in a targeted manner, model (4.1) was examined using a Relative Importance (RI) analysis. Because the lag term of carbon emissions ($L.PC0_2$) is used to study the path dependence of carbon emissions, and because the historical level of carbon emissions cannot be managed effectively, the lag term of carbon emissions is excluded from the relative importance analysis. The basic idea of relative importance is to standardize the variables and calculate each variable's elasticity according to its influence on the coefficient of determination, R^2. The estimated results are presented in Table 4.4.

The relative importance of the variables was analyzed and ranked in Table 4.4. The relative importance of the economic development level (PGDP) is 0.3000, making it the most important driving factor for carbon emissions. This is because the economic growth model is extensive and relies on high investments to maintain economic growth, which is not sustainable.

Therefore, it is necessary to transform the economic growth model to achieve high-quality growth in which economic growth is decoupled from carbon emissions. The coefficient of energy intensity (EI) ranks second only to the level of economic development (PGDP), indicating that energy intensity still requires improvement.

Table 4.4 Relative importance of factors influencing carbon emissions

Variable	Significance	Significance of standardization	Rank
People	0.0552	0.0979	3
FDI	0.0275	0.0487	4
IS	0.0093	0.0164	7
PGDP	0.3000	0.5318	1
RD	0.0128	0.0228	6
URBAN	0.0169	0.0299	5
EI	0.1425	0.2526	2

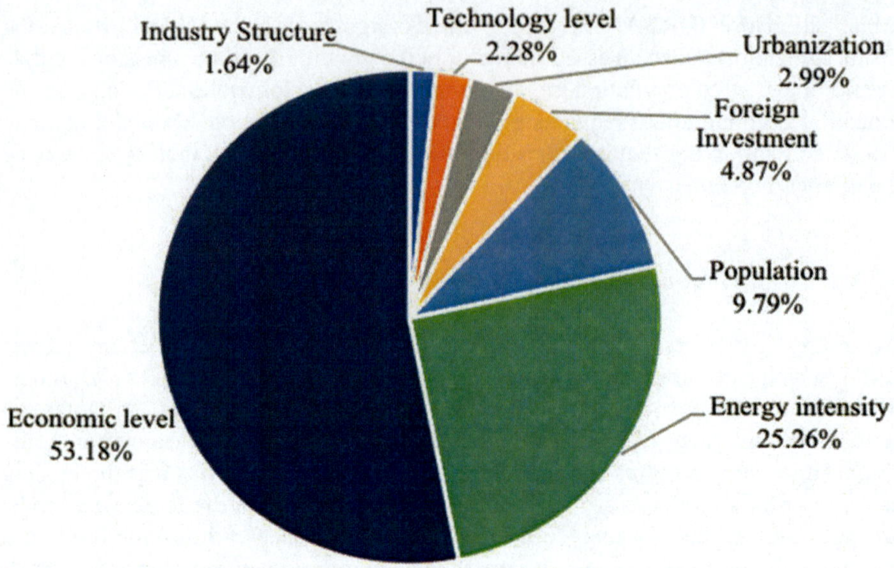

Fig. 4.8 Contribution of factors influencing carbon emissions

Fossil fuels, especially coal, are the primary source of energy in China. This is another major cause of emissions. Population growth is the third leading cause of carbon emissions. China has a large population; therefore, its total carbon emissions are among the highest in the world. However, the per-capita carbon emission level is still lower than those of developed countries. Some developed countries use this to suppress the economy and is one of the reasons China proposes to reach peak carbon and carbon neutrality.

In Fig. 4.8, after standardizing the washing of each driving factor, the relative contribution of each variable is obtained. The level of economic development has the highest contribution to carbon emissions, accounting for 53.18% of the total carbon emissions, followed by energy intensity contributed 25.26%; together, they account for 78.44% of the causes of carbon emissions. Therefore, to achieve peak carbon and carbon neutrality, governance should be carried out from economic growth and energy intensity.

4.3.2 Prediction of Carbon Emissions

4.3.2.1 Prediction Model

Predicting carbon emissions is an essential part of academic research. Predicting carbon emissions and seeking to achieve peak carbon and carbon neutrality requires a scientific approach. This study used a support vector regression (SVR) to predict carbon emissions using the latest machine learning methods. An SVR is a branch of support vector machine (SVM). SVM itself has only a two-class classification model, similar to the logistic model. The basic idea is to linearly classify the maximum interval of the model in the feature space and classify the space as a learning method for the learning strategy. SVM can transform the data and model into a convex quadratic programming problem. SVR inherits the spatial processing method of SVM, using the maximum spatial interval as the optimization strategy, and adding constraints, so that the model finds a bar area instead of a line for classifying data. The constraints are unique to SVR, enabling it to learn and predict carbon emissions repeatedly based on univariate or multivariate data. SVR uses the kernel function of SVM to adapt to different linear or nonlinear relationships. Compared with ordinary nonlinear regression models, SVR cannot calculate the loss within the calculated data isolation band and only calculates when the interval between the data is a substantial loss; it optimizes the model by maximizing the width of the interval and minimizing the total loss. Compared with a linear regression, which calculates the loss when the observed and actual values are not equal, the fitting effect of the problem and the prediction accuracy are better.

Linear regression equation for space:

$$F(x_i) = w^T \cdot \varphi(x_i) + g. \tag{4.3}$$

The SVR problem is formalized as:

$$\min L = \min_{w,b} \frac{1}{2}\|w\|^2 + C\sum_{i=1}^{m} \Gamma_\epsilon\big(f(x_i) - y_i\big), \tag{4.4}$$

where C is the normalization constant and Γ_ϵ is the ϵ-insensitive loss function; the slack variable is introduced:

$$\min L = \min_{w,b,\xi_I,\widehat{\xi_i}} \frac{1}{2}\|w\|^2 + C\sum_{i=1}^{m}\left(\xi_i + \widehat{\xi_i}\right) \tag{4.5}$$

$$\text{s.t}\begin{cases} f(x_i) - y \le \epsilon + \xi_i \\ y - f(x_i) \le \epsilon + \widehat{\xi_i} \\ \xi_i \ge 0, \widehat{\xi_i} \ge 0, i = 0, 1, 2, 3 \ldots .m \end{cases}. \tag{4.6}$$

Applying the Lagrangian function and converting it to a dual function to solve:

$$F(x) = \sum_{i=1}^{m} (\widehat{\alpha_i} - \alpha_i) K(X_t, X) + g\#, \tag{4.7}$$

where $\widehat{\alpha_i}$, α_i are two coefficients in the Lagrangian objective function.

The kernel function uses the Gaussian kernel function:

$$K(X_t, X) = e^{-\frac{\|x_i - x_j\|^2}{\sigma^2}}, \tag{4.8}$$

where $K(X_t, X) = \varphi(x_i)^T \cdot \varphi(x_j)$ is the kernel function, and the dual function is further transformed into:

$$F(x) = \sum_{i=1}^{m} c \cdot K(X_t, X) + g. \tag{4.9}$$

Through training, the parameter values c, g were obtained for the prediction model.

4.3.2.2 Prediction Steps

First, when making predictions, machine learning methods, such as SVR, usually improve the accuracy of the calculation by normalizing the data and slightly improving the speed of the calculation. Normalized data can eliminate the influence of outliers. Therefore, to standardize the data, the normalization function used was:

$$V_i' = \frac{v_i - \text{mediam}}{IQR}, \tag{4.10}$$

where v_i' is the normalized sample value, v_i is the sample value, mediam is the median of the sample, and IQR is the interquartile range of the sample.

Second, model parameter selection was performed. To obtain a reasonable prediction result for SVR, the parameters c and gamma were optimized by cross-validation. In general, the value of the penalty parameter c for the error term is in the range of $[10^{-5}, 10^5]$. The larger the value of c, the more training samples for the selected SVR model. The generalization ability of outliers is poor; the smaller the c value, the larger the interval of the decision boundary. Although the outliers can be effectively processed, the prediction accuracy of the model decreases. The gamma value represents the degree of the influence of a single dataset. When gamma is small, the influence of a single variable on the whole is small, and the information on the sample is less; that is, the data have outliers, which also have less impact on the overall sample. In contrast, the larger the gamma, the larger the influence of a single sample and the more the measurement of the information of a single sample, but the more obvious the influence of outliers. Therefore, it is necessary to find a balance between parameters c and gamma for cross-validation.

Third, after the parameters are determined, the historical data are tested, and for the SVR, the sample is divided into training and prediction samples to test the accuracy of the sample prediction. Whether a model achieves accurate prediction results depends on its prediction accuracy. In addition to the goodness of fit R^2, this study also calculates the mean absolute error MAE, and the root mean square error RMSE to test the prediction accuracy of the model. The formula is

$$\mathrm{MAE} = \frac{\sum_{t=1}^{n}\left|y_t - \widehat{y}_t\right|}{n} \tag{4.11}$$

$$\mathrm{RMSE} = \sqrt{\frac{1}{n}\sum_{t=1}^{n}\left|\frac{y_t - \widehat{y}_t}{y_t}\right|^2}, \tag{4.12}$$

where y_t is the sample value and $\widehat{y}_t$ is the predicted value.

4.3.2.3 Analysis of Prediction Results

Due to unbalanced economic development among the provinces, there are apparent differences in carbon dioxide emissions. The model parameters were optimized using comprehensive data from 30 provinces to train a model to predict carbon dioxide emissions. Therefore, this study chose provinces, autonomous regions, and municipalities as units to iterate and optimize the model parameters using the historical data of units to train a prediction model with a higher degree of fit and minor error. To verify the degree of fit of the model, this study first used national overall carbon emissions from 2000 to 2019 for training; the results are shown in Fig. 4.9.

In Fig. 4.9, the overall national carbon emissions showed an upward trend from 2000 to 2019, but the nonlinear characteristics were pronounced. If a linear method

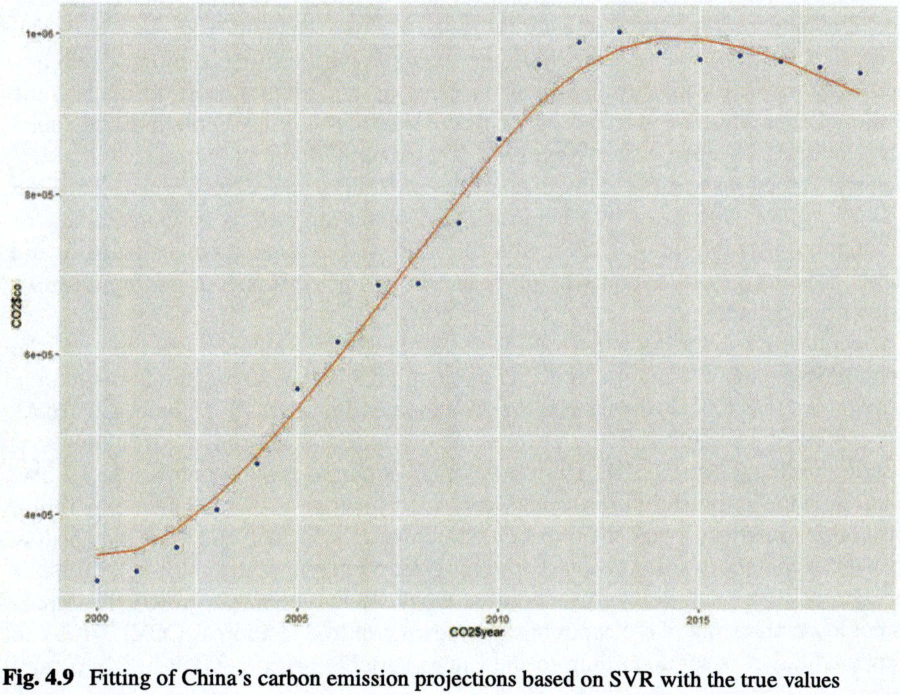

Fig. 4.9 Fitting of China's carbon emission projections based on SVR with the true values

is used for prediction, the gap between the actual and fitted values will be large, which illustrates the advantage of SVR in forecasting. The forecast in Fig. 4.9 is not only based on the historical trend of carbon emissions and control variables, including foreign direct investment (FDI), population (people), industrial structure (IS), economic level (PGDP), technology level (RD), urbanization (URBAN), and energy intensity (EI).

4.3.3 Carbon Emissions Scenario Simulation

4.3.3.1 Scenario Setting

According to the factors influencing carbon emissions in Fig. 4.8, the combination of economic level, energy intensity, and population contributes 88.23% of carbon emissions. The timing focuses on assumptions about GDP, energy, and population. The first is the setting of the population. According to the seventh census, the population is currently experiencing a cliff-like decline owing to rapid population decline and aging. Therefore, the fertility rate of the population should continue to decline. From the perspective of the regional distribution of the population, the seventh census shows that the resident population in Heilongjiang, Inner Mongolia, Jilin, Liaoning, Shanxi, and Gansu declined, and a large number of people migrated to other provinces. In contrast, the migration rates of Shandong, Guangdong, Jiangsu, and Zhejiang continued to increase along the eastern seaboard; therefore, different growth expectations were set for different regions. Furthermore, Beijing, Shanghai, and other megacities have specific plans to restrict the total population, such as the Beijing Urban Master Plan (2016–2035) and Shanghai National Economic and Social Development Thirteenth Five-Year Plan. The Annual Planning Outline (Draft) stipulates that the upper limits of the populations of Beijing and Shanghai are 23 million and 25 million, respectively. Therefore, the population policy was integrated, the population growth rate of each region over the past ten years was calculated, and the population growth target was set based on this growth rate, assuming negative population growth.

Second, the GDP growth rate is set. Since the reform and opening-up, the economic growth rate maintained a relatively high level. In 2016, China's economic growth rate was 6.85%, lower than the growth rate of more than 7% before 2016. As the economic structure shifted, it fell below 6% for the first time in 2019. However, due to the impact of COVID-19 in 2020, the economic growth rate reached 2.3%, making it the only economy in the world with positive growth. Therefore, future economic growth will face the impact of structural adjustment, and there may be low growth in the short term; however, the long-term economic growth rate still has a good trend. Even if a short-term economic shock affects economic growth, its impact is not lower than that of the economic shutdown required to address COVID-19. The expected target is set according to the future variable price of GDP in China, and

Table 4.5 Scenario settings for each indicator

Scenario	Year	Population (%)	GDP (%)	Energy intensity (%)
Baseline	2020–2025	−0.40	6	−3
	2026–2030	−0.60	5.5	−3
	2031–2035	−0.80	5	−3
Low-carbon	2020–2025	−0.50	6	−5
	2026–2030	−0.80	4	−5
	2031–2035	−1.00	2	−5

after the corresponding value is predicted, the carbon emissions forecast is converted into a constant price with 2000 as the base period.

The third scenario is the energy intensity scenario. Currently, energy intensity is calculated at the current price. According to the energy intensity forecast data released by the National Bureau of Statistics, China's energy intensity dropped by more than 4% annually since its reform and opening. During the 13th Five-Year Plan period, the energy intensity goal was to achieve a 15% reduction in unit GDP performance in 2020 compared to that in 2015. In the "14th Five-Year Plan," the target for reducing energy consumption per unit of GDP is 13.5%, so the average annual decrease in energy intensity should be in the range of 3%-5% (Table 4.5).

4.3.3.2 Scenario Analysis

To study the changing trend of total carbon emissions in each region from 2020 to 2035 under the low-carbon economy model and draw concise visualization graphs, the regions were divided into six geographic regions: North China, Northeast China, East China, Central South, Southwest, and Northwest China, as shown in Figs. 4.10, 4.11, 4.12, 4.13, 4.14 and 4.15. Figure 4.10 shows the carbon emissions forecast for the low-carbon scenario. In this case, the carbon emissions of Hebei and Inner Mongolia still show a relatively apparent upward trend from 2020 to 2035, indicating that Hebei Province and Inner Mongolia, according to the existing low-carbon path, peak carbon cannot be achieved, and targeted emission reduction policies need to be formulated. The growth trend of carbon emissions in Shanxi is very flat, indicating that the low-carbon development model affects carbon emissions reduction in Shanxi. However, the total carbon emissions in Shanxi are still relatively large. Under the low-carbon economy model, Beijing and Tianjin have a pronounced reduction in their total carbon emissions. Beijing, in particular, has shown a gentle downward trend, indicating that this low-carbon path can promote peak carbon.

Figure 4.11 shows the predicted carbon emissions for Northeast China. The total carbon emissions of Liaoning Province are much higher than those of Jilin and Heilongjiang, which is related to the fact that Liaoning is a coastal province, and its economic level is higher than that of Heilongjiang and Jilin. Under the low-carbon economic development model, the growth rate of total carbon emissions in Liaoning

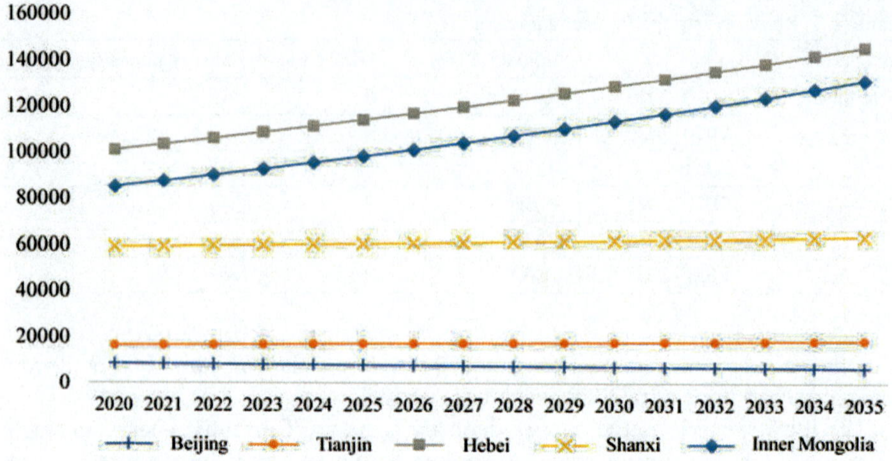

Fig. 4.10 Carbon emissions in the low-carbon scenario in North China

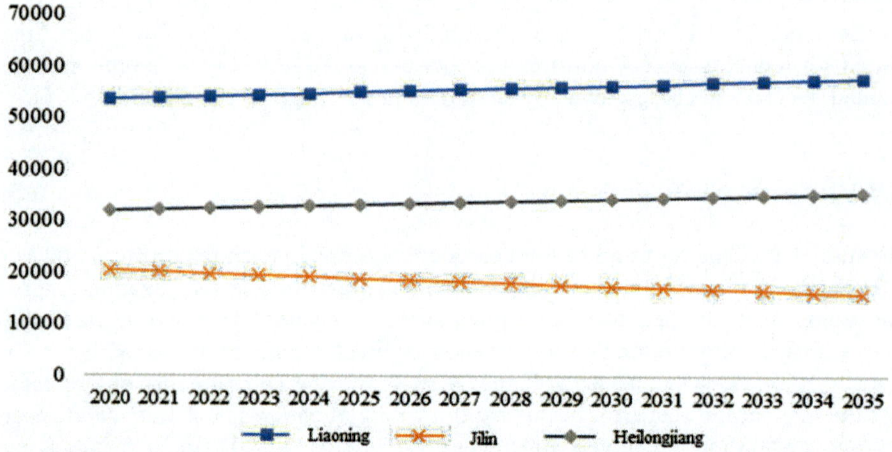

Fig. 4.11 Carbon emissions in the low-carbon scenario in Northeast China

Province slows, but is still rising. The total carbon emissions of Jilin Province are lower than those of Liaoning Province but higher than those of Heilongjiang Province, and its carbon emission trend is relatively flat. After 2020, carbon emissions show a downward trend under the set low-carbon path, ignoring the effects of population outflow and reduced economic growth, indicating the feasibility of a carbon peak on the low-carbon path in Heilongjiang.

Figure 4.12 shows the forecast of carbon emissions under low-carbon scenarios in East China. As the total amount of carbon emissions in Shandong Province is relatively large and the upward trend is pronounced, it is not included in the figure to improve readability. The total amount of carbon emissions in Jiangsu Province

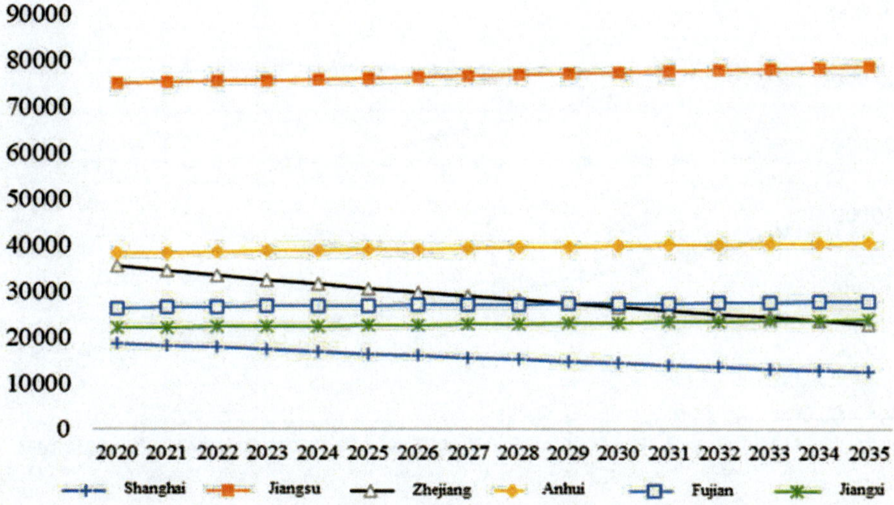

Fig. 4.12 Carbon emissions in the low-carbon scenarios in East China (except Shandong)

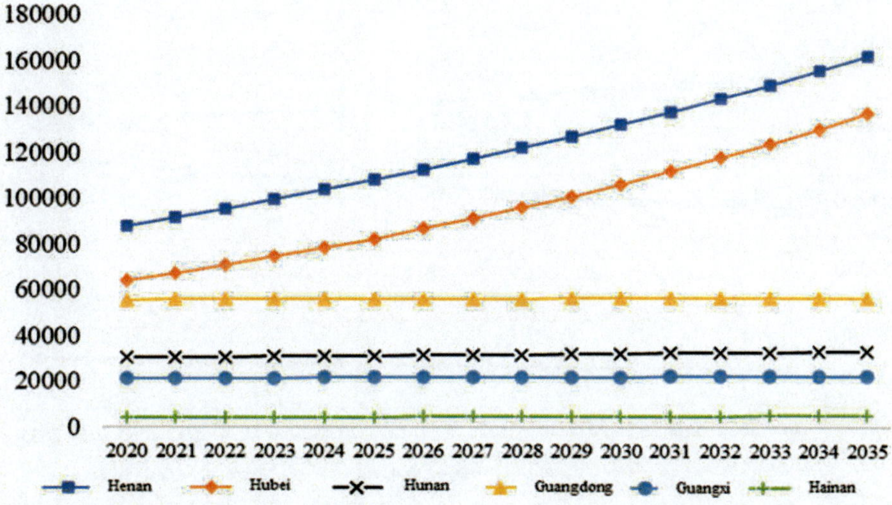

Fig. 4.13 Carbon emissions in the low-carbon scenario in South Central China

is relatively high, and there is a gentle upward trend from 2020 to 2035. This area has relatively high pressure to reduce carbon emissions. Carbon emissions in Anhui Province are also increasing, which is closely related to economic and population growth. Fujian Province experienced a phased increase in 2020–2028, and the trend was relatively flat in 2028–2035. Carbon emissions have an apparent downward trend in Zhejiang and Shanghai, and the total carbon emissions are forecasted to decline rapidly.

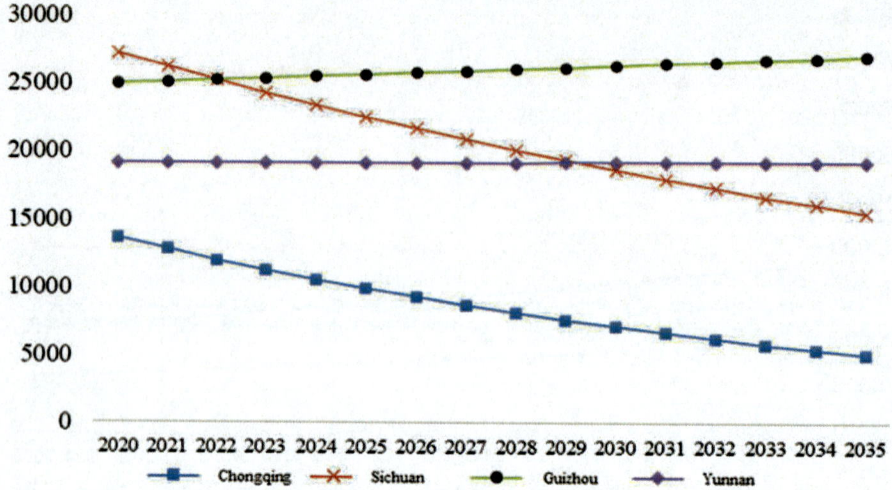

Fig. 4.14 Carbon emissions in the low-carbon scenario in Southwest China

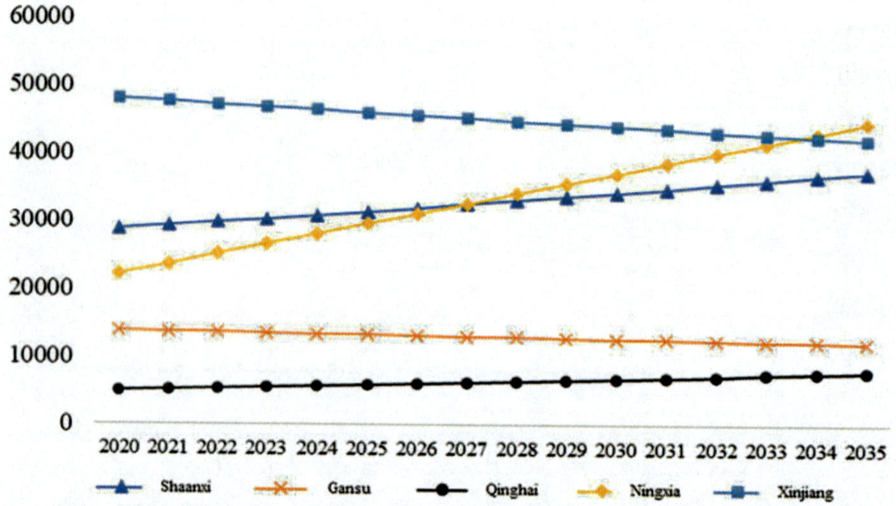

Fig. 4.15 Carbon emissions in the low-carbon scenario in Northwest China

Figure 4.13 shows the forecast of carbon emissions in the central and southern regions. The Henan and Hubei provinces, the most populous provinces in the central region, have a relatively fast growth rate in carbon emissions. Henan province also had the largest total carbon emissions among the central and southern regions. Some low-carbon pathways had less obvious effects on Henan and Hubei provinces. The carbon emissions of Guangdong, Guangxi, Hunan, and Hainan showed no obvious growth trends; however, the total carbon emissions of Guangdong were relatively large.

Figure 4.14 shows the change trend of total carbon emissions under the low-carbon path in Southwest China. Guizhou and Yunnan have an obvious growth trend of carbon emissions, and the total carbon emissions of Sichuan province and Chongqing City declined rapidly. In 2020, the carbon emissions of Sichuan will be higher than those of Guizhou. However, after 2022, the total carbon emissions of Guizhou will gradually surpass those of Sichuan, becoming the region with the largest total carbon emissions in the southwest.

Figure 4.15 shows the changing trend in total carbon emissions under the low-carbon path in Northwest China. The total carbon emissions in Ningxia will not decrease, but will increase along the set low-carbon development path. In 2020, the total carbon emissions in Ningxia will be lower than that of Xinjiang and Shaanxi provinces. However, after 2034, Ningxia will have the greatest total carbon emissions in the Northwest region. Although the total carbon emissions in Xinjiang were relatively high, they exhibited a clear downward trend. The total carbon emissions in Shaanxi were relatively high, and the upward trend was more prominent, indicating the failure of the low-carbon path in Shanxi's economic development model. The total carbon emissions in the Qinghai Province were the lowest in the Northwest, but there was also an upward trend. This indicates that a significant gap exists between Northwest China and other regions. Owing to industrial transfer and industrialization, the impact of carbon emissions in the Northwest region on the ecological environment, peak carbon, and carbon neutrality is increasing. The ecological environment in the northwestern region is fragile. If the relationship between environmental protection and economic development is not handled correctly, the consequences will be catastrophic, and policymakers should remain alert.

Figure 4.16 compares and analyzes the baseline and low-carbon carbon emissions scenarios in 2035 and calculates the rate of change in carbon emissions under the low-carbon path relative to the baseline scenario, reflecting the reduction in total carbon emissions. First, in the overall trend in the existing situation, the forecast of carbon emissions in various provinces in 2035 indicates that the total carbon emissions in Shandong will still be the largest, while Henan will surpass Hebei and rank second. Inner Mongolia and Hubei are among the top five carbon emissions.

Under the low-carbon scenario, total carbon emissions from each province exhibit a significant downward trend. The gap between the low-carbon development scenario and the existing development path was the largest in Chongqing, reaching 61.05%. This is followed by Jiangxi, with a rate of change of 53.42%. Moreover, Yunnan province and Beijing have the lowest rates of change at 0.42% and 3.33%, respectively. Regions with relatively high total carbon emissions such as Shandong, Henan, and Hebei fluctuated by approximately 30%. Hence, in the case of changing energy intensity, population, and economic growth rate, it is possible to reduce total carbon emissions, but the current pattern of national carbon emissions cannot be changed.

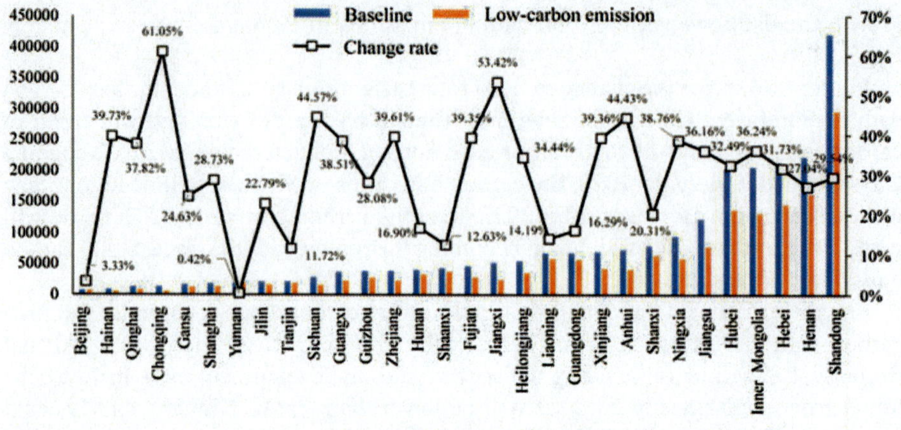

Fig. 4.16 Carbon emissions projections for 2035 for the base case and low-carbon case

4.4 Analysis of Carbon Emissions Reduction Paths

Based on the above analysis, the development path of a low-carbon economy is set only in terms of energy intensity, population, and GDP. Although it can help most regions reduce carbon emissions, carbon emissions in some regions still show a growing trend. Therefore, in addition to the above policies, it is necessary to introduce further supporting systems to form a complete policy system from technical, economic, and administrative aspects to help achieve peak carbon and carbon neutrality.

4.4.1 Technical Means

4.4.1.1 Reaching Carbon Neutrality Depends Ultimately on Scientific and Technological Progress

In the short term, the goals of peak carbon and carbon neutrality are superimposed on the impact of COVID-19 and the transformation of the economic structure. Here, scientific and technological support is urgently needed. China's economic growth is in a critical period for transforming its development model, optimizing its economic structure, and transforming its growth drivers. Structural, institutional, and cyclical issues are intertwined. The impact of the "three-phase superposition" will continue to deepen in the short term and maintain the downward pressure on the economy. With the continuous development of urbanization and industrialization, the consumption of energy, especially fossil energy, cannot be reduced in a short period, and carbon emissions reduction cannot be achieved overnight. Excessive adjustment of energy and industrial structures in the short term will cause economic fluctuations and affect economic security. Therefore, to achieve the goals of peak carbon and carbon neutrality, science and technology are required to promote renewable and clean energy, improve energy utilization, and reduce pollution without affecting residents' quality of life.

In the medium-term, the sustainability of a low-carbon economy depends on technological progress. The proposed goals of peak carbon and carbon neutrality mean that China must leave the traditional high-pollution, high-energy-consumption development path and rely on low carbon or decarbonization of the economy to achieve sustainable development. Low-carbon or decarbonization transformation requires technology-intensive industries to replace labor- and resource-intensive ones. The upgrading and transformation of traditional industries and seizing the commanding heights of global science and technology must rely on technology for support.

In the long term, low-carbon technological innovation is a crucial area of global technological innovation. Peak carbon and carbon neutrality are global activities related to the survival and development of all humans, international political patterns, and national security. In the context of global low-carbon development, low-carbon technology is a core technology and a point of economic growth. It is necessary to plan and deploy a path to achieve carbon neutrality, promote scientific and technological innovation from a higher starting point, and ensure competition in future international markets.

4.4.1.2 Peak Carbon and Carbon–Neutral Technology Breakthroughs Must Be a Priority

To achieve carbon peak and carbon neutrality, China needs to create plans for the required technology this year, focusing on power technology. Electricity is a commonly used clean and renewable energy source; however, China's electricity consumption is overly dependent on thermal power. Achieving clean power generation and promoting low-carbon power generation technology are essential technical directions. In addition to producing electricity, the transportation, storage, and distribution must be low-carbon and sustainable. Additionally, smart grids must be strengthened and the cost of electricity storage must be reduced.

Energy-saving materials and resource-recycling technologies are required. In addition to direct consumption, fossil fuels are an essential source of materials for production and life. For example, plastic and cement products originate from the development of fossil fuels. Therefore, the promotion of new materials and technologies to replace fossil fuel energy products is a technical direction that requires significant breakthroughs. Additionally, it is necessary to strengthen resource conservation and intensive recycling, promote the recycling and reuse of industries and products, and improve resource utilization efficiency.

Advanced deployment of sink-enhancement and negative-emission technologies is another area requiring development. Achieving peak carbon and carbon neutrality must start from the energy and emission sides to develop high-efficiency emission reduction equipment and technologies and air capture technology to reduce carbon stocks for the GHG that have been emitted. This requires the development of technologies to reduce emissions and increase sinks in agriculture, forestry, and grasslands; studies of carbon storage technologies such as oceans and soils; and development technologies such as marine blue carbon represented by mangroves, seagrass beds, and salina.

4.4.2 *Economic Means*

4.4.2.1 Fiscal Subsidies

Finance is the foundation and essential pillar of national governance. Peak carbon and carbon neutrality cannot be separated from support for fiscal policy. Fiscal policy includes financial support, taxation systems, and policy procurement, which are essential policy tools for promoting carbon emissions reduction. From 2016 to 2020, the central government has allocated 4,421.2 billion yuan of financial funds for the field of ecological and environmental protection, ensuring that the average annual growth rate of monetary funds for ecological and environmental protection exceeds 8%, focusing on the development of new energy vehicles, clean heating, and high-speed rail industry production capacity. Support is also required in other areas.

Under the current conflict of financial guidance in China, the use of ecological and environmental protection funds for peak carbon and carbon neutralization faces significant pressure. Solving energy conservation and emissions reduction problems requires substantial capital investment and a scientific capital allocation system. Optimizing the expenditure structure will play an essential role in reducing carbon emissions and in ecological protection. On this basis, the overall responsibilities of the central and local governments are divided to ensure the clarity and implementation of financial funding. In addition to direct financial subsidies, monetary funds should play the role of a "water pump" to attract diversified capital to the field of peak carbon and carbon emissions reduction, such as through the creation of public–private partnerships (PPPs) to promote the development of low-carbon projects.

A government procurement system is essential for financial macro control. Green procurement of low-carbon technologies and products can guide the flow of social funding. To expand greenness, the government should establish a sound priority procurement and emergency procurement system for low-carbon or decarbonized products. The scope of procurement standardizes green procurement and plays a guiding role.

Establishing a sound green tax system is also a necessity. A green tax system includes environmental, resource, consumption, vehicle and vessel, and vehicle purchase taxes. It is necessary to establish a complete taxation system to meet the requirements of the carbon neutralization target and for industries and enterprises with high energy consumption and high pollution through taxation of products to subsidize the development of low-carbon industries, giving play to the negative and positive incentives of taxation.

4.4.2.2 Green Financial System

Fiscal policy plays a more important role in macro control. The realization of peak carbon and carbon neutrality also requires green finance support. Under the current policy incentives, the scale of green finance in China grew rapidly. Green finance is indispensable for promoting the development of low-carbon industries, promoting changes in the energy structure, and improving energy efficiency. Green credit and claims are the primary components of green financing. According to the data on the green credit balance of central banks in China released by the Banking and Insurance Regulatory Commission, green credit is an integral part of financial institutions' loans, accounting for more than 7% of their total loans of financial institutions. This totals 10.2 trillion yuan, ranking first worldwide.

Green credit can limit loans to high-polluting and high-energy-consuming enterprises, provide loan support to low-carbon and clean enterprises, and ensure that green and low-carbon industries receive sufficient financial support. Green credit plays a vital role in promoting energy transformation in China. It plays an essential role in transforming future energy structures, thereby helping to achieve carbon neutrality. The ultimate purpose of green credit is to promote the flow of funds from high-polluting, low-efficiency enterprises or projects to environmentally friendly enterprises and projects. By combining green loans with peak carbon and carbon neutrality goals, it is necessary to ensure that low-energy-efficiency, high-efficiency, and high-efficiency enterprises have high-interest financing restrictions on polluting enterprises to effectively promote the allocation of credit funds, cut off bank support for high-pollution high-energy-consuming capital chains from the source, and force outdated production capacity to withdraw or transform from the market. Then, guide social resources to the green and clean field where carbon peaks and carbon neutrality need to be managed, guide social capital to enter the green industry field, and help energy-efficiency-improving, resource-saving, and environment-friendly enterprises develop green credit. Compared with traditional credit, the essential difference is whether banks consider the environmental impact of projects or enterprises when conducting a credit business. Green credit reflects that commercial banks consider their profits and take social responsibility when making loans and financing, and expand financing channels to achieve peak carbon and carbon neutrality. When approving loans, companies must consider environmental issues and social responsibility. With the development of the international carbon neutrality process, more and more financial institutions have participated in green credit, established the principles of responsible investment and the "Equator Principles," and incorporated green credit into banks' risk management without further improvement. This reduces the environmental risks and benefits of banks, strengthens their supervision, reduces the probability of financial risks, and enables commercial banks to transform industrial greening and sustainable economic development.

4.4.3 Administrative Measures

Although countries worldwide have taken positive action against global climate change and carbon neutrality, the results are currently unsatisfactory. One important reason is that climate issues have negative externalities, whereby individuals enjoy the benefits of economic activity, but society bears the costs. All countries hope to enjoy the benefits of the carbon neutrality, but are unwilling to pay the cost of carbon emission control and the negative externalities of GHG emissions. China is also facing the negative externality of carbon emissions. Owing to negative externalities, the market cannot regulate the allocation of resources, and the government needs to intervene in policies to correct market failures.

First, establishing a cross-departmental coordination mechanism for carbon emissions is necessary. Peak carbon and carbon neutrality involve all aspects of social and economic operations, including resources, energy, environment, industry, construction, transportation, and other fields. In turn, it is necessary to strengthen top-level design, formulate a departmental coordination mechanism for carbon emissions reduction, divide central and local responsibilities, and divide government and corporate responsibilities. According to China's goals of reaching peak carbon by 2030 and carbon neutrality before 2060, China will do an excellent job of short-term and long-term coordination mechanisms for economic transformation and overall planning.

Second, the guarantee mechanism should be improved. This requires designating a development roadmap toward peak carbon and carbon neutrality, focusing on managing industries with significant carbon emissions, clarifying emission reductions and reduction paths, strengthening financial and technical support, and coordinating the market and industry in carbon reduction roles. The government should promote the coordinated development of carbon emission reduction tools, such as capital and technology, pay attention to the quality of economic transformation, and improve the convergence of economic transformation.

Finally, international cooperation should be strengthened. Carbon neutrality is a common goal of human development set by the international community and is a fundamental issue in sustainable human development. Therefore, it is necessary to intensify cooperation with developed countries, especially regarding carbon neutrality, low-carbon technology, industrial structure upgrades, international trade, and other issues. In addition, it is also necessary to strengthen cooperation with developing countries and promote fairness in global carbon responsibility with the help of the "Belt and Road" and "South-South Cooperation." Finally, it is necessary to bolster cooperation with international organizations and establish a technical and industrial cooperation platform for carbon emissions reduction through participation in multilateral cooperation organizations.

References

Eggleston, H. S., Buendia, L., Miwa, K., Ngara, T., & Tanabe, K. (2006). *2006 IPCC guidelines for national greenhouse gas inventories.*

Houghton, J. T., Jenkins, G. J., & Ephraums, J. J. (1990). Climate change: The IPCC scientific assessment. *American Scientist (United States), 80*(6).

Liu, Z., Deng, Z., He, G., Wang, H., Zhang, X., Lin, J., & Liang, X. (2022). Challenges and opportunities for carbon neutrality in China. *Nature Reviews Earth & Environment, 3*(2), 141–155.

Zhao, X., Ma, X., Chen, B., Shang, Y., & Song, M. (2022). Challenges toward carbon neutrality in China: Strategies and countermeasures. *Resources, Conservation and Recycling, 176*, 105959.

Chapter 5
Replenishable but Depletable Resource: Water

5.1 Introduction

The International Terminology of Hydrology (Third Edition) was published by the World Meteorological Organization (WMO) and the United Nations Educational, Scientific, and Cultural Organization (UNESCO) in 2012 and defines water resources as follows: water sources that can be used or have the potential to be used, sufficient in quantity and quality, and able to meet the needs of a certain region for a certain period (WMO and UNESCO, 2012). The "Water Conservancy Science and Technology Terms 1997" issued by the National Science and Technology Terminology Approval Committee also defines water resources as those with a certain amount and usable quality that can be supplemented from nature and is available on Earth (Aba et al., 2022).

Water is an important and active factor in both natural and social environments. Chinese idioms such as green mountains and green waters, good weather, the same color in the sky, and flowing peaches describe the beauty that human beings yearn for. Water constantly moves through the human body, animals, plants, rivers, lakes, and seas. When humans enjoy nature, they should remember that it is the power provided by water. Water is the source of life that fills the world with vitality.

5.1.1 The Importance of Water Resources

5.1.1.1 The Importance of Water Resources to Human Beings

Humans are inseparable from water, one of the most important resources for their survival. Water accounts for approximately 65% of the human body weight. Particularly, the brain marrow contains 75% water, blood contains 83% water, muscles

X. Deng et al., *Environmental and Natural Resources Economics*,
https://doi.org/10.1007/978-981-99-9923-1_5

contain 76% water, and hard bones contain 22% water. Water is the main component of the human body and has several physiological functions.

Water is indispensable for human metabolism. Many substances are soluble in water and dissociate into ionic states that play important roles. For example, substances such as fat and protein are insoluble in water but can be suspended in water to form colloids or emulsions, which are beneficial for human digestion, absorption, and utilization. Metabolic products and toxic substances are transported and excreted through the interstitial fluid. Water is the internal lubricant of the human body and can moisturize the skin and keep it elastic; tears and saliva are the lubricants of the corresponding organs, preventing different organs from being damaged by friction. The specific heat capacity of water is large, which can adjust the temperature of the human body to maintain it at about 37 °C. At high temperatures, sweat evaporation helps dissipate heat and lowers the body temperature. When the weather is cold, the water in the body prevents it from dropping rapidly owing to the low external temperature.

Without water, it is difficult for people to absorb nutrients from food, the body cannot excrete waste, and medicines cannot reach the site of action. The dehydration of the human body can have serious consequences. For example, if the human body is dehydrated by 1–2%, people will feel thirsty; if it reaches 5%, the mouth will be dry, the skin will be wrinkled, consciousness will be unclear, and severe hallucinations will occur. Without water, people can only survive for approximately a week at most (Vestal et al., 2017). Without food, however, people can survive for a relatively long period. Water plays an important role in maintaining human health.

5.1.1.2 Importance of Water Resources to Flora and Fauna

Animals and plants cannot thrive without water. Crops need rain, and fish cannot survive without water. Water is also the basic unit that constitutes the animal and plant body and is the basic pre- requisite for the development and reproduction of animals and plants. Different locations have different water resources and ecosystem environments. Animals and plants have suitable living environments, and different ecosystems are often suitable only for the growth of specific animals and plants. Wetland ecosystems are created under high-water and excess humidity conditions. A swamp is a representative wetland ecosystem with many kinds of swamp plants and animals.

5.1.1.3 The Importance of Water Resources to Environmental Systems

Water is one way through which the Earth's surface is shaped. Rainfall and the flow of water form rivers, lakes, and seas, which scour the Earth's surface, forming landforms, surface landscapes, and deltas, and play an important role in the physical weathering of rocks. Many wildlife habitats have formed along riverbanks, and many species have been nurtured. Life exists when fresh water is present.

Water is an important component of the atmosphere that regulates it. Although atmospheric moisture is relatively low relative to the amount of water worldwide, it can play a significant role. The atmosphere and water interact, contributing to the movement of the Earth's water cycle and climate. Water in the atmosphere can coordinate Earth's energy balance, and the transmission of energy between different regions can be achieved through the circulation of water. Particulate matter in the atmosphere can settle in the water, which is then transported and dispersed as pollutants to farther and wider areas. Pollutants are removed from the water, diluted, and purified naturally.

5.1.1.4 The Importance of Water Resources in Economic and Social Development

Water is a special resource crucial for a region's sustainable economic and social development. The value of water resources lies in their ability to serve humans and meet their production and living requirements. In addition to being used in our daily lives, such as in cooking and washing, water is also used in agriculture, industry, shipping, hydroelectric power generation, freshwater aquaculture, creating aesthetic environments, and building places for entertainment and rest.

Any industry must use water resources during production. Many industries, such as steel plants, use water to cool equipment or products. Any enterprise needs to use multiple water functions to maintain normal production and operational activities, and usually, every production link requires water participation. Water is often used to rinse products, clean equipment, and as detergents for floors.

With the continuous development of science and technology, the understanding of water continues to deepen; thus, the utilization of water is becoming increasingly widespread. The more developed the production of human society, the higher the technical level and the higher the degree of development and utilization of water resources. The water in a specific area is limited. If the water consumption by people for production and living reaches a certain proportion of the available water resources in the local area, it will be difficult for natural water to meet the needs of economic and social development, which will restrict the development of the human economy and society.

5.1.2 Sources of Water Resources

Water for long has been considered a precious wealth that nature provides to human beings. It is a resource that humans understand more comprehensively. Its quantity and distribution will not change owing to discoveries such as coal, oil, natural gas, and other resources. People sailing into the sea, even if thirsty, do not drink seawater because direct consumption of seawater can cause the human body to lose more water and even endanger life.

Natural water resources include rivers, lakes, seawater, snow and glaciers, groundwater, and swamp water. Water can be classified as fresh water and saltwater. 97% of the Earth's water is undrinkable. Usually, what people call water resources refer to freshwater resources on land, which only account for approximately 2.5% of the earth's water. Water vapor rises from the sea surface and is entrained inland by air currents. As the altitude rises, it gathers into clouds and rain, the basic sources of fresh water. Life on land depends on freshwater, which affects the distribution of life worldwide. There are three sources of freshwater: surface, groundwater, and seawater.

5.1.2.1 Surface Water

Surface waters include rivers, lakes, and freshwater wetlands. The accumulation of long-term natural precipitation and snow forms it. Some naturally flow into the ocean, some evaporate and disappear, and some seep into the ground. The natural water of all surface water systems is derived solely from precipitation in the area; however, many factors affect the total amount of water in the system. These factors include surface runoff characteristics, seepage properties of soils, and the storage capacity of wetlands, lakes, and reservoirs. Human activities significantly influence these characteristics. For example, people can build reservoirs to increase the amount of water they can store and reduce the amount of water they drain from wetlands. The volume and intensity of runoff increased through reclamation activities and newly excavated ditches.

There are times when water requirements are temporary, such as when many farms need water in spring and very little in winter. Water must be collected for an entire year to supply farms with sufficient water to meet their daily needs. In this manner, a large amount of water is stored first and can be extracted and used in a short period when needed. There are times when water requirements recur, such as when cooling water is used in power plants. In such times, the surface water system must have a certain storage capacity that can be replenished when the power plant lacks it.

5.1.2.2 Groundwater

Groundwater is water buried in the crevices of underground rocks, and its narrow meaning is the water in saturated aquifers below the groundwater level. In the national standard "Hydrogeological Terminology" (GB/T 14157-93), groundwater refers to different forms of gravitational water under the surface.

Groundwater has stable volume and high quality and is an important water source for agricultural irrigation, urban production, and life. They can be classified according to certain groundwater characteristics or by comprehensively considering several groundwater characteristics. Based on the differences in origin, it is classified into buried water, primary water, infiltration water, and condensed water. Buried water is the groundwater produced with sediments or formed by the infiltration of seawater into the voids of primary sediments; primary water is the groundwater formed by the condensation of gases separated from magma, and condensed water is the infiltration of precipitation into the ground. Based on the differences in the properties of aquifers, they are classified as karst water, fissure water, and pore water. Karst water exists in karst voids, fissure water is gravity water in the cracks of hard and semi-hard bedrock, and pore water is water in the pores of loose rock. According to the difference in mineralization levels, water is classified as brine, saltwater, saltwater, brackish water, freshwater, salinity >50, 10–50, 3–10, 1–3, and <1 g/L respectively. Based on the differences in burial conditions, it can be classified as upper-layer stagnant water, diving, and confined water. Gravity water is buried near the surface and above the local aquifer in the vadose zone, buried below the surface and above the first stable aquifer, and gravity water with a free water surface, and is buried and full of two gravity water in an aquifer between two stable aquifers.

5.1.2.3 Desalination

Desalination, the process of converting seawater into freshwater, is also a source of freshwater. Most of the world's population lives within 120 km of oceans, and desalination effectively increases the total amount of freshwater. It is influenced by time, space, and climate and can ensure a stable supply of drinking, agricultural, and enterprise production water for coastal people.

Seawater desalination is an alternative and incremental technology for freshwater resources, and many scientific research institutions are engaged in this. There are more than 20 kinds of desalination technologies in Shanghai, among which the most mainstream methods are distillation and reverse osmosis. Large-scale modern desalination plants can produce tens of thousands or nearly one million tons of fresh water daily. With continuous improvements at the technical level, the cost of water has gradually decreased, and some countries can price this water close to that of tap water. Many desalination plants are co-located with power plants such that steam and electricity from the power plants can be used to power the desalination equipment, thereby reducing the cost of desalination. The extraction of freshwater from saline ocean water has become a booming industry. The cost of desalination is higher than that of other methods, and the supply of desalinated water can only meet the needs of a small number of people. With the reduction in treatment costs, obtaining fresh water for use through desalination will be promoted.

5.1.3 Characteristics of Water Resources

Unlike many other natural resources, water resources are renewable and can be reused multiple times. Water resources have changed over the years, showing certain cycles in accordance to some laws. Natural geographical factors and human activities affect movement processes and storage forms.

5.1.3.1 Inevitability and Contingency

Water resources follow their own basic laws, and their conditions (including surface water, groundwater and atmospheric water) in a certain period have their own objective reasons, which are all inevitable phenomena under specific conditions. Among the many changes in hydrology and water resources, those that can be explained or predicted are inevitability called. For example, the river's annual flood and dry periods, inter-annual dry year, and wet year, and the change in groundwater level exhibits a similar phenomenon. The year, month, etc. changes presented by this inevitability in time are called periodicity, and correspondingly, they are called multi-year periods, monthly periods, etc. Parts that cannot be explained or predicted are regarded as manifestations of hydrological phenomena or the contingency of water resources. The flow and process of any river cannot be the same each year; the changes in groundwater location each year will also be different, as will the changes in spring water flow. These laws need to be analyzed using countless statistical or long-term observational data.

5.1.3.2 Similarity and Particularity

Water resources in different locations have similarities and particularities. Similarity refers to watersheds with similar climatic and geographical conditions, hydrology, and water resources that are often similar. For example, the distribution of river runoff in humid regions was relatively uniform over the year, whereas there were large differences in arid regions. In particular, the changing laws of different hydrology and water resources formed under different underlying surface conditions differ. For example, although they are in the same climate zone, the floods of plain and mountain rivers have different changing characteristics. Although they are in the same semiarid conditions, the rules of occurrence of groundwater in different landforms are also different.

5.1.3.3 Circularity and Finiteness

Water is a highly active element constantly in motion and is found extensively in many biological, chemical, and physical processes in the natural environment. The fundamental difference between it and other solid resources lies in its fluidity, which is a dynamic resource produced in the water cycle process; that is, it is cyclic. Water resource systems are complex natural circulation systems. After the development and use of water resources, they can be supplemented by atmospheric precipitation and are in a cycle of exploitation, replenishment, consumption, and recovery. Therefore, it can be continuously provided to people to achieve an ecological balance. The recovery of water resources is critical and inexhaustible. However, the storage of freshwater resources on Earth is limited. It accounts for less than 3% of the global water, most of which is stored in polar ice caps and glaciers. From the perspective of the dynamic balance of water resources, if the consumption of water resources in a certain period is not greater than the amount of water supplied in the same period, the amount of water supply is appropriate. This maintains a dynamic balance that would otherwise lead to undesirable consequences. In summary, the water cycle process is constantly ongoing, and the storage quantity of water resources is limited, not inexhaustible.

5.1.3.4 Difference and Duality

Water resources also differ. Natural water resources have a specific spatial and temporal distribution. The distribution of water resources in different countries or regions varies, and the distribution of water resources in the same country or region at different times is also different. The difference is not only in quantity but also in quality.

The duality of water resources means that they benefit and harm human beings. When the quantity and quality of water resources are suitable, they contribute significantly to regional economic development, ecological environment construction, and social progress. In contrast, when natural disasters occur, or people use them improperly, they cause damage to the ecological environment and restrict economic and social development. For example, natural disasters often occur in regions with too much or too little precipitation. If the construction quality of water conservancy projects does not meet the standard or management is not in place, dam collapse accidents may occur, threatening the safety of life and property.

5.2 Water Resources

5.2.1 The Total Amount of Water Resources

The total amount of water resources refers to the amount of surface and underground water produced by precipitation in a specific area within a certain period and is equal to the sum of the surface runoff (excluding water inflow outside the area) and the amount of precipitation infiltration recharge. This is not equal to the simple sum of the surface water and groundwater resources, but the repetition of the two needs to be subtracted. This is because there is a time interval in the water resource statistics during which changes in the water resource categories occur.

The Earth has become a big water polo. The proverb "Thirty percent of mountains, sixty percent of water, and ten percent of fields" vividly describes the earth's appearance and shows that water resources are an important part of it. Three-quarters of the Earth's surface is covered by water, and water resources cover approximately 1.386 billion cubic kilometers. However, approximately 97% of this is saltwater, and freshwater accounts for a small proportion of the water found on Earth. More than two-thirds of the freshwater is trapped in glaciers, ice caps, and permanent snow, and less than 1% is not clean enough for drinking. Seven billion people worldwide consume less than 1% of freshwater.

With the world's population growth and the continuous development of economies, the disparity between the supply and demand of water resources continues to intensify. If a place's surface water and groundwater extraction accounts for more than 80% of the total available water, it is considered "extremely water-deficient"; if the proportion is between 40 and 80%, it is considered "highly water-deficient." According to a report released by the World Resources Institute in 2019, in 189 countries and regions, about one-third of the population (2.6 billion people) lives in countries with high water scarcity, of which 17 countries, 1.7 billion people live in "extreme water scarcity." China also has very scarce water resources and is one of more than ten water-poor countries worldwide (Dąbrowska et al., 2023). Its per capita water resources are only a quarter of the world average, but it consumes the largest amount of water in the world (Hartmann et al., 2014. Among the 669 Chinese cities, 400 have an insufficient water supply, and 110 have serious water shortages).

According to the 2020 China Water Resources Bulletin issued by the Ministry of Water Resources of the People's Republic of China in 2021 (Ministry of Water Resources of China, 2021a, 2021b), the average annual precipitation in China in 2020 will reach 706.5 mm, which is higher than the multiyear average (the multiyear average in this chapter). This refers to the average hydrological series from 1956 to 2000, which was 10.0 and 8.5% higher than the previous year. That year, China's total water resources were 3,160.52 billion cubic meters, 14.0% higher than the multiyear average and 8.8% higher than the previous year. It accounts for 47.2% of the total precipitation, and the average water production per unit area is 334,000 cubic meters per square kilometer. Among them, the number of surface water and groundwater resources in 2020 will be 3,040.70 billion cubic meters and 855.35 billion cubic meters, respectively, and the unique amount of the two will be 119.82 billion cubic meters. The amount of surface water resources was equivalent to an annual runoff depth of 321.1 mm, 13.9% higher than the multiyear average and 8.6% higher than the previous year. The groundwater resources were 6.1% higher than the multiyear average. Compared to the multiyear average, China's annual change in total water resources was not large; it was 3.9% higher from 1990 to 1999, 3.9% lower from 2000 to 2009, and 3.7% higher since 2010. In 2020, the amount of water flowing into China was 18.51 billion cubic meters, and the amount of water flowing out of China was 574.47 billion cubic meters. The amount of water flowing into the boundary rivers and the water entering the sea was 187.69 billion cubic meters and 1,907.10 billion cubic meters respectively.

In 2020, according to the statistics of 3,729 medium-sized and 705 large-scale reservoirs in China, the total water storage capacity of reservoirs at the end of the year was 435.87 billion cubic meters, 23.75 billion cubic greater than that at the beginning of the year. Among them, the year-end water storage of large and medium-sized reservoirs was 388.79 and 47.08 billion cubic meters, respectively, an increase of 19.06 and 4.69 billion cubic meters, compared with the beginning of the year. According to the data from 62 lakes monitored that year, the total water storage at the end of the year was 142.36, 4.75 billion cubic meters higher than that at the beginning of the year (Ministry of Water Resources of China, 2021a, 2021b).

5.2.2 *Spatial Distribution of Water Resources*

5.2.2.1 Spatial Distribution of Water Resources in the World

There are huge amounts of water resources on Earth, but very little can be used to sustain life. Its distribution in different continents, countries, and regions is considerably different. Regarding the distribution of water resources across continents, Asia has the largest annual runoff, followed by South America, North America, Africa, Europe, and Oceania. The per-capita runoff was approximately 10,000 cubic meters per person. Among the continents, Oceania has the largest per capita runoff, followed by South America, North America, Africa, Europe, and Asia.

According to the World Water Development Report 2021 released by the United Nations, a more in-depth understanding of the water situation in each continent can be obtained (UNESCO, 2021). Africa accounts for approximately 9% of the world's freshwater resources. However, it is unevenly distributed, with the six most water-rich countries in Central and Western Africa accounting for 54% of Africa's total water resources and the 27 least water-poor countries accounting for only 7%. Lake Victoria (spread across Kenya, Tanzania, and Uganda) is the second-largest freshwater lake in the world in terms of surface area, whereas Lake Tanganyika (shared by Burundi, the DRC, Tanzania, and Zambia) is the second-largest freshwater lake in the world. Africa is the world's second driest continent after Australia, with arid and semiarid regions accounting for approximately two-thirds. In 2017, approximately 73% of the total population in sub-Saharan Africa did not use safely managed drinking water services. An estimated 14% of Africa's population (approximately 160 million people) currently lives in water-stressed conditions, partly because of the uneven distribution of water resources and unequal access to clean and drinking water services.

Latin America and the Caribbean are water-rich regions. According to the latest regional estimates, the per capita water supply for residents in the region is approximately 28,000 cubic meters per year, more than four times the world average of 6,000 cubic meters per person per year. Similarly, it has the world's largest wetland, the Pantanal, with an area of 200,000 square kilometers, regulating the hydrology of most of the continent, whereas the Amazon has the largest flow in the world, and its water content is far greater than the sum of the Nile, Yangtze, and Mississippi Rivers. These facts often lead to the misconception that Latin America and the Caribbean water is easily and equally accessible to all citizens. However, this is far from true. Although these gifts account for one-third of the world's renewable freshwater resources, they are unequally distributed. Water is mainly found in rural and natural areas of the Amazon Basin and large cities in arid or semi-desert regions (such as Lima, Santiago, or Buenos Aires) or high-altitude areas with smaller catchment areas. Regions (such as Bogotá, Mexico City, and Quito) face greater challenges in ensuring stable water, as do the small island states in the Caribbean.

Asia and the Pacific account for 60% of the world's population but only 36% of the world's water resources; therefore, their per capita water availability is the lowest globally. Unsustainable extraction is a major problem, with some countries taking over half their water resources. Seven of the world's 15 largest groundwater extractors are in Asia and the Pacific.

In the Arab region, nearly 86% of the population (nearly 362 million people) lives underwater or in absolute scarcity conditions. Fourteen countries in the region use more than 100% of the available freshwater resources, increasing their reliance on transboundary waters, non-renewable groundwater resources, and unconventional water resources. The amount of freshwater that can be extracted sustainably may even be lower if water quality is considered.

At the national level, there are large differences in the water resource status of countries. The freshwater resources of nine countries, including the United States, Russia, Brazil, China, Canada, Colombia, Indonesia, the Congo, and India, account for 60% of global freshwater resources. In addition, 80 countries and regions accounting for approximately 40% of the world's total population seriously lack water resources. Today, approximately 1.5 billion people in more than 80 countries worldwide have insufficient freshwater, and 300 million people in 26 countries are in extreme water scarcity. Over time, the world's water-stressed population will continue to increase, affecting more countries and regions.

5.2.2.2 Spatial Distribution of Water Resources in China

In general, the spatial distribution of water resources in China was significantly higher in the south and lower in the north, higher in the east and lower in the west, and higher in summer and autumn. The 2020 China Water Resources Bulletin recognizes the spatial distribution of China's water resources (Ministry of Water Resources, 2021). Based on this, the following section first introduces the number of water resources and their changes in the first-level water resource districts and each provincial-level administrative region from the total amount of water resources, precipitation, surface water resources, and groundwater resources, and then from the large- and medium-sized reservoirs, lake water storage dynamics, and groundwater dynamics to introduce the water storage dynamics of each region.

Water resources in the first-level area of water resources

The total amount of water resources: Table 5.1 shows the total water resources for each first-level water resource area in 2020. Northwest River District, Huaihe District, Yellow River District, Haihe District, Liaohe District, and Songhua River District, the six first-level water resource areas referred to as the North Six Districts, have a total water resource of 664.50 billion cubic meters. The four first-level water resource areas, namely the Southwest River District, Pearl River District, Southeast Zhujiang District, and Yangtze River District (including the Taihu Lake Basin), are referred to as the South 4th District, with a total water resource of 2,496.02 billion cubic meters. They accounted for 21.03 and 78.97% of the total water resources, respectively, with the latter being 3.76 times that of the former. Among the six northern districts, the Songhua River District has the largest water resource at 225.31 billion cubic meters; the Haihe District has the least total water resources at only 28.31 billion cubic meters. Among the four southern regions, the Yangtze River region has the largest amount of water resources, reaching 1,286.29 billion cubic meters; the southeastern river regions have the smallest total amount of water resources, 167.73 billion cubic meters.

In terms of the change process of total water resources, the total amount of the six northern districts from 1990 to 1999 was close to the multiyear average: 6.9% lower from 2000 to 2009, 5.8% higher after 2010, 4.8% higher in 1999, 3.2% lower in 2000–2009, and 3.2% higher post 2010.

Table 5.1 The number of water resources in the first-level districts of water resources in 2020

Water resources first-level area	Total water resources (billion cubic meters)	Precipitation (mm)	Surface water resources (billion cubic meters)	Groundwater resources (billion cubic meters)
Nationwide	31,605.2	706.5	30,407.0	8553.5
Northern District 6	6645.0	373.1	5594.0	2820.1
Songhua River District	2253.1	649.4	1950.5	647.3
Liaohe River District	565.0	589.4	470.3	200.0
Haihe River District	283.1	552.4	121.5	238.5
Yellow River District	917.4	507.3	796.2	451.6
Huaihe River District	1303.6	1060.9	1042.5	463.1
Northwest River District	1322.8	159.6	1213.1	819.6
South District 4	24,960.2	1297.0	24,813.0	5733.4
Yangtze River District	12,862.9	1282.0	12,741.7	2823.0
Southeast River District	1677.3	1582.3	1665.1	429.4
Pearl River District	4669.0	1540.5	4655.2	1068.7
Southwest River District	5751.1	1091.9	5751.1	1412.4

Note The data comes from the China Water Resources Bulletin, 2020

Precipitation: Table 5.2 compares precipitation in the first-level areas of water resources in 2020 with the previous year and multiyear averages. In 2020, the precipitation in the six regions of northern China was 373.1 mm, and that in the four southern regions was 1297.0 mm. Among the six northern districts, the Huaihe district had the highest precipitation, reaching 1060.9 mm; the northwest river districts had the lowest precipitation, only 159.6 mm. Among the four southern regions, the southeastern rivers have the highest precipitation, reaching 1582.3 mm; the southwestern rivers have the lowest rainfall (1091.9 mm).

Seven of the ten first-level water resource areas had more precipitation than the multiyear average, among which the Songhua River and Huaihe areas were 28.8 and 26.5% higher, respectively. The other three had less precipitation; the southeast River District was below the multiyear average (4.8%). Compared with the previous year, the precipitation in the same seven areas increased, among which the Huaihe, Haihe, and Yangtze River Districts increased by 73.9, 23.0, and 21.0%, respectively. The other three precipitations decreased, among which the southeastern River District,

Table 5.2 Comparison of precipitation in water resource first-level area

Water resources first-class area	Precipitation (mm)	Compared to 2019 (%)	Compared with the multiyear average (%)
Nationwide	706.5	8.5	10.0
Northern District 6	373.1	7.8	13.8
Songhua River District	649.4	7.6	28.8
Liaohe River District	589.4	5.6	8.1
Haihe River District	552.4	23.0	3.3
Yellow River District	507.3	2.1	13.8
Huaihe River District	1060.9	73.9	26.5
Northwest River District	159.6	−12.9	−0.8
South District 4	1297.0	8.8	8.1
Yangtze River District	1282.0	21.0	18.0
Southeast River District	1582.3	−14.2	−4.8
Pearl River District	1540.5	−5.3	−0.5
Southwest River District	1091.9	7.7	0.5

Note The data comes from the China Water Resources Bulletin, 2020

Northwest River District, and Pearl River District decreased by 14.2, 12.9, and 5.3%, respectively.

Surface water resources: Table 5.3 compares the surface water resources of each first-level water resource area in 2020 with the previous year's and multiyear averages. The surface water resources in the six northern districts are 559.40 billion cubic meters, and the four southern districts are 2481.30 billion cubic meters. They account for 18.40 and 81.60% of China, respectively, the latter being 4.43 times that of the former. Among the six northern districts, the Songhua River District has the largest surface water resources, reaching 195.05 billion cubic meters; Haihe District has the least, with only 12.15 billion cubic meters. Among the four southern regions, the Yangtze River region has the largest surface water resources, reaching 1,274.17 billion cubic meters; the southeastern river region has the least amount, with 166.51 billion cubic meters.

Six of the ten first-level water resource areas had more surface water resources than the average for many years, especially the Huaihe and Songhua River areas,

Table 5.3 Comparison of the surface water resources

Water resources first-class area	Surface water resources (billion cubic meters)	Compared to 2019 (%)	Compared with the multiyear average (%)
Nationwide	30,407.0	8.6	13.9
Northern District 6	5594.0	18.7	27.8
Songhua River District	1950.5	0.8	51.1
Liaohe River District	470.3	53.8	15.3
Haihe River District	121.5	16.2	−43.8
Yellow River District	796.2	15.4	30.2
Huaihe River District	1042.5	217.7	54.0
Northwest River District	1213.1	−10.1	3.5
South District 4	24,813.0	6.6	11.1
Yangtze River District	12,741.7	22.2	29.3
Southeast River District	1665.1	−32.7	−16.2
Pearl River District	4655.2	−8.1	−1.1
Southwest River District	5751.1	8.3	−0.4

Note The data comes from the China Water Resources Bulletin, 2020

which were 54.0 and 51.1% higher than the average, respectively. The other four surface water resources were lower than the multiyear average, among which the Haihe and Southeast Rivers were 43.8 and 16.2% lower, respectively. Compared to the previous year, surface water resources increased by 7. Among them, Huaihe and Liaohe Districts increased by 217.7 and 53.8%, respectively, whereas the other three surface water resources decreased, among which the Southeast River District decreased by 32.7%.

In 2020, China's groundwater resources (salinity $\leq$ 2 g/L) were 855.35 billion cubic meters, 6.1% higher than the multiyear average. Among them, the plain and the hill areas are 202.24 and 683.61 billion cubic meters, respectively, and the double-counting amount between them is 30.50 billion cubic meters. The total amount of shallow plain groundwater recharged in China was 209.32 billion cubic meters. The calculated area of plain shallow groundwater in the 4th area in the south accounts for 9%, and its total groundwater recharge is 38.58 billion cubic meters; the calculated area in the 6th area in the north accounts for 91%, and its total groundwater recharge is 170.74 billion cubic meters. Among them, Songhua River District, Liaohe District, Haihe District, Yellow River District, Huaihe District, and Northwest River District are 40.16, 12.91, 18.57, 16.65, 34.14 billion cubic meters, and Northwest River District covers 48.31 billion cubic meters.

Water resources in each provincial administrative region

The total water resources for each provincial administrative region in 2020 are listed in Table 5.4. Among the provinces and cities, Tibet ranked first in total water resources in 2020 with 459.73 billion cubic meters, greatly exceeding other provinces and cities. Sichuan followed closely with 323.73 billion cubic meters, and Hunan, Guangxi, and Yunnan ranked third, fourth, and fifth, respectively, in addition to Jiangxi, Guangdong, Heilongjiang, Zhejiang, Guizhou, Anhui, and Qinghai, with total water resources of more than 100 billion cubic meters. The three provinces and cities with the least total water resources that year were Ningxia (1.10 billion cubic meters), Tianjin (1.33 billion cubic meters), and Beijing (2.58 billion cubic meters).

Table 5.5 compares the precipitation of each provincial administrative region in 2020 with the previous year's sand multiyear averages. Among the provincial administrative regions, there are 24 provinces (autonomous regions and municipalities) with precipitation higher than the multiyear average, among which four provinces (municipalities), Shanghai, Anhui, Hubei, and Heilongjiang, are all higher by more than 30%. The other seven provinces (autonomous regions and municipalities directly under the Central Government) are less than the annual average, especially the three provinces of Fujian, Guangdong, and Yunnan, which are less than 14.2, 11.1, and 9.5%, respectively. Compared with the previous year, the precipitation in 22 provinces (autonomous regions and municipalities) increased, among which Hubei, Anhui, Henan, and Jiangsu were all over 50% higher; the other nine provinces (autonomous regions and municipalities) in the table when compared with the previous year are less, especially in Guangdong, Xinjiang, Fujian, Zhejiang, and Ningxia, which are more than 10% less.

Table 5.4 The number of water resources in China, 2020

Province	Total water resources (billion cubic meters)	Precipitation (mm)	Surface water resources (billion cubic meters)	Groundwater resources (billion cubic meters)
Nationwide	31,605.2	706.5	30,407.0	8553.5
Beijing	25.8	560.0	8.2	22.3
Tianjing	13.3	534.4	8.6	5.8
Hebei	146.3	546.7	55.7	130.3
Shanxi	115.2	561.3	72.2	85.9
Inner Mongolia	503.9	311.2	354.2	243.9
Liaoning	397.1	748.0	357.7	115.2
Jilin	586.2	769.1	504.8	169.4
Heilongjiang	1419.9	723.1	1221.5	406.5
Shanghai	58.6	1554.6	49.9	11.6
Jiangsu	543.4	1236.0	486.6	137.8
Zhejiang	1026.6	1701.0	1008.8	224.4
Anhui	1280.4	1665.6	1193.7	228.6
Fujian	760.3	1439.1	759.0	243.5
Jiangxi	1685.6	1853.1	1666.7	386.0
Shandong	375.3	838.1	259.8	201.8
Henan	408.6	874.3	294.8	185.8
Hubei	1754.7	1642.6	1735.0	381.6
Hunan	2118.9	1726.8	2111.2	466.1
Guangdong	1626.0	1574.1	1616.3	399.1
Guangxi	2114.8	1669.4	2113.7	445.4
Hainan	263.6	1641.1	260.6	74.6
Chongqing	766.9	1435.6	766.9	128.7
Sichuan	3237.3	1055.0	3236.2	649.1
Guizhou	1328.6	1417.4	1328.6	281.0
Yunnan	1799.2	1157.2	1799.2	619.8
Tibet	4597.3	600.6	4597.3	1045.7
Shaanxi	419.6	690.5	385.6	146.7
Gansu	408.0	334.4	396.0	158.2
Qinghai	1011.9	367.1	989.5	437.3
Ningxia	11.0	309.7	9.0	17.8
Xinjiang	801.0	141.7	759.6	503.5

Note The data comes from the China Water Resources Bulletin, 2020

Table 5.5 Comparison of precipitation in China

Province	Precipitation (mm)	Compared to 2019 (%)	Compared with the multiyear average (%)
Nationwide	706.5	8.5	10.0
Beijing	560	10.7	−4.1
Tianjing	534.4	22.5	−7.0
Hebei	546.7	23.5	2.8
Shanxi	561.3	22.5	10.3
Inner Mongolia	311.2	11.3	10.3
Liaoning	748	8.8	10.3
Jilin	769.1	13.2	26.3
Heilongjiang	723.1	−0.7	35.6
Shanghai	1554.6	11.9	42.7
Jiangsu	1236	54.8	24.3
Zhejiang	1701	−12.8	6.0
Anhui	1665.6	78.0	42.0
Fujian	1439.1	−16.9	−14.2
Jiangxi	1853.1	8.4	13.1
Shandong	838.1	49.9	23.3
Henan	874.3	65.2	13.4
Hubei	1642.6	83.8	39.2
Hunan	1726.8	15.2	19.1
Guangdong	1574.1	−21.0	−11.1
Guangxi	1669.4	4.2	8.6
Hainan	1641.1	2.9	−6.2
Chongqing	1435.6	29.7	21.2
Sichuan	1055	10.7	7.8
Guizhou	1417.4	13.7	20.3
Yunnan	1157.2	14.8	−9.5
Tibet	600.6	0.7	5.1
Shaanxi	690.5	−9.1	5.2
Gansu	334.4	−7.6	11.0
Qinghai	367.1	−1.8	26.4
Ningxia	309.7	−10.4	7.3
Xinjiang	141.7	−18.9	−8.4

Note The data comes from the China Water Resources Bulletin, 2020

In 2020, there were 18 provinces (autonomous regions and municipalities directly under the Central Government) with surface water resources higher than the multiyear average, among which Shanghai was 104.9% higher, and four provinces of Jiangsu, Anhui, Heilongjiang, and Hubei were more than 70% higher, while the other 13 provinces (autonomous regions and municipalities) were relatively low, of which two provinces (municipalities), Hebei and Beijing, both dropped by more than 50%.

Regional water storage dynamics

Dynamics of water storage in large and medium-sized reservoirs: In terms of water resources zoning, the Haihe District, the Northwest River District, and the Southeast River District have reduced water storage capacity at the end of the year by 190, 530 million cubic meters, and 1.60 billion cubic meters, respectively. The other seven areas increased by varying degrees; the Pearl, Huaihe, and Yangtze River areas increased by 4.17, 5.81, and 10.41 billion cubic meters, respectively. At the provincial level, the water storage capacity of reservoirs in 22 provinces (autonomous regions and municipalities), including Hunan, Henan, Hubei, Anhui, and Guangxi, increased by 31.39 billion cubic meters in total, including Sichuan, Fujian, Zhejiang, and Guangdong. The water storage capacity of reservoirs in 10 provinces (autonomous regions and municipalities directly under the Central Government) decreased by 7.64 billion cubic meters.

Dynamics of lake water storage: In 2020, according to the statistics of 62 monitored lakes, the total water storage of the lakes at the end of the year was 142.36 billion cubic meters, an increase of 4.75 billion cubic meters from the beginning of the year. Among them, the water storage capacity of Qinghai Lake and Hongze Lake increased by 1.43 and 1.68 billion cubic meters, respectively. In addition, water storage in Taihu Lake decreased by 370 million cubic meters.

Groundwater dynamics were evaluated at 10,131 National Groundwater Monitoring Project Sites of the Ministry of Water Resources and 4,279 National Groundwater Monitoring Project Sites of the Ministry of Natural Resources in 31 provinces (autonomous regions and municipalities) of China. Their monitoring area was approximately 3.5 million square kilometers, covering the main plains, basins, and karst mountains of China.

From the perspective of shallow groundwater, in 2020, the water level of shallow groundwater in the middle and lower reaches of the Yangtze River Plain, the Huanghuaihai Plain, and the Northeast Plain will generally rise, while the plains and basins in Shanxi and Northwest China will slightly decrease. In the middle and lower reaches of the Yangtze River plain, the average buried depth of groundwater in the Poyang Lake Plain, Jianghan Plain, Yangtze River Delta Plain, and general plain along the east coast of Zhejiang is 4.9, 4.5, 2.3, and 1.6 m; the groundwater levels in the plain area, the Yangtze River Delta Plain, and the general plains along the east coast of Zhejiang rose by 0.9, 0.5, and 0.1 m, and were the same. The average buried depth of groundwater in the Haihe Plain and the Huanghuai Plain is 14.0 and 4.9 m, respectively; compared with the end of the previous year, the Haihe Plain and the Huanghuai Plain increased by 0.1 and 0.6 m respectively. In the Northeast Plain, the average buried depths of groundwater in the Sanjiang Plain, Songnen Plain,

Liaohe Plain, and Muling Xingkai Plain are 7.8, 7.5, 5.1, and 4.9 m, respectively; The groundwater level rose by 0.3 m, the Songnen Plain rose by 0.2 m, and the Sanjiang Plain and the Liaohe Plain were the same. Among the plains and basins in Shanxi and Northwest China, the average burial depths of groundwater in the Guanzhong Plain, Hexi Corridor, Junggar Basin monitoring area, major basins in Shanxi, Tarim Basin, Hubao Plain, Qaidam Basin, Hetao Plain, and Weining, Yinchuan were different. Compared to the end of the previous year, the groundwater levels of the Hubao Plain, Tarim Basin Monitoring Area, Guanzhong Plain, Hexi Corridor Plain, and Yinchuan Weining Plain decreased by 0.9, 0.9, 0.5, 0.5, and 0.4 m, respectively, and the groundwater levels of the Qaidam Basin, Hetao Plain, Junggar Basin monitoring area, and major basins in Shanxi increased by 0.4, 0.3, 0.1 and 0.1 m respectively.

From the perspective of deep confined water, among the 3362 deep confined water groundwater level monitoring sites selected in 22 provinces (autonomous regions and municipalities) across China, there were a total of 1089 sites whose water levels fluctuated at the end of 2020 compared with the end of the previous year. The range is between ±0.5 m (inclusive), accounting for 32.4%; a total of 727 stations have a water level increase of more than 2 m, accounting for 21.6%; a total of 291 stations have a water level decrease of more than 2 m, and the proportion is 8.7%; in provinces (autonomous regions and municipalities) with more than 20 deep confined water monitoring sites, Xinjiang, Henan and Yunnan provinces (autonomous regions) have a larger proportion of sites with a water level drop of more than 0.5 m, Tianjin, Liaoning and Jiangsu provinces (municipalities directly under the central government) accounted for a relatively large proportion of sites where the water level rose more than 0.5 m.

5.3 Water Resource Allocation

5.3.1 Rational Allocation of Scarce Water Resources

5.3.1.1 The Concept of Rational Allocation of Water Resources

Water scarcity refers to the lack of water. The world's total water resources are limited, and the amount of freshwater available for drinking is even lower. With economic development, increased population, and mounting waste, this problem is becoming increasingly serious. Water resource shortages mainly include resource water shortages and water quality shortages. The former is caused by geographical differences in the distribution of water resources, resulting in less distribution of water sources in local areas, whereas the latter is caused by the physical form of water resources in the region or the deterioration of water quality, resulting in water shortages due to the difficulty in using water resources. This often occurs in high-water areas. Therefore, rational allocation of water resources is very important.

Rational allocation of water resources refers to the reasonable allocation of limited and different forms of water resources among demand users through engineering or non-engineering measures in a specific area or basin according to the principles of effectiveness, fairness, and sustainability distribution (Yang et al., 2013). Rational allocation of water resources relies on a comprehensive system composed of engineering and non-engineering measures. Its basic functions include the following two aspects: one is the demand side, that is, through the adjustment of the industrial structure, the development of a water-conserving society, and the adjustment of the layout of the productive forces to suppress the rising trend of water demand, to adapt to the relatively poor water resource conditions. On the supply side, by strengthening the management of water resources, coordinating competitive water use in all aspects, and using engineering measures to affect the natural temporal and spatial distribution of water resources to conform to the productivity layout. They complement each other and jointly promote sustainable development of the region.

The "reasonable" in the rational allocation of water resources is reflected in a relatively fair and acceptable allocation scheme in the allocation process. A series of complex relationships exist, such as economic and ecological water use efficiency and the mutual transformation of different water sources. This is a goal that people hope to achieve when allocating scarce water resources. Generally, the effect of rational allocation is not that the individual obtains the highest or best benefit, but as far as the entire water resource distribution system is concerned, its overall benefit is the highest or best. The optimal configuration involves the means and methods to explore a reasonable configuration scheme.

5.3.1.2 The Principle of Rational Allocation of Water Resources

Due to the scarcity of water resources, their effective use is extremely important. Economic benefits reflect the wealth created by utilizing existing water resources, whereas social equity reflects its distribution among various social sectors and groups. In the long term, it is necessary to ensure that people have access to the continuous use of water resources, that is, to follow the principles of sustainability. According to the economic principles of scarce resource allocation and sustainable economic and social development requirements, the basic principles of rational water resource allocation must be followed: effectiveness, fairness, and sustainability (Dogra et al., 2011). Correspondingly, effective, equitable, and sustainable water policy measures based on social objectives are required.

Principle of validity

Water resources determine effectiveness as commodity attributes in economic market activities. From an economic perspective, because of investment in water conservancy projects, the distribution of water resources in various economic sectors can be understood as follows: water is a limited resource or capital, and various economic sectors can generate returns after use. The economic efficiency of water resource allocation implies that the marginal benefits of using water resources are the same in all water-use sectors to obtain the greatest social benefits.

In other words, the benefits of the additional use of one unit of water resource in one sector are the same as those in any other sector. If the marginal benefit of a sector is lower than that of other sectors, this portion of the water resources can be provided to a sector that can generate greater benefits. The use of water resources must be based on its economic benefits as an important basis for economic sector cost accounting, in addition to its ecological environment protection function (or benefit) to measure the status of the healthy development of society so that it can achieve the goal of making the best use of it. It should be noted that the effectiveness here is not purely economic but includes environmental benefits with a marginal negative impact on the environment and the pursuit of social benefits that can increase the per capita income of the entire society. This can guarantee the comprehensive utilization benefit of the coordinated development of the economy, society, and environment. Therefore, from the aspect of the rational allocation of water resources, it is necessary to set corresponding economic, social, and environmental development goals and, at the same time, pay attention to the degree of competitiveness and coordinated development among various goals to realize the effective allocation of water resources.

Principle of fairness

The principle of fairness seeks to protect the interests of all parties in the distribution of water resources between regions and social classes and to ensure fair distribution. This may follow or conflict with the validity principle. It focuses on the coordinated development between different regions (eastern, central, and western, various provinces, upstream and downstream, left and right banks, etc.), as well as the fair distribution of development achievements or water resource use benefits among all social classes within the same region. In China, the state owns water resources, and everyone owns them. From the perspective of each family, regardless of whether they can purchase water resources, they have the basic right to use them, which is a fair distribution of domestic water. It is necessary to take certain measures to protect the rights of low-income families, such as using different water price structures to allocate water resources according to differences in income levels. In addition, everyone is obligated to protect their water resources. "Whoever develops it will protect it, and whoever pollutes it will control it," which reflects fairness.

Principles of sustainability

The principle of sustainability considers the fairness of water resource distribution between generations and the rationality of comparing the total amount of water resources consumed by society in a certain period with the amount that future generations can obtain. The principle of sustainability implies that for short- and long-term water resources, current and future generations should coordinate and use resources fairly. The sustainable use of water resources should be guaranteed. In other words, the development and use of water resources by the current generation should not hamper the basic usage of future generations. In contrast, the predatory development and utilization of water resources and their destruction can cause serious pollution and waste. Water resources are unique resources that can be renewed throughout the water cycle, and the renewal cycles of water resources in different occurrence states differ. Therefore, to achieve sustainable development, the current generation must protect the regeneration capacity of water resources and ensure that economic and social activities are limited within their carrying capacity; at the same time, the discharge of pollutants is not higher than the capacity of the water environment according to the water resources in different regions; and the conditions and the different characteristics of precipitation, surface water, and groundwater are rationally developed and utilized to meet the needs of the current generation.

5.3.1.3 Chinese Practice of Water Resource Allocation

Long ago, China had early water diversion projects, such as digging ditches to divert water for irrigation and digging canals to transport goods and develop trade. According to historical sources, the Hangou Project, built in 486 BC, diverted water from the Yangtze River into the Huaihe River, which became the main water transportation artery in the eastern plains of China and the earliest inter-basin water transfer project in China. The chasm dug in 361 BC connected the Yellow and Huai Rivers and created the main waterway and navigation route between them. Since then, the Dujiangyan Water Conservancy Project, Ling Canal, and Beijing-Hangzhou Grand Canal have enabled water resource allocation for economic and social development. After the founding of the People's Republic of China, the South-to-North Water Diversion Project, Guangdong's Dongshen Water Supply Project, Gansu's Water Diversion to Qin, Tianjin's Luan to Tianjin, Shandong's Yellow River to Jiqing, Shanxi's Yellow to Shanxi, Liaoning's Yellow River to Shanxi, Liaoning's Bi-Jiang Water Diversion, Jiangsu's River-to-North Water Diversion, Jilin Important water diversion projects, including the introduction of pine into Changjiang were undertaken. They provide stable and reliable water sources for receiving areas and promote the economic and social development of the region (Ministry of Water Resources, 2021).

The distribution of water resources in northern and southern China is quite different. In the north, especially in the Huanghuaihai region, the population is dense, the plain land is rich, the light and heat resources are rich, and the mineral resources are rich; however, the water resources are very scarce, which seriously hinders economic and social development. The demand for water resources in the northern region is increasing with development. Still, surface water is overused, and groundwater is overexploited, further exacerbating the water shortage problem. Taking the Beijing-Tianjin-Hebei region as an example, it covers an area of 217,000 square kilometers, accounting for 2.2% China; the total population is 110 million, accounting for 8.0% of China; the regional GDP accounts for more than 8.0% of China; and the effective irrigation area is 77 million mu, accounting for 8.2% of China. However, compared with the national average, there is a big gap in the annual average precipitation, total water resources, and per capita water resources. Before the construction of the South-to-North Water Diversion Project, the Beijing-Tianjin-Hebei region had to source water in a way that would destroy the sustainable development of water resources and maintain economic and social development. The contradiction between the carrying capacity of water resources, economic and social development, and the construction of ecological civilization has become increasingly prominent. The over-exploitation area of groundwater accounts for more than 90% of the total plain area, forming dozens of funnel areas where the groundwater level drops and land subsides in some areas. This situation is serious (Liang et al., 2012).

As the population and economic layout cannot be changed on a large scale in the short term, the use of cross-basin water transfer to transfer water resources from water-rich areas to water-deficient areas to meet their development needs is an effective measure for solving the shortage in China's water resources. The South-to-North Water Diversion Project is a typical example of water resource allocation. Since its opening to water, it has played an important role in assisting water supply safety, groundwater overexploitation, water environmental protection, and implementing major national strategies, greatly promoting the economic, social, and ecological development of water-receiving and water source areas.

5.3.2 *Water Resource Allocation System*

5.3.2.1 Water Transfer

With the rapid development of China's economy and society and the acceleration of industrialization, industrial and urban water use is increasing daily. The transfer of agricultural water to urban, industrial, and ecological water has become the main trend in water resource allocation use. Thus, structural contradictions have become increasingly prominent. The transfer and use of water resources between different water users and departments can solve new problems in developing other departments, promoting the efficient use of water resources, and improving allocation efficiency (Zheng, 2010). In addition, the cross-regional transfer of water resources can facilitate rational use and promote coordinated regional development. It is necessary to establish sound compensation, revenue sharing, and coordinated management mechanisms for the cross-regional transfer and use of water resources so that rules and regulations can govern the cross-regional use of water resources and the efficiency of water resource utilization can be improved. This is not only an economic issue (considering the economic efficiency of water use) but also a social (social equity in water use) and ecological (sustainable use of water) issue.

5.3.2.2 Market for Water Resources

Water is an important resource for human survival, and everyone has a basic right to water. When optimizing the allocation, development, and utilization of water resources, it is necessary to consider the basic water requirements upstream and downstream of a certain basin, left and right banks, different regions, and different groups. Therefore, it is a special commodity that follows the same pricing principle of the law of value as ordinary commodities but also has to weigh the purchasing power of users. The price is scientifically determined according to cost conditions, government regulations, and user negotiations. The allocation of water resources should consider technical feasibility, economic rationality, and political fairness. It is necessary to give full play to the adjustment function of the water resources market, coordinate the distribution of interests among different regions and groups, reduce regional differences and polarization between the rich and the poor, and realize the rational allocation of water resources.

Under the planned economic system, because the subject of water rights and ownership is unclear, the water resources market and transactions cannot be developed, the disorderly development of water resources further deteriorates, efficiency is low, and the conflict between supply and demand is becoming increasingly serious. The optimal allocation and sustainable use of water resources must establish, cultivate, and develop a water resources market based on water conservation so that it can be transferred from low-income areas and used in high-efficiency areas and truly reflect the commodity attributes of water and the value of water rights. The establishment and improvement of the socialist market economy system have created favorable conditions for promoting the rational flow of water resources and giving better play to the important role of the market in resource allocation. Establishing a water resource market should follow the following principles: 1. Water is a special commodity with a value of 2. Its ownership belongs to state 3. Water rights, which can be transferred and exchanged, are an important condition for developing the water-resource market, which can be transferred and exchanged; 4. The government can reasonably determine prices (Li et al., 2021).

5.3.2.3 Price of Water Resources

From the economic perspective, water resources belong to the category of economic resources, with production and consumption needs. There is a large gap between supply and demand, with a price. The users and beneficiaries of water resources in their natural state and after development should pay for their use. Water price is a tax levied on the main body of water resources to better manage and protect water resources and achieve sustainable development (Anuradha & Ambujam, 2010).

In the face of different water resource situations, the price is also different and is affected by factors such as water quantity, water quality, geographical location, use, supply, and demand conditions. The charging standard for water must comprehensively consider these factors to ensure the rationality of the price. For example, in general, the price of water is lower in periods of high water than in periods of poor water, the price of water with higher quality water resources is often higher than that of poor -quality water, and in areas with abundant water resources, the price of water is generally lower than that of areas with scarce water resources.

The new "Water Law of the People's Republic of China" (revised in July 2016) stipulates those units and individuals who directly draw water resources from rivers, lakes, and underground must pay to source water under the provisions of the national water drawing permit system and the system of paid use of water resources. Administrative departments and river basin management agencies apply for water licenses and pay water resource fees to obtain water rights. Household life, sporadic free-range, captive livestock, poultry drinking, and other small amounts of water were excluded. Water use adopts metered charging and an over-quota progressive price increase system (The Standing Committee of the National People's Congress of China, 2016). The charged items for water resources in China include water resource fees, water conservancy project water fees, urban tap water fees, urban sewage treatment fees, and sewage discharge fees of enterprises and institutions.

5.4 Water Resource Management

5.4.1 Water Rights Management

5.4.1.1 The Meaning of Water Rights Management

Water rights are the rights enjoyed by units or individuals following the law regarding the rated water resources of a certain place. Water rights in China include rights to use and benefit from water resources. Water rights management refers to the water administrative departments of governments at all levels, as representatives of the property rights of state-owned water resources, and uses economic, administrative, legal, and other means to control the acquisition, use, and fulfillment of obligations of water rights holders. Supervision and management activities were also conducted in other respects. Once water rights are damaged, they can be applied to suspend the damage or obtain the corresponding compensation according to the law. Water rights management aims to ensure fair and reasonable development, use, and protection of water resources, meet the water demand of society to the greatest extent, and achieve the greatest economic, social, and environmental benefits.

The ownership of water resources has three meanings. First, the water resources belong to the state; all the people jointly own them, and the relevant laws and regulations have stipulated this. This aspect is mainly reflected in the state's unified planning and dispatching of water resources and the issuance of water abstraction licenses. Second, the "Water Law of the People's Republic of China" stipulates that the water in the ponds of rural collective economic organizations and the reservoirs they built and managed should be used by the corresponding rural collective economic organizations. Third, the legal right to use water resources is protected by the units and individuals that develop and use water resources according to the law, which manifests the principle of the separation of ownership and uses rights of water resources. Water rights can be obtained according to law, canceled according to law, and adjusted by the state when necessary.

The main contents of water rights management are as follows: First, grant water rights following the law, that is, through the registration of water rights or the implementation of a water drawing permit system, and the water rights are granted to applicants after approval through legal procedures. The second is to promulgate laws and regulations related to water rights implementation, such as the rights and obligations of water rights holders, the implementation scope and methods of water-drawing licenses, and the adjustment of water rights. The third is supervising and managing the exercise of rights and the performance of water rights holders' obligations. Fourth, the right to use water resources is shared. Water resources have mobility and multiple functions and can be reused or used by different entities. The fifth is the timely adjustment of water rights. The government's water administrative department can adjust water rights according to the state's requirements for construction during different periods or in the case of severe drought.

5.4.1.2 The Rights and Obligations of Water Rights Holders

The rights that water rights holders can exercise include obtaining the right to use rated water resources according to the water abstraction permit and the right to construct water projects with rated water resources, which is based on compliance with statutory infrastructure procedures and approval requirements for construction projects within the scope of river management—conditions: the right to benefit from the production of commercial water, such as water fees. When water rights obtained according to the law are infringed upon, they can apply to the water administrative department for administrative protection or to the court for judicial protection to protect their legitimate rights and interests. The obligations that water rights holders must perform mainly include strictly implementing the water-drawing permit system, paying water resource fees in full and in time as required, accepting the supervision of administrative law enforcement departments, obeying the needs of flood control and water hazard prevention, and consciously and proactively doing a good job in water pollution prevention.

5.4.1.3 Water Rights Control System

Effective management of water rights requires an efficient control system. Building a perfect water rights trading market requires a "water bank," a water-related bank, which must be established—a configuration method used to operate rights. Like a normal bank, water is considered "currency," Some entities deposit excess water into it, and some entities withdraw it when they need it. Some entities choose to obtain water from loans, and interest on loans is higher than interest on deposits. The water depositors can get the deposit interest, and the "water bank" can also get the interest difference between the deposit and the loan.

The water bank acts as an intermediary between the buyers and sellers of water resources, providing market transaction information and trading platforms that can adjust the supply and demand of water resources, promote scientific pricing of water rights, reduce transaction costs, and facilitate the sustainable supply of water resources. It is generally a for-profit commercial organization, independent of government departments, and subject to effective government supervision. It can avoid negative influences, such as the non-marketization of water rights trading and the distortion of market prices caused by excessive government intervention. Transferring the right to use water resources generates tangible benefits and meets the needs of the water resource market. This will help people improve water conservation awareness and cultivate good water habits and enable the society to better distribute and use water resources. China can learn from developed countries (Papacharalampou et al., 2016).

5.4.2 *Development and Utilization of Water Resources*

With economic and societal development, there is an urgent need to change the development and utilization of water resources in China from extensive to intensive conservation. China has begun to control its entire water resource development and utilization process. According to the "2020 China Water Resources Bulletin" issued by the Ministry of Water Resources of China in 2021, we can understand the current status of water resource development and utilization in China (Ministry of Water Resources, 2021).

5.4.2.1 Water Supply

In 2020, the total water supply in China was 581.29 billion cubic meters, accounting for 18.4% of the total water resources that year. Water sources were 479.23, 89.25, and 12.81 billion cubic meters, respectively, accounting for 82.4, 15.4, and 2.2% of the total water supply, respectively. The total water supply decreased by 20.83 billion cubic meters compared to the previous year. Among them, the water supply from surface water sources and groundwater sources decreased by 19.02 and 4.18 billion cubic meters, respectively, while the water supply from other water sources increased by 2.37 billion cubic meters. In the water supply volume of surface water sources, the proportion of water supply in the first-level interval of water resources, the water supply volume of the water lifting project, the water supply volume of the water diversion project, and the water supply volume of the water storage project were 4.8, 31.0, 31.3, and 32.9%, respectively. Regarding the water supply of the groundwater sources, the proportions of brackish water, deep confined water, and shallow groundwater were 0.4, 3.9, and 95.7%, respectively. In addition, rainwater harvesting projects and recycled water account for 6.2 and 85.0% of the water supplied from other sources, respectively.

In 2020, the total water supply in the six regions of northern China was 268.13 billion cubic meters, and the total water supply in the four southern regions was 313.16 billion cubic meters. They account for 46.13 and 53.87% of the total national water supply, respectively, the latter being 1.17 times that of the former. Among the six northern districts, the Northwest Rivers District had the largest total water supply, reaching 67.57 billion cubic meters, while the Liaohe District had the least total water supply, 191 cubic meters. Among the four southern regions, the Yangtze River region has the largest total water supply, reaching 195.76 billion cubic meters; the southeastern river region has the smallest total water supply, 29.51 billion cubic meters.

Among the provinces and cities, the total water supply in 2020 ranked first in Jiangsu with 57.20 billion cubic meters, much higher than most other provinces and cities, followed by Xinjiang with 57.04 billion cubic meters, Guangdong, Heilongjiang, and Hunan ranked third, fourth, and fifth, respectively. In addition to Anhui, Jiangxi, Shandong, Henan, Hubei, Guangxi, and Sichuan, the total water supply is greater than 20 billion cubic meters. The three provinces and cities with the least total water supply that year were Qinghai (2.43 billion cubic meters), Tianjin (2.78 billion cubic meters), and Tibet (3.22 billion cubic meters).

5.4.2.2 Water Consumption

After 1997, total water consumption in China generally showed a slow growth trend and remained relatively stable after 2013. In 2020, total water consumption was 581.29 billion cubic meters. The agricultural, industrial, and domestic water amounts were 3612.4, 1030.4, and 863.1 billion cubic meters, respectively, and the proportions of the total water consumption were 62.1, 17.7, and 14.9%, respectively. The artificial ecological environment's water supply reached 30.70 billion cubic meters, accounting for 5.3% of total water consumption. Compared to the previous year, the total water consumption in China decreased by 20.83 billion cubic meters. Among them, agricultural, industrial, and domestic water use decreased by 6.99 billion, 18.72 billion, and 860 million cubic meters, respectively, while the artificial ecological environment water supply increased by 5.74 billion. Primary, secondary, and tertiary industries accounted for 62.2, 17.7, and 4.1% of total water consumption, respectively.

5.4.2.3 Water Use Indicator

In 2020, China's per capita total water consumption was 412 cubic meters. The national water consumption per 10,000 yuan of GDP (current-year price) was 57.2 cubic meters, and the water consumption per 10,000 yuan of industrial added value (current-year price) was 32.9 cubic meters, 84% lower than that in 1997 at comparable prices. The actual average water consumption per mu of farmland irrigation is 356 cubic meters, the effective utilization coefficient of farmland irrigation water is 0.565, the average domestic water consumption per person in urban areas (including public water) is 207 L/day, and the average domestic water consumption per person in rural areas is 100 L/day. Water-use efficiency has significantly improved since 1997. Water consumption per 10,000 yuan of GDP, water consumption per 10,000 yuan of industrial added value, and average water consumption per mu of actual irrigation of cultivated land declined significantly. Per capita, comprehensive water consumption was roughly maintained at 400–450 cubic meters.

In northern China, the per capita comprehensive water consumption in 2020, the water consumption per 10,000 yuan of GDP (current year price), the water consumption per 10,000 yuan of industrial added value (current year price), and the actual irrigation water consumption per mu of cultivated land. The average water consumption is 1962 cubic meters in the Northwest River District, 361.1 cubic meters in the Northwest River District, 44.5 cubic meters in the Songhua River District, and 522 cubic meters in the Northwest River District. The per capita domestic water consumption of urban residents (including public water) and that of rural residents in the Northwest River District, is 211 and 125 L/day, respectively. In contrast, Haihe District has the lowest per capita domestic water consumption: 138 and 80 L/day in the Yellow River area. In the four southern districts, the per capita comprehensive water consumption in 2020, the water consumption per 10,000 yuan of GDP (current year price), the water consumption per 10,000 yuan of industrial added value (current year price), and the average per mu of cultivated land irrigated. The highest water consumption is 478 cubic meters (Southwest River District), 114.1 cubic meters (Southwest River District), 52.9 cubic meters (Yangtze River District) and 679 cubic meters (Pearl River District), corresponding to the lowest value of 1.46, 3.46, 2.52 and 1.70 times.

Among the provinces and cities, the highest per capita total water consumption in 2020 was in Xinjiang (2,218 cubic meters), much higher than in other provinces and cities. The second, third, and fourth highest were Ningxia (977 cubic meters, Heilongjiang (976 cubic meters), and Tibet (886 cubic meters), respectively. There are 16 provinces and cities higher than the national level. The three provinces and cities with the lowest comprehensive water consumption per capita were Beijing (185 cubic meters), Tianjin (201 cubic meters), and Shanxi (208 cubic meters). In terms of water consumption per 10,000 yuan of GDP, the three highest provinces are Xinjiang with 413.4 cubic meters, Heilongjiang with 229.3 cubic meters, and Ningxia with 179.1 cubic meters, which are 36.91 times, 20.47 times, and 15.99 times that of Beijing, the province with the least per capita total water consumption, respectively. This shows that, like some provinces, there is still much room for improvement in water efficiency.

5.4.3 Water Pollution Control

With economic and societal development, China's efforts to prevent and control water pollution and protect water resources have been continuously strengthened. By applying legal, administrative, engineering, technical, and other means and measures, such as structural, management, and engineering emission reduction, water pollution control has been steadily promoted, and good control results have been achieved. According to the *2015 Bulletin on the State of China's Environment* (Ministry of Ecological Environment of China, 2016) and the *Bulletin on the State of China's Ecological Environment in 2020* (Ministry of Ecology and Environment, 2021), three aspects of freshwater pollution, marine pollution, and acid rain will be introduced.

First, freshwater pollution has greatly improved. In 2020, among the 1,937 water quality sections (points) monitored by the national surface water system, the proportion of water quality sections I–III (points) accounted for 83.4%, an increase of 8.5 percentage points compared to the previous year and 2015. 972 state-controlled sections of surface water increased by 18.9 percentage points; the proportion of the inferior V category was 0.6%, a decrease of 2.8 percentage points from the previous year and a decrease of 8.2 percentage points from 2015. The main pollution indicators were total phosphorus, chemical oxygen demand, and the permanganate index. 30 cities, including Zhangye, Liuzhou, and Guilin, have relatively good water environmental quality in the national surface water assessment section, while 30 cities, such as Xingtai, Tongchuan, and Cangzhou, have relatively poor water environmental quality.

In 2020, among the 1,614 water quality sections monitored in the seven major river basins of the Songhua River, Liaohe River, Pearl River, Huaihe River, Haihe River, Yellow River, and Yangtze River, as well as major rivers in Zhejiang and Fujian, southwest rivers, and northwest rivers, the Grade I–III water quality sections were 87.4%, which was 8.3 percentage points higher than that of the previous year, and 15.3 percentage points higher than that in 2015 (700 state-controlled sections); the proportion of the inferior V category was 0.2%, which was lower than that of the previous year—increased by 2.8 percentage points, a decrease of 8.7 percentage points from 2015. The main pollution indicators were chemical oxygen demand, permanganate index, and five-day BOD. Rivers in the northwest, southwest, Yangtze River Basin, Jiang-Fujian Rivers, and Pearl River Basin have excellent water quality. The Yellow, Huai, and Songhua River Basins have good water quality, whereas the Haihe and Liaohe River Basins are slightly polluted.

In 2020, among the 10,171 groundwater quality monitoring points in the natural resources department (7923, 1338, and 910 groundwater monitoring points in plain basins, bedrocks in hilly and mountainous areas, and mountainous karst areas, respectively), the water quality monitoring points of Class I–III, Class IV, and Class V will be The proportions are 13.6, 68.8 and 17.6% respectively; among the 10,242 groundwater quality monitoring points (mainly shallow groundwater) of the water conservancy department, the proportion of water quality monitoring points of Class

I–III, Class IV, and Class V were 22.7, 33.7, and 43.6%, respectively. The main excess indicators were Mn, total hardness, and total dissolved solids.

Secondly, the overall marine environment is stable and improving, and marine pollution requires continuous attention. In 2020, the area close to the sea with Class I water quality in China's jurisdictional waters accounted for 96.8%, which was roughly stable compared with the previous year, while the sea areas with water quality inferior to Class IV water was 30,070 square kilometers, an increase of 1,730 square kilometers over the previous year, a decrease of 9,950 square kilometers from 2015. Inorganic nitrogen and active phosphate were the main excess indicators. The proportion of monitoring sites with good sediment quality was 96.5%. The quality of coastal waters across China is generally stable and has improved. The sea areas with excellent (Class I, II) water quality accounted for 77.4%, an increase of 0.8 percentage points over the previous year and an increase of 6.9 percentage points over the year 2015; the proportion of the class was 9.4%, a decrease of 2.3 percentage points from the previous year, and a decrease of 8.9 percentage points from the year 2015. The main indicators that exceeded the standard were active phosphate and inorganic nitrogen. The water quality in the coastal waters of Guangxi, Shandong, Hainan, Liaoning, Hebei, Guangdong, and Fujian was good. The water quality in the coastal waters of Tianjin was average. The water quality in the coastal waters of Zhejiang and Jiangsu was poor, and the water quality in the coastal waters of Shanghai was poor. Sea water quality is extremely poor. In 2020, among the 24 typical marine ecosystems monitored, there were seven, 16, and one healthy, sub-healthy, and unhealthy states, respectively.

Third, the problem of acid rain was alleviated. In 2020, the acid rain area in China was approximately 466,000 square kilometers, accounting for 4.8% of national land area, which was 0.2 and 2.8 percentage points lower than in the previous year and 2015, respectively. The proportion of the national land area was 0.4%. In 2020, the average frequency of acid rain in 465 cities or districts monitoring precipitation was 10.3%, an increase of 0.1 percentage points compared with 2019 (469 cities or districts monitoring precipitation) and a decrease of 3.7 percentage points compared with 2015 (480 cities or districts monitoring precipitation). The proportion of cities experiencing acid rain was 34.0%, an increase of 0.7 percentage points from 2019 and a decrease of 6.4 percentage points from 2015; the proportion of cities with acid rain frequencies of 75% and above, 50% and above, and 25% and above, and 2.8, 7.5, and 16.3%, respectively, were 2.2 percentage points, 5.2 percentage, and 4.5 percentage points lower than that of 2015, respectively. The annual average pH of precipitation in China ranges from 4.39 to 8.43, with an average of 5.60. The proportion of acid rain, heavy acid rain, and cities with acid rain were 0.2, 2.8, and 15.7%, respectively, which were 0.8 percentage points, 5.7 percentage, and 6.8 percentage points lower than those in 2015, respectively. In general, the acid-rain type was sulfuric acid.

5.4.4 *Measures to Ensure Water Resources*

5.4.4.1 Water Resource Legislation

The government should formulate and improve relevant laws and regulations on water resources, take the lead in preventing and controlling water pollution, and focus on planning, supervising, and managing water pollution prevention and control. With economic and societal development, China's laws and regulations on water pollution prevention and control have continuously improved, and related policies and systems have become increasingly effective, thereby providing an important guarantee for the protection of water resources. On May 11, 1984, the Fifth Session of the Standing Committee of the Sixth National People's Congress passed the Law of the People's Republic of China on the Prevention and Control of Water Pollution. On January 21, 1988, the 24th meeting of the Standing Committee of the Sixth National People's Congress passed the "Water Law of the People's Republic of China," the first basic law regulating water affairs in China (Vice-Gao Erkun, 2003). In December 2010, the State Council promulgated the "National Main Function Zone Plan," which put forward requirements for the development and utilization of water resources (The State Council of China, 2011). In April 2015, the State Council issued the "Water Pollution Prevention and Control Action Plan," an action guide for China's water pollution prevention and control (The State Council of China, 2015).

Water pollution prevention and control in China has shifted from decentralized governance to a combination of centralized and decentralized governance, end-to-end governance to whole-process control, single-concentration control to total and concentration control, and regional management as the mainstay of area management combined with watershed management. Supervision, management, and law enforcement of water resources in China have been continuously strengthened to ensure the sustainable development of water resources.

5.4.4.2 Establish and Improve the Water Resources Market

The development of the water resource market is related to the security of water resources. Regarding water pollution control, aquatic environment control cannot be equated with compulsory administrative measures. It is necessary to strengthen supervision and law enforcement, strengthen the operational supervision of urban sewage treatment plant facilities, and establish a government accountability mechanism for water conservation and emission reduction. Contrastingly, economic means should be used mainly to fully mobilize the enthusiasm of water users and pollutant discharge units by establishing incentive and restraint mechanisms so that water resource conservation, pollutant discharge, and comprehensive treatment should become conscious actions of enterprises and the implementation of water resources laws and regulations should be improved. In April 2004, the General Office of the State Council issued a "Notice on Promoting Water Price Reform to Promote

Water Conservation and Protection of Water Resources," which promoted water price reform and better-leveraged market mechanisms and price levers for water resource allocation, water demand regulation, and water pollution. This can help build a water-saving society and promote the sustainable use of water resources (General Office of the State Council of China, 2004).

Increasing investment in scientific and technological research and development in water resources, and promoting the development of water resource conservation and emission reduction technologies are needed. Accelerate the construction of technological innovation systems; develop the water pollution control technology service industry; and rely on scientific and technological progress and advanced management levels to achieve low-cost, high-efficiency water resource development, and the operation of pollution control facilities. In addition, plants can be cultivated through market-oriented means, and the use of plants can effectively prevent soil erosion and have various purification effects, especially on forests, to promote the natural protection of water resources.

5.4.4.3 Government Macro-Control

The government must check the source well and conduct overall planning, coordination, and comprehensive decision-making. To do a good job in allocating water resources between different regions, it is necessary not only to maintain the water rights of each region but also to save water sources and protect the environment. In terms of environmental protection approval, a pollution one-vote veto system was adopted to approve new projects, and the water environment impact assessment of the projects was strictly carried out. It is strictly forbidden to approve projects contrary to the national industrial policy and development plan, phase-out projects prohibited by the state, and projects that do not meet the requirements of environmental standards.

Water pollution control must be combined with the restoration and improvement of the ecological environment. Build a certain number of modern sewage treatment plants, strengthen the application of treated water, and lead treated clean water to rivers or reuse it for other purposes. The development and utilization of water resources should be considered for the entire basin to promote a fundamental solution to aquatic environment problems. The basin nature of water pollution issues coordinates production, living, and ecological water use, coordinates the relationship between water use in various regions, and strengthens cooperation between regions along rivers and lakes.

5.4.4.4 Community Participation

Protecting water resources requires the extensive participation of all sectors of society, and everyone should be conscious of protecting water resources. However, there are still some phenomena that affect wastewater resources. For a long time, the construction and operational management of urban drainage facilities and sewage treatment plants have mainly been based on the investment of the state and local governments. The main reasons for serious water pollution in China are the wastage of water resources, the low cost of pollution, and the lack of awareness of water resource protection. Therefore, there is an urgent need to change the concept and implement "whoever pollutes will control" water pollution and collect water pollution control fees while charging water fees.

Carry out publicity and education on protecting water resources through various means to improve the public's awareness of saving and protecting the environment. For example, warning signs are set up on the seaside of rivers and lakes, calling on people to protect water resources jointly, and encouraging the public to participate actively in environmental management. Relevant management departments should establish channels for communication with surrounding residents, publicize reporting methods, and allow residents to join teams that monitor pollution sources to identify problems and deal with them in a timely manner, effectively exerting great enthusiasm among people to protect the water environment and carrying out comprehensive supervision and protection of the water environment.

5.5 Summary

This chapter first introduces the importance of water resources from the perspectives of human beings, animals, plants, environmental systems, and economic and social development. It then describes the concept and principle of rational allocation of water resources, as well as China's great practices in water resource allocation, and analyzes its allocation system regarding water resource transfer, market, and price. It then introduces the meaning of water rights management, the rights and obligations of water rights holders, and the water rights control system; it discusses the development and utilization of water resources in China from the perspective of water supply, water consumption, and water use indicators; and discusses water resource pollution control. Finally, safeguard measures are discussed from the aspects of water resource legislation, establishment and improvement of its market, government macro-control, and participation of all sectors of society. Through the study in this chapter, you can gain a more comprehensive and in-depth understanding of the water resource field, which will help you understand the government's policies and measures on water resources and guide people's specific practices in daily life.

References

Aba, M. M., Parente, V., & dos Santos, E. M. (2022). Estimation of water demand of the three major Brazilian shale-gas basins: Implications for water availability. *Energy Policy, 168*, 113170. https://doi.org/10.1016/j.enpol.2022.113170.

Anuradha, B., & Ambujam, N. K. (2010). Impact of water resources protection on local ground water market. *Journal of Water Resource and Protection, 02*(8).

Dąbrowska, J., Menéndez Orellana, A. E., Kilian, W., Moryl, A., Cielecka, N., Michałowska, K., et al. (2023). Between flood and drought: How cities are facing water surplus and scarcity. *Journal of Environmental Management, 345*, 118557. https://doi.org/10.1016/j.jenvman.2023.118557.

Dogra, P., Sharda, V. N., Ojasvi, P. R., Prasher, S. O., & Patel, R. M. (2011). Optimal allocation of water for enhanced food production in a mid-Himalayan watershed. Ceska Slov Farm.

General Office of the State Council of China. (2004). Notice on promoting water price reform, promoting water conservation and protecting water resources [EB/OL]. (2004-04-19) [2021-09-21].

Hartmann, A., Goldscheider, N., Wagener, T., Lange, J., & Weiler, M. (2014). Karst water resources in a changing world: Review of hydrological modeling approaches. *Reviews of Geophysics, 52*(3), 218–242.

Li, J. Y., Cui, L. B., Dou, M., & Ali, A. (2021). Water resources allocation model based on ecological priority in the arid region. *Environmental Research.*

Liang, Y. S., Wang, W., Li, H. J., Shen, X. H., Xu, Y. L., & Dai, J. R. (2012). The south-to-north water diversion project: Effect of the water diversion pattern on transmission of Oncomelania hupensis, the intermediate host of Schistosoma japonicum in China. *Parasites & Vectors, 5*(1), 1–6.

Ministry of Ecology and Environment of China. (2016). Ecological and Environmental Bulletin [EB/OL]. (2016-05-20) [2021-09-21].

Ministry of Ecology and Environment of China. (2021). Bureau of Ecology and Environment [EB/OL]. (2021-05-24) [2021-09-21].

Ministry of Water Resources of China (2021a). 2020 China Water Resources Bulletin. [EB/OL]. (2021-07-09) [2021-09-21]. 2020 China Water Resources Bulletin.

Ministry of Water Resources of China. (2021b). What are the major water transfer projects in China. [EB/OL]. (2021-09-01) [2021-09-21].

Ministry of Water Resources of the People's Republic of China: What are the major water-transfer projects in China?

Papacharalampou, C., Mcmanus, M., Newnes, L. B., & Green, D. (2016). Catchment metabolism: Integrating natural capital in the asset management portfolio of the water sector. *Journal of Cleaner Production*, S0959652616319291.

The State Council of China. (2011). National main functional area planning. (2011-06-09) [2021-09-21]. Circular of the State Council on Printing and Distributing the National Plan for Major Functional Areas.

The State Council of China. (2015). Water pollution prevention action plan [EB/OL]. (2015-04-16) [2021-09-21]. Circular of the State Council on printing and distributing the action plan for the prevention and control of water pollution. http://www.gov.cn/zhengce/content/2015-04/16/content_9613.htm

The Standing Committee of the National People's Congress of China (2016). Water law of the people's Republic of China [EB/OL]. (2016-07-02) [2021-09-21].

UNESCO. (2021). The United Nations World Water Development Report 2021: Valuing Water. [EB/OL]. (2021-03-21) [2021-09-21]. United Nations Educational, Scientific and Cultural Organization, Paris.

Vestal, M. K., Guerrero, B. L., Golden, B. B., & Harkey, L. D. (2017). The impact of discount rate and price on intertemporal groundwater models in grant county, Kansas. *Water Resources and Protection, 9*(7), 15.

WMO, UNESCO. (2012). International glossary of hydrology. WMO-No.385.
Yang, J., Wang, Z., & Zhou, Z. L. (2013). Research on water resources allocation in the west Anhui based on analytic hierarchy process. *Journal of West Anhui University*.
Zheng, X. W. (2010). Calculation of water resources value of inter-basin water transfer projects. *Journal of Economics of Water Resources*.

Chapter 6
A Fixed and Versatile Resource: Land Resources

This chapter focuses on analyzing the current situation of land resource utilization, characteristics of land use change, economic analysis of land resource allocation, causes of ineffective land use and utilization change, and remedies based on market innovation policies. On this basis, the Malmquist index is used to evaluate the total factor productivity, technical progress, technical efficiency, pure technical efficiency, and scale efficiency of land use in 31 provinces (municipalities) in China, and to analyze their spatial and temporal evolution characteristics.

6.1 Introduction

Since the Industrial Revolution, the world has experienced rapid economic development and urban expansion. According to UN estimates, the urbanization rate of developed countries will reach 86% in 2050, whereas China's urbanization rate will reach 71.2% in 2050. An increasing urbanization rate means more people are shifting from rural to urban areas (Deng et al., 2021). Cities need to build more residential houses to accommodate the large newly transferred population, and the demand for residential bases is increasing. In addition, as the scale of nonagricultural industrial development gradually expands, the demand for urban construction land, as the main carrier of nonagricultural production activities, increases. As the total land resources are limited, the area of urban construction land increases, and the area of other types of land is bound to decrease. To ensure a continuous supply of urban construction land, many cities convert agricultural lands around the city, such as arable land, forest land, and grassland into urban construction land, which causes a significant decrease in the number of agricultural land resources. Once arable land resources are converted into urban construction land, they are difficult to effectively replenish in a short period of time, leading to decreased food production and threatening food security (Si et al., 2021). The significant reduction of agricultural land resources, such

X. Deng et al., *Environmental and Natural Resources Economics*,
https://doi.org/10.1007/978-981-99-9923-1_6

as woodlands and grasslands, may lead to ecological and environmental problems, such as the sharp decline of biological species, soil erosion, and land desertification, threatening ecological security.

The 2030 Agenda for Sustainable Development was adopted by 193 UN member states in September 2015 (Jain et al., 2020). It sets out 17 Sustainable Development Goals (SDGs) that present the global challenges we face, including those related to poverty, inequality, climate, environmental degradation, prosperity, peace, and justice. Interconnected and designed to leave no one behind, these goals are a blueprint for a better and more sustainable future for all (Meinshausen et al., 2015). Therefore, under the reality of limited total land and serious ecological damage, to achieve the sustainable development goals, both economic output and environmental protection must be considered in the process of land resource utilization. This brings a series of problems such as how to optimize land allocation, improve land use efficiency, and how to apply scientific research methods to measure land resource use efficiency. To solve the above problems, this chapter first depicts the current situation of land use in China from two aspects: the quantitative and structural characteristics of land use and its spatial distribution analyzes the characteristics of land use changes in China since 1990, and the problems of land resource use in China. Second, it examines the economic principles of land use, land use transformation, and optimal land allocation. Furthermore, the problems associated with inefficient land use in industrialized countries and the special problems of inefficient land use in developing countries are analyzed, and remedial methods based on market innovation policies are proposed. Finally, the Malmquist index is used to measure the total factor productivity, technical progress, technical efficiency, pure technical efficiency, and scale efficiency of agricultural and construction land use in 31 provinces (municipalities) of China from 2006 to 2019, and to analyze their spatial and temporal evolution characteristics.

6.2 Land Use

6.2.1 Analysis of Current Land Use

6.2.1.1 Land Use Quantity Structure Characteristics

China is located in the eastern part of Asia and on the west coast of the Pacific Ocean, with a land area of about 9.6 million square kilometers, a mainland coastline of more than 18,000 km, an island coastline of more than 14,000 km, and a water area of more than 4.7 million square kilometers in the inland and border seas. In 2020, China's land use structure was divided into: 1.78 million square kilometers of arable land, accounting for 18.76% of the total land area; 2.26 million square kilometers of forest land, accounting for 23.76% of the total area; 2.71 million square kilometers of grassland, accounting for 28.51% of the total land area; 0.30 million

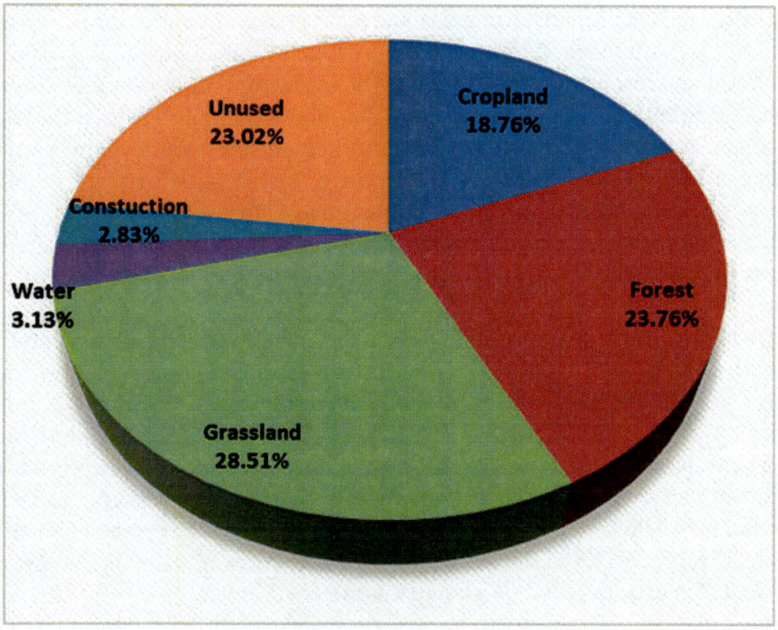

Fig. 6.1 China's land use structure in 2020

square kilometers of water area, accounting for 3.13% of the total land area; 0.27 million square kilometers of construction land, accounting for 2.83% of the total land area; and 2.19 million square kilometers of unused land, accounting for 23.02% of the total land area. In terms of the structure of land resource utilization, China has a high proportion of arable land, a low proportion of construction land, and a high proportion of unused land, indicating that there is more room for future land development and utilization. The proportion of ecologically functional land, such as grassland and forest land, is high, with rich ecological resources and a high ecological value (Fig. 6.1).

6.2.1.2 Spatial Distribution of Land Use

As observed in Fig. 6.2, arable and forest land are mainly distributed in the northeast, central, and southeast coastal regions of China; grassland is mainly distributed in the central and western regions of China; watershed and construction land are more scattered; and unused land is mainly distributed in the northwest of China. The 2020 land use types in China are mainly grassland, followed by forest land and unused land, arable land, watershed and construction land, and each type of land. The distribution of each type of land is uneven in space. Among them, grassland resources are concentrated in four provinces—Tibet, Inner Mongolia, Xinjiang, and Qinghai—with grassland areas of 555,562, 526,253, 486,456, and 396,639 square kilometers, respectively; and the proportion of grassland areas in these four provinces is as high as 72.46%, while the number of grassland resources in the remaining provinces (cities) is small. Woodland resources are widely distributed, except in six provinces (cities)—Shanghai, Beijing, Tianjin, Jiangsu, Ningxia, and Shandong—where the number of woodland resources is low. The remaining regions are more evenly distributed. The largest area of unused land is in Xinjiang, with 994,666 square kilometers, followed by Tibet, Inner Mongolia, Qinghai, and Gansu, with 395,327, 308,654, 248,432, and 152,676 square kilometers, respectively, while the remaining areas have less unused land and are relatively evenly distributed. Arable land resources are widely distributed; the largest amount of arable land is in Heilongjiang, with 173,297square kilometers, followed by Sichuan, Inner Mongolia, Henan, and Shandong, with 117,935, 113,543, 103,101, and 101,328 square kilometers, respectively. The least amount of arable land is in Shanghai, with only 3,215 square kilometers, followed by Tianjin and Taiwan, with 5,864 and 6,679 square kilometers, respectively. The water area is relatively small, concentrated mainly in Tibet, Xinjiang, and Qinghai, with an area of 80,238, 35,349, and 33,246 square kilometers, respectively, accounting for 49.95% of the total water area of the country. The area of construction land is relatively small, and the least area of construction land among the provinces (cities) is in Tibet, with only 557 square kilometers (Table 6.1).

6.2.2 *Land Use/Cover Change*

The ArcGIS10.6 software was used to classify the land use raster data in China for 1990, 2000, 2010, and 2020. The area data of arable land, forest land, grassland, water area, construction land, and unused land in China for these four years were obtained (see Table 6.2), and the raster map of land use types for these four years was drawn (see Fig. 6.2). On this basis, the changes in the area of each land use type in different periods were further calculated according to Table 6.2, as shown in Table 6.3. It can be seen that the area of the watershed, construction land, and unused land increased, the wetland area decreased, and the area of cultivated land and forest land did not change significantly from 1990–2020. Specifically, the arable land area shows a trend of increasing before decreasing, with an overall increase of 12,797 square kilometers (0.72%); forest land area shows a trend of decreasing before increasing, with an overall increase of 9345 square kilometers (0.42%); grassland area shows a continuous decrease, with a total decrease of 333,366 square kilometers (10.96%); water area shows a continuous increase, with a total increase of 28,207square kilometers (10.49%); construction land shows an increasing trend with an increase of 70.79%; and unused land shows a trend of decreasing and then increasing, with an overall increase of 188,079 square kilometers (9.41%).

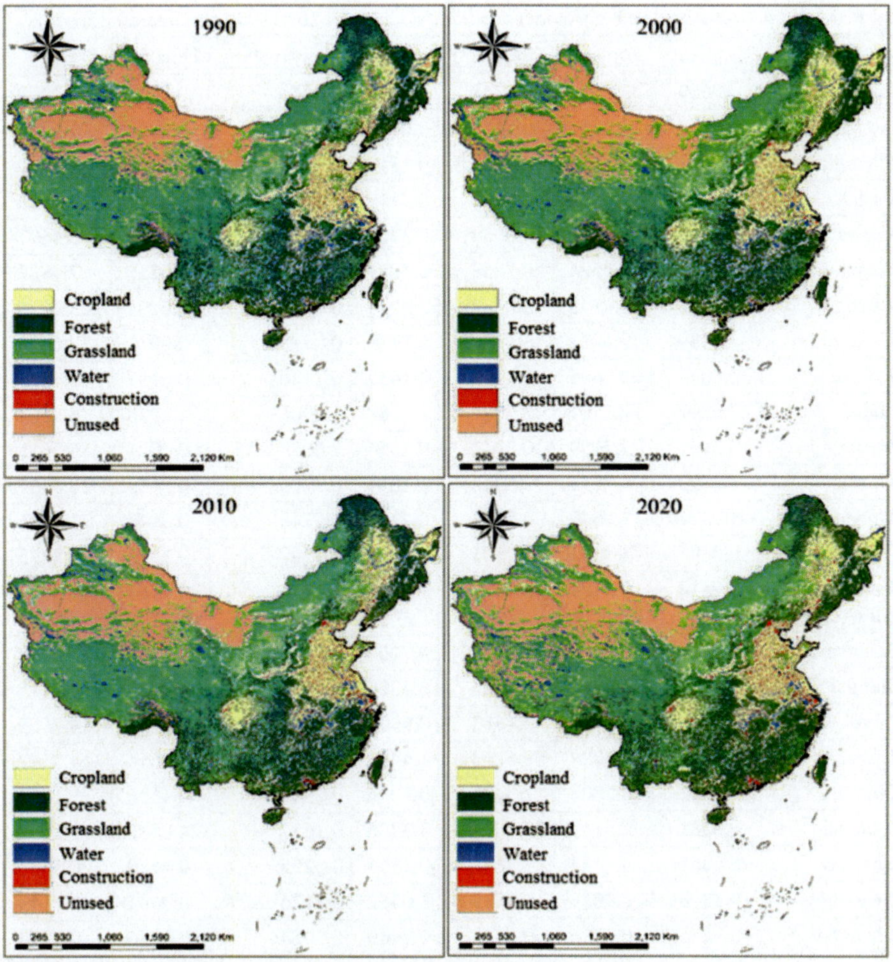

Fig. 6.2 China land use type map from 1990 to 2020

6.2.3 Problems of Land Resource Use in China

The quantity of agricultural land is gradually decreasing, and the situation is serious. After the reform and opening up, China's rapid economic development and urbanization process are accelerating. The demand for construction land is also increasing, which will lead to a decrease in the area of arable land, grassland, and other agricultural land. This contradiction is difficult to resolve over a long period of time. In addition, the growing industry, catering industry, feed industry, and so on, all need to consume crops. Therefore, the demand for crops is growing, further driving the demand for agricultural land. This contradiction between supply and demand for agricultural land continues to expand. Although China has a large area of unused land,

Table 6.1 Land use structure by provinces (cities) in China in 2020 (million square kilometers)

Province	Cropland	Forest	Grassland	Water	Construction	Unused	Area
Anhui	7.7609	3.2113	0.8348	0.7380	1.4665	0.0025	14.0140
Beijing	0.8557	0.2557	0.1278	0.0438	0.3542	0.0016	1.6388
Chongqing	3.7504	3.3575	0.7608	0.1312	0.2399	0.0012	8.2410
Fujian	2.0646	7.5752	1.8486	0.2116	0.5582	0.0143	12.2725
Gansu	6.4048	3.9031	14.0054	0.3326	0.5542	15.2676	40.4677
Guangdong	4.2371	10.7624	0.7858	0.8087	1.3684	0.0258	17.9882
Guangxi	5.6678	14.9143	2.0482	0.3911	0.6463	0.0299	23.6976
Guizhou	4.8289	9.3027	3.1184	0.1189	0.2378	0.0031	17.6098
Hainan	0.8624	2.1445	0.1177	0.1437	0.1446	0.0157	3.4286
Hebei	17.3297	19.1761	2.2097	1.1588	1.0901	4.2920	45.2564
Heilongjiang	10.3101	2.7159	0.8942	0.4354	2.2038	0.0021	16.5615
Henan	6.6893	9.2293	0.6948	1.1644	0.7770	0.0403	18.5951
Hubei	5.9371	13.1647	0.6837	0.7216	0.5746	0.0978	21.1795
Hunan	9.0387	3.7551	3.2501	0.4974	2.1347	0.1515	18.8275
Inner Mongolia	6.2514	0.3107	0.1079	1.5463	2.1430	0.0181	10.3774
Jiangsu	4.4100	10.2503	0.7194	0.7202	0.5424	0.0535	16.6958
Jiangxi	7.6487	8.4285	0.6722	0.4326	0.7970	1.1325	19.1115
Jilin	6.0961	6.0973	0.4683	0.5906	1.2892	0.1742	14.7157
Liaoning	11.3543	16.5224	52.6253	1.4463	1.5210	30.8654	114.3347
Ningxia	1.7463	0.2808	2.3471	0.1036	0.2342	0.4696	5.1816
Qinghai	0.8581	2.8117	39.6639	3.3246	0.1617	24.8432	71.6632
Shaanxi	6.6861	4.8684	7.8641	0.1866	0.5255	0.4420	20.5727
Shandong	10.1328	0.9027	0.8555	1.0462	2.7126	0.0645	15.7143
Shanghai	0.3215	0.0086	0.0113	0.2449	0.3054	0.0356	0.9273
Shanxi	5.7878	4.4489	4.4080	0.1431	0.8594	0.0101	15.6573
Sichuan	11.7935	16.8879	16.8389	0.4749	0.6229	1.7540	48.3721
Tianjin	0.6679	2.4834	0.1115	0.1653	0.2195	0.0031	3.6507
Yunnan	0.5864	0.0476	0.0302	0.1808	0.3252	0.0292	1.1994
Zhejiang	6.7105	21.9823	8.5995	0.3838	0.4775	0.1540	38.3076
Xinjiang	2.3877	6.5246	0.2483	0.3531	0.9087	0.0062	10.4286
Tibet	8.9804	2.7536	48.6456	3.5349	0.9015	99.4666	164.2826
Shanxi	0.7602	16.5070	55.5562	8.0238	0.0557	39.5327	120.4356

Note Given that the Data Center for Resource and Environmental Sciences of the Chinese Academy of Sciences does not publish land use data for Hong Kong and Macau, this chapter does not consider land use in Hong Kong and Macau

Table 6.2 Area and share of each land use type in China, 1990–2020

Land types	Indicators	1990	2000	2010	2020
Cropland	Area (10^4 km^2)	176.9858	179.8170	178.537	178.2655
	Proportion (%)	18.66	18.95	18.82	18.76
Forest	Area (10^4 km^2)	224.8647	223.8449	224.1859	225.7992
	Proportion (%)	23.70	23.60	23.63	23.76
Grassland	Area (10^4 km^2)	304.2457	300.7944	299.5769	270.9091
	Proportion (%)	32.07	31.71	31.58	28.51
Water	Area (10^4 km^2)	26.8899	27.1622	27.3952	29.7106
	Proportion (%)	2.83	2.86	2.89	3.13
Construction	Area (10^4 km^2)	15.7719	17.2051	19.7343	26.9367
	Proportion (%)	1.66	1.81	2.08	2.83
Unused	Area (10^4 km^2)	199.9189	199.8495	199.2713	218.7268
	Proportion (%)	21.07	21.07	21.00	23.02
Total	Area (10^4 km^2)	948.6769	948.6731	948.7006	950.3479
	Proportion (%)	100.00	100.00	100.00	100.00

Table 6.3 Changes in area by land use type in China, 1990–2020 (%)

Period	Cropland	Forest	Grassland	Water	Construction	Unused
1990–2000	1.60	−0.45	−1.13	1.01	9.09	−0.03
2000–2010	−0.71	0.15	−0.40	0.86	14.70	−0.29
2010–2020	−0.15	0.72	−9.57	8.45	36.50	9.76
1990–2020	0.72	0.42	−10.96	10.49	70.79	9.41

the area of land that can be developed into agricultural land resources is limited. As the quality of the ecological environment continues to deteriorate, resource constraints continue to increase, severely limiting the ability to replenish agricultural land resources.

The total amount of urban construction land is out of control, and the structure is unbalanced (Peng et al., 2021). Rapid economic development, urbanization, and industrialization have led to excessive growth in construction land and urban and rural construction, and production and construction scales have been expanding (Tao et al., 2018). The area of construction land has increased by 70.79% from 1990 to 2020. In addition, despite the high levels of economic development and urbanization in China's eastern coastal cities, the space for future expansion of built-up areas is limited. However, the strategy of the rise of central China and the western plan will certainly pull the central and western urban construction and infrastructure development, indicating that the demand for construction land will remain high for a considerable period in the future and the area of construction land will further increase. The unreasonable land use structure is manifested in the unreasonable ratio of industrial, commercial, and residential land and the unreasonable ratio within the industry. The limited land cannot be effectively allocated, resulting in the waste of land resources.

Ecological and environmental problems caused by the development and utilization of land resources are becoming increasingly serious. Since the industrial revolution, excessive deforestation and overutilization of land owing to rapid economic development have considerably damaged the ecological environment, leading to ecological and environmental problems, such as land sanding, soil erosion, and ground strength reduction (Yang et al., 2018). The ecological environment of land is facing a situation wherein the deterioration is increasing and intensifying, which is manifested in serious soil erosion and salinization, continuous degradation of grassland and forests, shrinking of wetlands, and loss of biodiversity. China is a country with one of the largest desertification areas, largest population impacted, and greatest wind and sand hazards. The country's total area of desertified land is 2,611,600 square kilometers, accounting for 27.2% of the national land area, and the area of stone desertification in karst areas is 10.07 million hectares. The destruction of the ecological environment will, in turn, restrict economic development. The 17th National Congress of the Communist Party of China proposed building an ecological civilization, which requires China to form an industrial structure, growth mode, and consumption pattern that conserves energy resources, protects the ecological environment, and promotes the concept of an ecological civilization to be firmly established in society in the future development and construction. Therefore, China should build an ecologically safe and environment-friendly land use model by strictly protecting the land, such as forest land, grasslands, and wetlands, and improving the efficiency of land use for construction, based on an understanding of the current status of existing land use.

6.3 Economic Analysis of Land Resource Allocation

6.3.1 Land Use

Market allocation of resources refers to the process of allocating, combining, reallocating, and recombining resources in the process of economic operation by the market mechanism according to the changes in demand and supply, resulting in price changes. Both theoretical and practical evidences show that market-based resource allocation is highly efficient.

In any given location, there is always one use of land that pays a higher rent than any other use, and that is always the most efficient use. Competitive rent theory states that different land users will bid against each other to obtain land close to the city center, such that those who demand land are willing to pay higher rents for land close to the city center and only lower rents for land far from the city center. The rent competition principle can explain the allocation process of land use in and around a typical urban area.

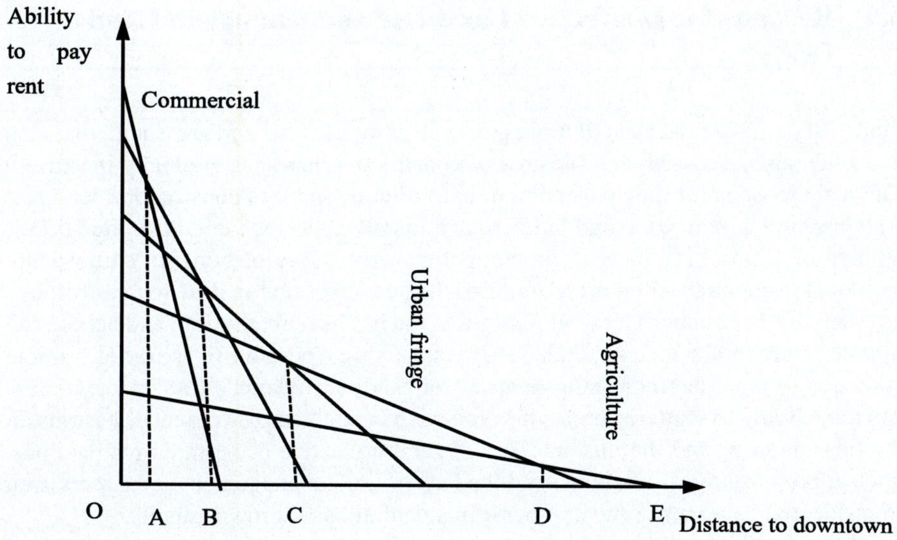

Fig. 6.3 Land configuration

In the 1960s, American land economist W. Alonso introduced the spatial economics theory of marginal equilibrium and marginal returns to location, proposed the competitive rent theory, and drew the urban rent gradient curve (Fig. 6.3). He argued that competitors (e.g., business services) who are more sensitive to location and have a higher ability to pay land rent would gain access to land use in the downtown area, with land use for other activities pushing out in turn. According to Fig. 6.3, as land rent gradually decreases from the city center to the suburbs, the land-use functions from the city center to the suburbs are, in order, commercial (OA segment), industrial (AB segment), residential (BC segment), urban fringe (CD segment), and agricultural (DE segment).

6.3.2 Land Use Conversion and Optimal Land Allocation

When the competitive rent curve in Fig. 6.3 shifts, the land-use pattern is transformed. When the competitive rent curve of commercial land moves upward and the competitive rent curve of industrial land does not move or moves downward, industrial land may be transformed into commercial land. If the competitive rent curve of commercial land moves upward to a greater extent, residential, marginal, and even agricultural land may be transformed into commercial land.

6.4 Reasons for Ineffective Land Use and Change in Land Use

Since 1990, the area of agricultural land, such as arable land and grassland, in China has gradually decreased, and the area of construction land has gradually increased. The main reasons for the conversion of agricultural land into construction land are: (i) increasing urbanization and industrialization owing to the rapid economic development of China have caused the competitive rent curve of nonagricultural land, including residential, commercial, industrial, and transportation land, to rapidly move upward; (ii) the productivity of agricultural land has been increasing, and people can produce more food with less arable land area, making it possible to use the less arable land area to meet the increasing demand for food; (iii) China's rural population is large and living in scattered areas; the arable land occupied by residential bases is on the rise annually, and the unsoundness of rural collective economic organizations, unclear property rights relations, and the lack of strict management and supervision of residential bases make the number of residential bases grow annually.

The optimal allocation of land resources refers to the adjustment of the internal structure of land resources, that is, the conversion of different land use types. As the total amount of land resources is certain, with the development of the economy and society, the internal structure of land resources changes to meet social development needs (Xiong et al., 2017). For example, as cities continue to expand and basic public service facilities continue to improve, arable land, forest land, and grassland are transformed into urban residential land, land for public administration and public services, land for commercial service facilities, industrial land, land for logistics and warehousing, land for transportation facilities, land for public utilities, and so on. Land has multiple uses, but the actual use of each unit of land resource can only choose a certain use; it is difficult and costly to change the use of land. Therefore, the optimal allocation of land resources is of great importance. The means to achieve the optimal allocation of land resources mainly include the following aspects.

Land creation around the sea to increase total land area: The Earth's surface comprises two parts—land and sea—of which the sea accounts for 70.8% of the total area of the Earth's surface. Therefore, using modern science and technology, appropriate sea-enclosing land effectively increases the total land area.

Transformation of unused land: In 2020, China had 2.19 million square kilometers of unused land, such as arid land, high mountains, and steep slopes, which accounted for 23.02% of the total land area—a relatively high percentage. Therefore, to increase the total area of usable land and alleviate the problem of an insufficient supply of land resources owing to population growth, funds should be invested to transform unused land into usable land.

Adjusting the structure of utilized land resources according to the needs of social and economic development: The adjustment of the structure of utilized land resources refers to the transformation of land from one type of use to another to meet the demands of socioeconomic development (Bai et al., 2021). This includes the transformation of arable and forest land into residential, commercial, and transportation land, and the mutual transformation of arable land, forest land, and grassland, specifically adjusting the proportion of each type of land resource in the total land area and adjusting the spatial distribution of each type of land resource in the region.

6.4.1 Urban Sprawl and Leapfrog Development

Urban sprawl refers to the spread of urban activities, mainly concentrated in the central area, to the periphery of the city. The urban form is characterized by dispersion, low density, a single regional function, and dependence on automobile traffic. Owing to urban sprawl, people living on the periphery of the city need to travel long distances to reach the central area for work. Transportation increases energy consumption, increases pollutants, and pressurizes the ecological environment. Urban sprawl is mainly influenced by market, policy, and socio-cultural factors. For example, urban sprawl in the United States is the result of spontaneous emergence under the market economy. The dominant driver of economic development and urban construction in the United States is the market force, and policies follow the market to promote urban sprawl.

Frog-hopping development refers to selecting new development not at the edge of current development but further away. This results in discontinuous land development, possible vacant lots, and new development zones far from the center of economic activity. The main factors that promote frog-hopping development include topographic characteristics, land use externalities, and government policies.

6.4.2 Incompatible Land Use Practices

Land resources have externalities, and the value of a piece of land is influenced not only by its location but also by the characteristics of its neighboring land. Negative externalities are prevalent in land transactions, where the landowner may not bear some of the costs of using a piece of land that the neighboring landowner bears. For example, toxic emissions from chemical plants can harm the health of nearby residents; excessive use of chemical fertilizers and pesticides on a particular piece of arable land can also lead to the destruction of the soil structure of the surrounding arable land; and noise from airports can disturb nearby residents. The negative externality of land use can lead to incompatible private and social benefits (Li et al., 2018); that is, the allocation solution that maximizes the private net benefit decision may not be the allocation solution that maximizes the net social benefit, resulting in inefficient or even ineffective land use.

One traditional approach to solving this incompatible land-use problem is to adopt land-use zoning. Land use zoning refers to an area divided based on different basic uses of land as per its natural and economic attributes and the regional variability of natural and socioeconomic conditions, land use directions, and the degree of land use. Through land-use zoning, the direction of land use and the focus of development and management in different areas are determined, which is conducive to the realization of localization, making use of the strengths and avoiding the weaknesses, and achieving the best use of land and materials (Asamoah et al., 2021). Based on the problems and characteristics of land use in each district, the classification guidance to achieve scientific management of land use through zoning to determine the adjustment indicators of the structure and layout of land use in different regions, the establishment of various production bases and industrial concentration areas in a planned and systematic manner. However, land use zoning has drawbacks, such as increasing urban sprawl and inefficient land use.

6.4.3 Underestimating the Value of Environmental Comfort

Negative externalities of land use can lead to incompatible land use problems, resulting in inefficient land use. Positive externalities may also arise from land use; that is, some ecological goods or services produced by a particular land use benefit the landowner and others. Often, when landowners conduct land use decision analysis, they ignore the positive externalities generated by land use and underestimate land use patterns. For example, if an area is rich in ecological resources, the developer may develop an area with ecological management and environmental protection to make it a scenic spot. The beautiful ecological environment not only attracts tourists and generates economic income but also benefits the health of nearby residents. However, whether the developer is willing to convert the area into a scenic spot will depend on whether creating a scenic spot can bring net benefits to the developer, not the contribution that benefits others. If the economic benefits of creating a scenic spot cannot cover the cost of developing and protecting the area, the developer will not choose this strategy, even if the benefits of creating a scenic spot are greater from a social perspective.

6.4.4 The Impact of Taxation on Land Use Change

Taxes not only raise government revenue but also incentivize land use change, even if the change is not efficient. In the United States, for example, real estate tax comprises two components: tax rate and tax base. The tax base is determined either by the market value of recent sales or by an appraisal by a professional appraiser. As appraisals are based on observed market values, these can differ significantly from current land-use values (Dong et al., 2019). This discrepancy can lead to the conversion of agricultural land to other types of land because developable land is worth more than the cultivation value, and the real estate taxes levied on developable land for growing crops are significantly higher than the cost of cultivation, thus reducing net income.

6.4.5 Market Power

Market power is one of the more common phenomena of market failure. This refers to the ability of an economically active person or a small group of economically active persons to unduly influence market prices. Market power can prevent the market from allocating resources efficiently, leading to market inefficiency because it can cause prices and quantities to deviate from the equilibrium between supply and demand. This is also true for land resources, and when market forces intervene in the allocation of land resources, it can lead to high and inefficient land sale prices that are not conducive to adding social value. This phenomenon is most likely to occur during the market demolition process. When the government or a developer decides to demolish an area, some landowners, recognizing that the government or developer is bound to buy their land, may insist on not giving in to gain more revenue from the demolition until the government or developer pays the monopoly price they expect to obtain the land. If this happens, inefficient land use occurs owing to land cost rising to an inefficiently high level.

6.4.6 Special Problems of Developing Countries

Developing countries have a lower degree of economy, technology, and standard of living than developed countries. Land property rights in many developing countries are informal or nonexistent. Land use is based on a first-come, first-served basis, and many land occupants have acquired land not by purchase or lease but by direct occupation. In such a case the land user does not have land ownership and may be evicted from that land. Consequently, those who occupy the land have a weak awareness of protecting the value of the land, resulting in inefficient land use. In addition, as there is no land title, it is difficult for those who have used a certain piece of land for a long time and have worked and created value on that land to protect their interests when the government asks them to move out.

Many developing countries with low levels of economic development have serious poverty problems. Poverty largely constrains people's choice of how to use land, and they use it primarily for survival. The main capital owned by poor farmers is land and unskilled labor, and they often face financial constraints in transforming their land. Owing to the lack of investment, the land is not transformed, and its overuse leads to its reduced fertility and infertility. Farmers have to migrate from degraded to other marginal land to produce more crops. After the newly migrated land is degraded, they continue to search for new land, contributing to the inefficient transformation of land to agricultural land.

In addition to the absence of property rights and poverty, government action can also result in inefficient land use. Government failure occurs when public policies distort land use allocation.

6.5 Remedies Based on Market Innovation Policies

This section examines policy remedies to improve the efficiency of land allocation.

6.5.1 Establishing Property Rights

Land use based on a first-come, first-served basis can result in inefficient land use and even insecure interests of land users. Establishing an effective property rights system can alleviate or even avoid these problems. The establishment of a secure, enforceable, transferable, and adequate property rights system can stimulate the effective transfer of property rights and maintenance of property values and can promote the alignment of individual interests and social incentives.

6.5.2 Transferable Development Rights

Some land should be protected based on the principle of maximizing social benefits, but its owners usually oppose zoning ordinances that favor land protection because they bear all the costs of protection, while the society, as a whole, benefits. This inefficient land allocation owing to positive land-use externalities is addressed by creating transferable development rights (TDRs) (Shen & Sun 2016). A TDR is the transfer of residential development rights from a portion of a community where the owner of the conservation land does not have to bear all the costs. The local government delineates areas into the area from which development rights can be transferred and the area to which these rights can be transferred.

6.5.3 Grazing Rights

Grazing rights refer to the ability of ranchers to allow livestock to graze on land that does not belong to them. Owing to the competitive and non-exclusive nature of "public resources" such as shared grazing land, it is easy for ranchers to overuse such land and degrade the ecosystem. Grazing rights can effectively solve this problem.

6.5.4 Conservation Easement

A conservation easement is a popular method of protecting land. It is a legal agreement between a landowner and the private or public sector that restricts how the land can be used to protect its environmental value. Once established, conservation easements can be sold or donated. Conservation easements allow a bundle of powers associated with land ownership to be broken down into separate, transferable units. Divesting land for development rights and allowing it to be transferred to the highest value use protects the land while increasing the value of the entire bundle of rights.

6.5.5 Land Conservation Trust

A land conservation trust is a trust in which the owner (the principal) gives the land to a trustee for the effective use of the land and to improve the efficiency of the development and operation of the real estate. Most land trusts have a term of 30–50 years.

6.5.6 Development Impact Fee

Development impact fees are levied on developers to offset the cost of new public services brought about by development, and are used to build public services, such as justification, schools, and hospitals, that meet the needs of the new residents. These costs are passed on by developers to home buyers through higher property prices, which internalize the cost of new public services, and residents take the initiative to protect these public services.

6.5.7 Regulating Real Estate Taxes

Some states in the United States use real estate tax incentives to promote the allocation of land resources in a manner consistent with current social expectations. For example, when real estate taxes are based on market value rather than its current use, the tax structure puts pressure on landowners to convert their land use. To avoid situations where arable or forest land, for example, is forced to be converted to construction land owing to high real estate taxes that are detrimental to sustainable social development, eligible landowners may receive a certain amount of tax relief based on the assessed value of the land.

6.6 Exploring the Efficiency of Land Resource Use in China

6.6.1 Evaluation Index System Construction

Article 4 of the Land Management Law stipulates that "the State shall prepare a general land use plan, specify land use, and divide land into agricultural, construction, and unused land." As the use of agricultural and construction land is different in each region of China, and the problems are also different, it is difficult to objectively evaluate the actual situation of land use efficiency in each province of China if only one set of index systems is used. Therefore, it is necessary to establish two sets of evaluation index systems for agricultural and urban construction land.

Input indicators. In the evaluation indexes of agricultural land use efficiency, four types of input indicators are selected in this chapter—labor, land, mechanical power, and fertilizer input. Labor input is expressed as the number of employees in the primary industry; land input is expressed as the area of agricultural land; mechanical power input is expressed as the total power of agricultural machinery; and fertilizer input is expressed as the discounted amount of fertilizer used in agriculture. Regarding the selection of urban construction land use efficiency evaluation indicators, this chapter intends to select input indicators from land, capital, and labor, that is, urban construction land area, total investment in urban fixed assets, and the number of employees in secondary and tertiary industries in the city.

Output indicators. The total agricultural output value is selected to represent the output of agricultural land, and the added value of urban secondary and tertiary industries is selected to represent the output of construction land.

In summary, this chapter constructs two sets of land use efficiency evaluation index systems to evaluate the utilization efficiency of agricultural and urban construction land in each province (city) of China. The number of employees in the primary industry, area of agricultural land, total power of agricultural machinery, and discounted amount of agricultural fertilizer use are used as input indicators, and the total agricultural output value is used as an output indicator for agricultural land. The total investment in urban fixed assets, number of employees in urban secondary and tertiary industries, and area of urban construction land are used as input indicators, and the added value of urban secondary and tertiary industries is used as an output indicator for urban construction land.

6.6.2 Methodology and Data Sources

6.6.2.1 Malmquist

The Malmquist index is defined using a distance function, which is the inverse of Farrell's technical efficiency. The distance function for the output indicator variable in period t is as follows:

$$d_0^t(x^t, y^t) = \inf\{\theta : (x^t, y^t/\theta) \in L^t\} = \left(\sup\{\theta : (x^t, y^t) \in L^t\}\right)^{-1} \tag{6.1}$$

where x^t and y^t are the matrices of the input and output variables for period t, θ is an output-oriented efficiency indicator, and L^t is also known as the set of production possibilities, where the maximum output subset for each given input is also known as the frontier of production technology. Malmquist is calculated using the ratio of distance functions based on period t and the output-based Malmquist index defined as

$$M_0^t(X^t, Y^t, X^{t+1}, Y^{t+1}) = \frac{D_0^t(X^{t+1}, Y^{t+1})}{D_0^t(X^t, Y^t)} \tag{6.2}$$

where $D_0^t(X^t, Y^t)$ represents the current period's technical efficiency level based on the technical representation of period t, and $D_0^t(X^{t+1}, Y^{t+1})$ represents the technical efficiency level of period $t + 1$. To avoid the effect of different base periods on the results, the geometric mean of the Malmquist index for periods t and $t + 1$ was used to calculate the output TFP index for total factor productivity by Fare et al.

$$\begin{aligned} M_0^{t,t+1}(X^t, Y^t, X^{t+1}, Y^{t+1}) &= \left[\frac{D_0^t(X^{t+1}, Y^{t+1})}{D_0^t(X^t, Y^t)} \times \frac{D_0^{t+1}(X^{t+1}, Y^{t+1})}{D_0^{t+1}(X^t, Y^t)}\right]^{1/2} \\ &= \frac{D_0^{t+1}(X^{t+1}, Y^{t+1})}{D_0^t(X^t, Y^t)} \\ &\quad \times \left[\frac{D_0^t(X^{t+1}, Y^{t+1})}{D_0^{t+1}(X^{t+1}, Y^{t+1})} \times \frac{D_0^t(X^t, Y^t)}{D_0^t(X^t, Y^t)}\right]^{1/2} \\ &= Ech \times Tch \end{aligned} \tag{6.3}$$

where Ech represents technical efficiency, which measures the catch-up of each observation from period t to period $t + 1$ to the optimal production frontier; and

Tch represents technical progress, which measures the observation transfer from period t to period $t+1$. The above total factor productivity decomposition is based on the assumption of constant returns to scale. Under the variable returns to scale assumption, technical efficiency can be further decomposed into pure technical and scale efficiency. Equations (6.2–6.5) can be further decomposed as

$$\begin{aligned} M_0^{t,t+1}(X^t, Y^t, X^{t+1}, Y^{t+1}) &= \frac{D_0^{t+1}(X^{t+1}, Y^{t+1}, C)/D_0^{t+1}(X^{t+1}, Y^{t+1}, V)}{D_0^t(X^t, Y^t, C)/D_0^t(X^t, Y^t, V)} \\ &\times \frac{D_0^{t+1}(X^{t+1}, Y^{t+1}, V)}{D_0^t(X^t, Y^t, V)} \\ &\times \left[\frac{D_0^t(X^{t+1}, Y^{t+1})}{D_0^{t+1}(X^{t+1}, Y^{t+1})} \times \frac{D_0^t(X^t, Y^t)}{D_0^{t+1}(X^t, Y^t)}\right]^{\frac{1}{2}} \\ &= PEch \times SEch \times Tch \end{aligned} \tag{6.4}$$

where C represents constant payoffs to scale, V represents changes in payoffs to scale, *PEch* represents pure technical efficiency, and *SEch* represents scale efficiency.

6.6.2.2 Data Sources

Based on existing studies, this chapter constructs two sets of evaluation indicators to measure the efficiency of agricultural and urban construction land use in each province (city) of China. In this chapter, the data of primary industry employees, agricultural land area, total power of agricultural machinery, amount of fertilizer used in agriculture, total agricultural output value, total investment in urban fixed assets, number of urban employees in secondary and tertiary industries, urban construction land area, and value-added by secondary and tertiary industries in urban areas are obtained from the China Urban Statistical Yearbook and the statistical yearbooks of each province (city) in the corresponding years.

6.6.3 Analysis of Land Use Efficiency Measurement Results

This chapter adopts the data envelopment analysis software DEAP Version 2.1 to measure the total factor productivity of agricultural and urban construction land in 31 provinces (cities) of China from 2006 to 2019 based on the constructed evaluation index system of agricultural land and construction land use efficiency. Subsequently, it decomposes total factor productivity into technical efficiency and technical progress, further decomposes technical efficiency into pure technical efficiency and scale efficiency, and analyzes the spatial and temporal evolution characteristics of total factor productivity, technical efficiency, technical progress, pure technical efficiency, and scale efficiency of land in China.

6.6.3.1 Time-Series Evolutionary Characteristics of Land Use Efficiency

Table 6.4 shows that the total factor productivity of agricultural land in China generally showed a fluctuating upward trend with an average annual growth rate of 0.72% from 2006 to 2019. Only in 2007, 2008, 2010, 2014, and 2019 did agricultural land use efficiency increase, while in the remaining eight years it decreased. In 2010, agricultural land use efficiency increased the fastest, up 17.86%, followed by 2019, up 12.15%. In 2011, agricultural land use efficiency decreased the fastest, down 14.93%, followed by 2009 with a decline of 4.44%.

From the decomposition of total factor productivity, the average annual growth rate of technical progress is 1.02%. Only in 2007, 2010, 2011, 2014, and 2016 was technical progress up, and in the remaining eight years it was down. The fastest technical progress occurred in 2019, up 74.64% from the previous year, followed by 2007, up 17.21% from the previous year. Technical progress declined the fastest in 2018, with a 33.44% decline, followed by 2015, with a 14.04% decline. Technical efficiency declined by 0.29% per year, with 2008, 2010, 2012, 2013, 2015, 2017, and 2018 being up, and the remaining six years being down. The fastest increase in technical efficiency was seen in 2018, up 46.22%, followed by 2010, up 17.72%;

Table 6.4 Results of the analysis of agricultural land use efficiency in China from 2006 to 2019

Year	Technical efficiency	Pure technical efficiency	Scale efficiency	Technological progress	Total factor productivity
2006	1.042	1.465	0.711	0.976	1.017
2007	0.970	0.985	0.985	1.144	1.110
2008	1.025	1.032	0.993	1.120	1.148
2009	1.016	0.944	1.077	1.079	1.097
2010	1.196	0.900	1.329	1.081	1.293
2011	0.960	1.188	0.808	1.145	1.100
2012	1.014	1.019	0.995	1.040	1.054
2013	1.024	0.908	1.128	1.003	1.027
2014	0.992	0.889	1.116	1.097	1.088
2015	1.108	1.007	1.100	0.943	1.044
2016	1.047	0.953	1.098	0.983	1.029
2017	1.071	1.181	0.907	0.948	1.015
2018	1.566	1.138	1.376	0.631	0.988
2019	1.006	0.997	1.009	1.102	1.108
Average	1.066	1.034	1.031	1.011	1.078

2019 saw the fastest decline in technical efficiency, down 35.76%, followed by 2011, down 19.73%.

From the decomposition of technical efficiency, pure technical efficiency declined by 3.16% per year, of which only the pure technical efficiency in 2008, 2011, 2015, and 2017 was up, and the remaining nine years were down. The fastest rise in pure technical efficiency occurred in 2011, up 32.00%, followed by 2017, up 23.92%; 2012 saw the fastest decline in pure technical efficiency, down 14.23%, followed by 2019, down 12.39%. The average annual increase in scale efficiency is 2.96%, of which scale efficiency increased in 2007, 2008, 2009, 2010, 2012, 2013, and 2019 and decreased in the remaining six years. Scale efficiency rose the fastest in 2018, up 51.71%, followed by 2007, up 38.54%; scale efficiency fell the fastest in 2011, down 39.20%, followed by 2019, down 26.68%.

To further explore the time-series evolution characteristics of agricultural land use efficiency in China, the trend graphs of total factor productivity, technical efficiency, technical progress, pure technical efficiency, and scale efficiency of agricultural land are shown in Fig. 6.4, based on the data in Table 6.4. Among them, the overall total factor productivity of agricultural land in China shows a wave-like change, and the magnitude of change is relatively small. Technical progress of agricultural land in China shows a fluctuating change trend from 2006 to 2017, but the magnitude of fluctuation is not obvious. It first decreases more sharply and then increases more sharply in 2018–2019. The technical efficiency of agricultural land in China showed a fluctuating change pattern from 2006 to 2017, with only a small increase in 2010, a small decrease in 2011, insignificant fluctuations in the remaining years, and a larger increase followed by a larger decrease in 2018–2019. Overall, the total factor productivity of agricultural land in China is determined by both technical efficiency and technological progress in agricultural land use, but the role of technological progress is more obvious than that of technical efficiency. It can also be seen from Fig. 6.4 that during the period 2007–2011, the pace of change of total factor productivity of agricultural land and the pace of change of technical efficiency are the same, but the trend of technical efficiency is steeper, owing to the improvement of the technical level of agricultural production, which makes the magnitude of change of total factor productivity of agricultural land more moderate compared with technical efficiency. During the period 2012–2019, the pace of change of total factor productivity of agricultural land is consistent with the pace of technological progress, but the trend of technological progress is steeper, owing to the driving and hindering effects of technical efficiency of agricultural land use, moderating the magnitude of change in total factor productivity of agricultural land.

The scale efficiency of agricultural land in China continued to rise in 2006–2010, declined more sharply in 2011, continued to rise in 2012–2013, fluctuated insignificantly in 2014–2017, and showed a decline, then a rise, then a decline in 2018–2019. The pure technical efficiency of agricultural land in China fluctuated and decreased in 2006–2010, showed a small increase in 2011, continued to decrease in 2012–2014, fluctuated and increased in 2015–2017, and decreased in 2018–2019. The technical efficiency of agricultural land in China is determined by a combination of pure technical efficiency and scale efficiency of agricultural land use. As can be

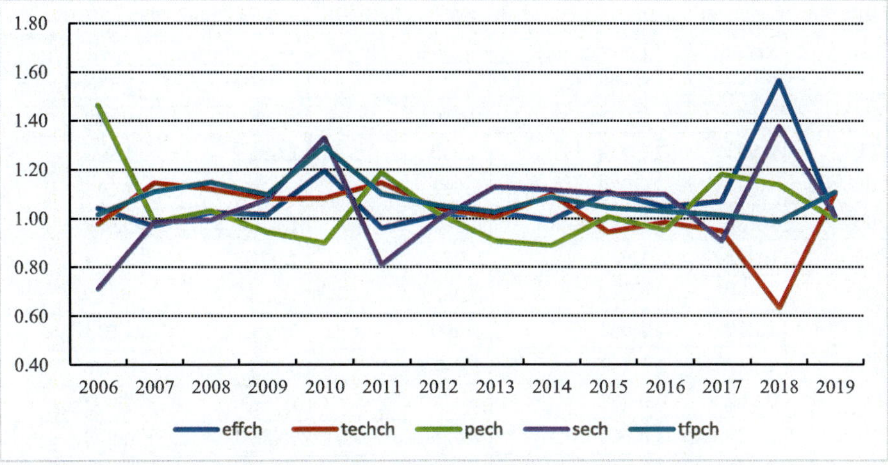

Fig. 6.4 Time series evolution of agricultural land use efficiency in China, 2006–2019[1]

seen in Fig. 6.4, the pace of change of technical efficiency is consistent with that of pure technical efficiency in the period 2006–2010 and 2011–2019. The pace of change of technical efficiency of grain is consistent with that of scale efficiency in the period 2011–2012. In the period 2013–2017, the pace of change of scale efficiency, technical efficiency, and pure technical efficiency are consistent, and those of food technical efficiency and scale efficiency are consistent for the period 2018–2019.

Table 6.5 shows that during the period 2006–2019, the total factor productivity of construction land in China showed an overall fluctuating upward trend with an average annual growth rate of 0.42%. Construction land use efficiency was up only 2007, 2010, 2011, 2014, 2016, and 2018, and it was down in the remaining seven years. In 2014, construction land use efficiency rose the fastest, up 24.66%, followed by 2010, up 17.82%. In 2013, construction land use efficiency fell the fastest, down 15.85%, followed by 2015 with a decline of 10.71%.

From the decomposition of total factor productivity, the average annual growth rate of technical progress is 0.0007%, among which technical progress increased in 2007, 2010, 2011, 2014, 2016, 2017, and 2018 and decreased in the remaining six years. Technical progress was the fastest in 2014, up 57.12% from the previous year, followed by 2011, up 23.12% from the previous year. Technical progress declined the fastest in 2013 by 27.04%, followed by 2012 with a decline of 25.26%. Technical efficiency increased by 0.42% per year, of which 2008, 2010, 2012, 2013, 2015, 2016, and 2019 were up, and the remaining six years were down. The fastest increase in technical efficiency occurred in 2012, up 19.32%, followed by 2013, up 15.43%. The fastest decrease in technical efficiency occurred in 2014, down 20.63%, followed by 2011, down 11.56%.

[1] In Fig. 6.4, effch, techch, pech, sech, and tfpch denote technical efficiency, technical progress, pure technical efficiency, scale efficiency, and total factor productivity, respectively.

Table 6.5 Results of the analysis of construction land use efficiency in China, 2006–2019

Year	Technical efficiency	Pure technical efficiency	Scale efficiency	Technological progress	Total factor productivity
2006	0.942	0.955	0.986	1.048	0.987
2007	0.937	0.938	1.000	1.093	1.025
2008	0.976	0.979	0.997	1.040	1.015
2009	0.886	0.917	0.967	1.026	0.909
2010	0.995	1.018	0.977	1.077	1.071
2011	0.880	0.896	0.982	1.326	1.167
2012	1.050	1.061	0.990	0.991	1.041
2013	1.212	1.157	1.047	0.723	0.876
2014	0.962	0.988	0.973	1.136	1.092
2015	0.989	0.966	1.025	0.986	0.975
2016	1.063	1.055	1.008	1.065	1.132
2017	0.994	1.001	0.993	1.100	1.093
2018	0.950	1.000	0.950	1.178	1.119
2019	0.990	0.975	1.015	1.049	1.038
Average	0.985	0.991	0.993	1.052	1.036

From the decomposition of technical efficiency, the pure technical efficiency increased by 0.17% annually, among which only the pure technical efficiency increased in 2008, 2010, 2012, 2013, and 2016 and decreased in the remaining eight years. The fastest increase in pure technical efficiency occurred in 2012, up by 18.42%, followed by 2010, up by 11.01%. The fastest decrease in pure technical efficiency occurred in 2014, by 14.61%, followed by 2011 with a decline of 11.98%. Scale efficiency increased by 0.24% annually, which increased in 2007, 2010, 2011, 2012, 2013, 2015, and 2019, and decreased in the remaining six years. Scale efficiency increased fastest in 2019, up 6.84%, followed by 2013, up 5.76%. Scale efficiency decreased fastest in 2014, down 7.07%, followed by 2018, down 4.33%.

To further explore the time-series evolution characteristics of construction land use efficiency in China, the trend graphs of total factor productivity, technical efficiency, technical progress, pure technical efficiency, and scale efficiency of construction land are shown in Fig. 6.5 based on the data in Table 6.5. Among them, the total factor productivity of construction land in China shows a wave-like change of rising and then falling several times in a row, with a large increase in 2009–2011 and a large decrease in 2013–2013, while the fluctuations are smaller in the rest of the years. The technical progress of China's construction land shows a fluctuating change, with a rising, then falling, and then continuing to rise trend in 2010–2014, and a larger fluctuation and a smaller fluctuation in the rest of the years. The technical efficiency of China's construction land shows a fluctuating trend, with an obvious change of rising and then declining from 2011 to 2014, and insignificant fluctuations in the

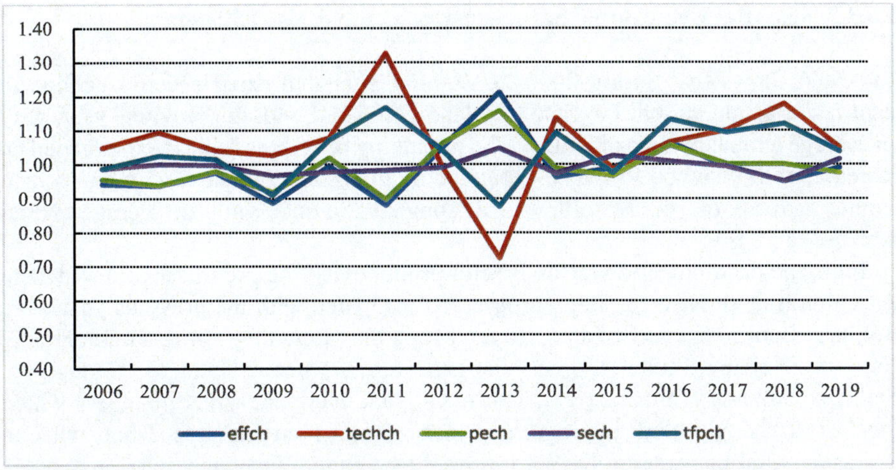

Fig. 6.5 Time series evolution of construction land use efficiency in China, 2006–2019

remaining years. Overall, the total factor productivity of construction land in China is mainly determined by the technical progress of construction land. It can also be seen from Fig. 6.4 that the pace of change in construction land total factor productivity and technological progress are consistent during 2006–2019. However, the trend of technological progress is steeper. This is because the pace of change of total factor productivity of construction land is consistent with that of technical efficiency only in 2008–2010 and 2015–2017. The remaining years are opposite to the pace of change of technical efficiency. The reverse movement of technical efficiency makes total factor productivity change more moderate relative to technical efficiency.

The scale efficiency of construction land only rose slightly in 2012–2014 and then declined, while the remaining years showed a stable change with insignificant fluctuations. The pure technical efficiency of construction land shows an overall fluctuating trend, with an obvious upward and downward trend in 2011–2014 and insignificant fluctuations in the remaining years. The technical efficiency of construction land in China is primarily determined by the pure technical efficiency of construction land use. As can be seen in Fig. 6.4, the pace of change of technical efficiency is the same as that of pure technical efficiency in 2006–2019, and the two curves almost overlap in the remaining years, except that the technical efficiency curve is slightly higher than the pure technical efficiency curve in 2012–2014. This is because scale efficiency in 2012–2014 first rose and then declined, suggesting that the magnitude of change in technical efficiency is more obvious relative to pure technical efficiency.

6.6.3.2 Spatial Evolution Characteristics of Land Use Efficiency

Table 6.6 shows that during 2006–2019, only Shanghai experienced a decline in total factor productivity of agricultural land compared with the previous year, with an average annual decline of 0.40%. The remaining provinces (cities) experienced an increase, with Guizhou Province experiencing the fastest increase, with an average annual increase of 14.2%, followed by Qinghai Province with an average annual increase of 13.0%.

From the decomposition of total factor productivity, the provinces (cities) where agricultural land use technology progressed compared with the previous year were Beijing, Tianjin, Hebei, Shanxi, Inner Mongolia, Liaoning, Jilin, Heilongjiang, Jiangsu, Zhejiang, Anhui, Fujian, Jiangxi, Hubei, Hunan, Guangxi, Chongqing, Sichuan, Guizhou, Yunnan, Tibet, Gansu, Qinghai, Ningxia, and Xinjiang, of which the fastest rising technical progress in agricultural land use was Tibet, with an average annual increase of 4.40%, followed by Qinghai Province, with an average annual increase of 4.30%. The technical progress of agricultural land use in Fujian Province remained unchanged compared with the previous year, while the technical progress of agricultural land use in the remaining provinces (cities) declined compared with the previous year, with the fastest decline being in Shaanxi Province, with an average annual decline of 1.50%, followed by Shandong Province, with an average annual decline of 0.80%. The technical efficiency of agricultural land use in Shanghai remained unchanged compared with the previous year, it increased in all the remaining provinces (cities) compared with the previous year, with the fastest increase being in Shaanxi Province, with an average annual increase of 10.70%, followed by Ningxia, with an average annual increase of 10.40%.

Regarding the decomposition of technical efficiency, only Jilin Province in China experienced a decrease in the pure technical efficiency of agricultural land use compared with the previous year, with an average annual decrease of 2.20%, while Shanghai, Jiangsu Province, Shandong Province, Guangdong Province, Hainan Province, Tibet, and Qinghai Province all experienced an increase compared with the previous year. Shaanxi Province experienced the fastest increase, with an average annual increase of 8.20%, followed by Shaanxi Province, with an average annual increase of 8.20%, and Guizhou Province, with an average annual increase of 7.80%. Only the scale efficiency of agricultural land use in Beijing decreased compared with the previous year, with an annual average decrease of 2.50%. The scale efficiency of agricultural land use in Shanghai remained unchanged compared with the previous year. In contrast, the scale efficiency in the remaining provinces (cities) increased, among which Jiangsu Province experienced the fastest increase, with an annual average increase of 8.70%, followed by Qinghai Province, with an annual average increase of 8.40%.

As seen in Table 6.8, only Jiangsu, Anhui, Hubei, Guangxi, Sichuan, and Guizhou provinces showed an increase in total factor productivity of agricultural land in the study interval, while the remaining provinces (cities) showed both increasing and decreasing trends. However, the number of years in which total factor productivity increased was more than the number of years in which it decreased in all

Table 6.6 Total factor productivity of agricultural land and its composition in 31 provinces (cities) in China

Region	Technical efficiency	Pure technical efficiency	Scale efficiency	Technological progress	Total factor productivity
Beijing	1.043	1.016	1.070	0.975	1.060
Tianjin	1.068	1.011	1.059	1.008	1.079
Hebei	1.064	1.003	1.019	1.044	1.067
Shanxi	1.066	1.011	1.047	1.018	1.078
Inner Mongolia	1.056	1.009	1.008	1.047	1.065
Liaoning	1.056	1.007	1.030	1.026	1.063
Jilin	1.034	1.016	0.978	1.057	1.050
Heilongjiang	1.094	1.023	1.046	1.046	1.118
Shanghai	1.000	0.996	1.000	1.000	0.996
Jiangsu	1.087	1.012	1.000	1.087	1.100
Zhejiang	1.064	1.004	1.042	1.022	1.068
Anhui	1.072	1.005	1.022	1.048	1.077
Fujian	1.095	1.000	1.057	1.036	1.094
Jiangxi	1.074	1.016	1.053	1.020	1.092
Shandong	1.077	0.992	1.000	1.077	1.068
Henan	1.083	0.998	1.012	1.070	1.080
Hubei	1.093	1.006	1.056	1.035	1.099
Hunan	1.062	1.017	1.054	1.007	1.080
Guangdong	1.022	0.994	1.000	1.022	1.015
Guangxi	1.079	1.010	1.074	1.005	1.090
Hainan	1.019	0.995	1.000	1.019	1.013
Chongqing	1.080	1.009	1.067	1.012	1.090
Sichuan	1.082	1.022	1.063	1.017	1.105
Guizhou	1.101	1.037	1.078	1.021	1.142
Yunnan	1.029	1.012	1.016	1.013	1.042
Tibet	1.026	1.044	1.000	1.026	1.072
Shaanxi	1.107	0.985	1.082	1.023	1.091
Gansu	1.054	1.034	1.033	1.020	1.090
Qinghai	1.084	1.043	1.000	1.084	1.130
Ningxia	1.104	1.007	1.069	1.033	1.112
Xinjiang	1.071	1.019	1.018	1.053	1.092
Average	1.066	1.011	1.034	1.031	1.078

provinces (cities). As seen in Table 6.9, the technical efficiency of agricultural land in Shanghai remained unchanged, while the technical efficiency of agricultural land in the remaining provinces (cities) showed both upward and downward trends, and the number of years in which technical efficiency increases is more than that in which it decreases in most provinces (cities). As seen in Table 6.10, the technical progress of agricultural land in all Chinese provinces (municipalities) showed both upward and downward trends, and the technical progress of agricultural land in all provinces (municipalities) increased in 2007, 2008, 2009, 2010, 2011 and 2019, and decreased in 2015 and 2018. As seen in Table 6.11, the pure technical efficiency of agricultural land in Shanghai, Jiangsu, and Tibet remained unchanged, and the remaining provinces (cities) showed both rising and falling trends in the pure technical efficiency of agricultural land, and both trends appeared more frequently. As seen in Table 6.12, the scale efficiency of agricultural land in Shanghai remained unchanged, while agricultural land in the remaining provinces (municipalities) showed upward and downward trends. Most provinces (municipalities) have more years of upward-than downward-scale efficiency.

Table 6.7 shows that during 2006–2019, only Heilongjiang and Jiangxi provinces in China experienced a decline in total factor productivity of construction land compared with the previous year, with an average annual decline of 3.70 and 0.40%, respectively. The remaining provinces (cities) experienced an increase, with Shanxi Province experiencing the fastest increase, with an average annual increase of 8.70%, followed by Inner Mongolia, with an average annual increase of 8.10%.

Regarding the decomposition of total factor productivity, all provinces (cities) in China experienced an increase in technical progress of construction land use compared with the previous year, among which the fastest increase was in Inner Mongolia, with an average annual increase of 7.30%, followed by Tianjin, with an average annual increase of 7.20%. The provinces (cities) where the technical efficiency of construction land use in China increased compared with the previous year are Beijing, Shanxi, Inner Mongolia, Liaoning, Shanghai, Jiangsu, Chongqing, Shaanxi, Gansu, and Ningxia. Among them, Shanxi Province experienced the fastest increase, with an average annual increase of 2.50%, followed by Jiangsu Province, with an average annual increase of 0.90%. The technical efficiency of construction land use in Shanghai remained unchanged in Shanghai, while it decreased in the remaining provinces (cities). Heilongjiang Province experienced the fastest fall, with an average annual decrease of 8.20%, followed by Jiangxi Province, with an average annual decrease of 6.30%.

Regarding the decomposition of technical efficiency, the provinces (cities) which experienced an increase in pure technical efficiency of construction land use compared with the previous year are Beijing, Shanxi, Inner Mongolia, Liaoning, Shanghai, Jiangsu, Guangdong, Chongqing, Sichuan, Tibet, Shaanxi, Gansu, Qinghai, and Ningxia. Among them, Ningxia experienced the fastest increase, with an average annual increase of 3.10%, followed by Gansu Province, with an average annual increase of 2.70% The pure technical efficiency of construction land use in Shanghai, Guangdong, Sichuan, and Tibet remained unchanged compared with the previous year, while the remaining provinces (cities) experienced a decrease.

Table 6.7 Total factor productivity of construction land and its composition changes in 31 provinces (cities) in China

Region	Technical efficiency	Pure technical efficiency	Scale efficiency	Technological progress	Total factor productivity
Beijing	1.004	1.065	1.003	1.001	1.069
Tianjin	0.964	1.072	0.964	1.000	1.033
Hebei	0.967	1.053	0.978	0.989	1.018
Shanxi	1.025	1.061	1.021	1.003	1.087
Inner Mongolia	1.008	1.073	1.009	0.998	1.081
Liaoning	1.001	1.062	1.004	0.997	1.063
Jilin	0.970	1.065	0.969	1.001	1.033
Heilongjiang	0.918	1.048	0.915	1.003	0.963
Shanghai	1.000	1.067	1.000	1.000	1.067
Jiangsu	1.009	1.052	1.015	0.994	1.061
Zhejiang	0.985	1.052	0.994	0.991	1.036
Anhui	0.965	1.047	0.980	0.985	1.011
Fujian	0.979	1.053	0.989	0.990	1.031
Jiangxi	0.937	1.063	0.944	0.993	0.996
Shandong	0.986	1.051	0.988	0.998	1.037
Henan	0.964	1.049	0.984	0.980	1.012
Hubei	0.989	1.040	0.997	0.992	1.029
Hunan	0.984	1.043	0.992	0.993	1.027
Guangdong	0.965	1.051	1.000	0.965	1.015
Guangxi	0.977	1.039	0.976	1.001	1.015
Hainan	0.982	1.048	0.980	1.003	1.029
Chongqing	1.008	1.057	1.006	1.002	1.065
Sichuan	0.985	1.046	1.000	0.984	1.030
Guizhou	0.996	1.041	0.991	1.006	1.037
Yunnan	0.983	1.049	0.984	0.998	1.031
Tibet	0.985	1.055	1.000	0.985	1.039
Shaanxi	1.003	1.023	1.001	1.002	1.026
Gansu	1.001	1.046	1.027	0.975	1.047
Qinghai	0.993	1.038	1.008	0.985	1.031
Ningxia	1.008	1.049	1.031	0.978	1.058
Xinjiang	0.987	1.052	0.986	1.001	1.038
Average	0.985	1.052	0.991	0.993	1.036

The pure technical efficiency of construction land use in Heilongjiang Province decreased the fastest, with an average annual decrease of 8.50%, followed by Jiangxi Province, with an average annual decrease of 5.60%. The provinces (cities) which experienced an increase in the scale efficiency of construction land use compared with the previous year were Beijing, Tianjin, Shanxi, Jilin, Heilongjiang, Shanghai, Guangxi, Hainan, Chongqing, Guizhou, Yunnan, Tibet, Shaanxi, and Xinjiang (Liao et al., 2019). Among them, Guizhou Province experienced the fastest increase, with an average annual increase of 0.60%, followed by Shanxi Province, with an average annual increase of 0.30%. Tianjin and Shanghai's construction land use scale efficiency remained unchanged compared with the previous year, and the remaining provinces (cities) experienced a decline. Guangdong Province showed the fastest decline, with an average annual decline of 3.50%, followed by Gansu Province, with an average annual decline of 2.50%.

6.7 Summary

This chapter considers land resources, a fixed location, and multi-purpose resources as the object of study. First, it depicts the current situation of land use in China from two aspects—the quantitative structural characteristics of land use and spatial distribution—analyzes the characteristics of land use changes in China since 1990, and highlights the problems of land resource use in China. Second, it analyzes the economic principles of land use, land use transformation, and optimal land allocation. The problems associated with inefficient land use in industrialized countries and the special problems of inefficient land use in developing countries are analyzed, and remedial methods based on market innovation policies are proposed. Finally, the Malmquist index is used to measure the total factor productivity, technical progress, technical efficiency, pure technical efficiency, and scale efficiency of agricultural and construction land use in 31 provinces (municipalities) of China from 2006 to 2019, to analyze their spatial and temporal evolution characteristics. The findings of this study are as follows.

Regarding the quantitative structure of land use, China has a large proportion of arable land, small proportion of construction land, and large proportion of unused land, indicating more room for future land development and utilization (Göpel et al., 2018). The proportion of ecologically functional land, such as grassland and forest land, is high, with rich ecological resources and high ecological value. Regarding the spatial distribution of land, arable land and forest land are mainly distributed in the northeast, central, and southeast coastal regions of China; grassland is mainly distributed in the central and western regions of China; water area and construction land are scattered; and unused land is mainly distributed in the northwest of China. Regarding land use changes since 1990, the area of watershed, construction land, and unused land have increased, that of wetlands has decreased, and changes in the area of arable and forest land are not obvious. China's land resource use has problems such as a gradual decrease in agricultural land, uncontrolled total urban construction land, structural imbalance, and increasingly serious ecological and environmental problems caused by the development and use of land resources.

The principle of competitive rent can easily elucidate the allocation process of land use in and around a typical urban area. When the competitive rent curve shifts, the land use pattern is transformed. The means to achieve the optimal allocation of land resources include creating land around the sea to increase the total land area, transforming unused land, and adjusting the structure of utilized land resources according to the needs of socioeconomic development (Strauss et al., 2021).

Problems related to ineffective land use commonly found in industrialized countries include urban sprawl and frog-jumping development, incompatible land use patterns, underestimation of environmental amenity values, and the influence of taxes on land use patterns and market forces. Poverty is a serious problem in many developing countries. It largely constrains people's choice of land use patterns and contributes to the inefficient conversion of land to agricultural land. In addition, land property rights in many developing countries are informal or nonexistent, and people who occupy the land have a weak sense of protection of land values as they may lose their land ownership at any time, resulting in inefficient land use. Policy remedies to improve the efficiency of land allocation include establishing property rights, transferable development rights, grazing rights, conservation easements, land conservation trusts, development impact fees, and the regulation of real estate taxes.

The overall total factor productivity of agricultural land in China from 2006 to 2019 shows a wave-like change of a small magnitude. The total factor productivity of agricultural land in China is jointly determined by the technical efficiency of agricultural land use and technical progress, but the role of technical progress is more obvious than that of technical efficiency. The technical efficiency of agricultural land in China is determined by the pure technical efficiency and scale efficiency of agricultural land use. In 2006–2019, the total factor productivity of construction land in China shows a wave-like trend of rising and falling several times in a row, with a relatively large rise in 2009–2011 and a relatively large decline in 2013–2013. The fluctuations in the remaining years are smaller. Total factor productivity of construction land in China is mainly determined by the technical progress of construction land. The technical efficiency is mainly determined by the pure technical efficiency of construction land utilization.

Between 2006 and 2019, only the total factor productivity of agricultural land in Shanghai decreased compared with the previous year, while the remaining provinces (cities) experienced an increase. Compared with the previous year the technical progress of agricultural land use in China rose in 25 provinces (cities), including Beijing, remained unchanged in Fujian, and decreased in the remaining provinces (cities). The technical efficiency of agricultural land use in Shanghai remained unchanged compared with the previous year, and remained unchanged in all the remaining provinces (cities). The pure technical efficiency of agricultural land use declined in Jilin Province compared with the previous year, remained unchanged in seven provinces (cities), including Shanghai, and increased in the remaining provinces (cities). The scale efficiency of agricultural land use in Beijing decreased compared with the previous year, remained unchanged in Shanghai, and increased in all the remaining provinces (cities).

During 2006–2019, the total factor productivity of construction land in Heilongjiang and Jiangxi provinces decreased compared with the previous year, and increased in the remaining provinces (cities). All provinces (cities) in China experienced increased in technical progress of construction land use compared with the previous year. Compared with the previous year, technical efficiency of construction land use increased in 10 provinces (cities), including Beijing, remained unchanged in Shanghai, and decreased in the remaining provinces (cities). Pure technical efficiency of construction land use increased in 14 provinces (cities), including Beijing, Shanghai, Guangdong, Sichuan, and Tibet, while the remaining provinces (cities) experienced a decrease compared with the previous year. Compared with the previous year, the scale efficiency of construction land use increased in 14 provinces (cities), including Beijing, remained unchanged in Tianjin and Shanghai, and decreased in the remaining provinces (cities).

Appendix

See Tables 6.8, 6.9, 6.10, 6.11, and 6.12.

Table 6.8 Total factor productivity of agricultural land in 31 Chinese provinces

Region	2006	2007	2008	2009	2010	2011	2012	2013	2014	2015	2016	2017	2018	2019
Beijing	1.166	1.168	1.142	1.124	2.259	1.103	1.098	0.468	0.905	1.100	1.017	1.020	1.004	0.973
Tianjin	1.138	1.126	1.098	1.101	2.308	1.115	1.102	0.594	1.078	1.060	0.777	1.045	1.128	1.065
Hebei	1.090	1.161	1.074	1.098	1.234	0.997	1.080	1.103	0.976	0.982	0.989	1.053	1.086	1.042
Shanxi	1.050	1.099	0.894	1.754	1.070	1.088	1.153	0.968	1.052	0.979	0.986	1.128	1.035	1.020
Inner Mongolia	1.038	1.059	1.086	1.011	1.201	1.180	1.033	1.156	1.063	1.029	0.971	0.973	1.070	1.072
Liaoning	1.105	1.102	1.038	0.965	1.176	1.088	1.257	0.953	1.017	1.173	0.880	1.023	1.082	1.089
Jilin	1.117	1.071	1.126	1.029	1.092	1.128	1.150	1.084	1.100	1.060	0.701	0.976	1.136	1.027
Heilongjiang	1.060	1.101	1.137	0.964	1.056	1.249	1.241	1.232	1.059	0.969	0.997	1.175	1.102	1.400
Shanghai	1.030	1.112	1.152	1.035	1.112	1.118	0.971	1.046	1.045	0.942	0.957	0.961	0.559	1.089
Jiangsu	1.103	1.090	1.133	1.116	1.165	1.164	1.133	1.090	1.084	1.160	1.036	1.082	1.013	1.044
Zhejiang	1.045	1.088	1.105	1.070	1.210	3.185	1.082	0.300	1.060	1.059	1.111	1.037	1.035	1.118
Anhui	1.077	1.120	1.129	1.021	1.160	1.077	1.078	1.088	1.099	1.040	1.037	1.064	1.021	1.073
Fujian	1.099	1.126	1.100	1.043	1.155	1.139	1.233	0.950	1.071	1.027	1.162	1.074	1.094	1.074
Jiangxi	1.059	1.115	1.118	1.028	2.567	0.473	1.091	2.557	1.050	1.041	1.071	1.069	0.546	1.092
Shandong	1.121	1.142	1.113	1.114	1.139	1.048	1.031	1.141	1.054	1.036	0.943	0.994	1.027	1.077
Henan	1.124	1.121	1.136	1.114	1.242	1.017	1.100	1.028	1.067	1.028	0.994	1.052	1.034	1.088
Hubei	1.046	1.122	1.104	1.038	1.222	1.177	1.099	1.060	1.080	1.068	1.097	1.110	1.067	1.108
Hunan	1.087	1.123	1.114	1.036	1.228	1.104	1.235	0.854	1.031	1.006	1.069	1.050	1.039	1.201
Guangdong	0.488	1.036	1.081	1.020	2.340	0.543	1.064	1.094	2.016	1.024	1.227	0.551	0.949	1.131
Guangxi	1.081	1.147	1.130	1.002	1.134	1.164	1.050	1.054	1.055	1.072	1.089	1.073	1.085	1.142
Hainan	0.428	0.970	1.116	2.573	1.153	1.180	0.461	2.317	1.057	1.093	1.082	0.989	0.499	1.128
Chongqing	0.886	1.198	1.096	1.079	1.953	1.190	1.103	0.606	1.053	1.057	1.179	1.053	1.125	1.102

(continued)

Table 6.8 (continued)

Region	2006	2007	2008	2009	2010	2011	2012	2013	2014	2015	2016	2017	2018	2019
Sichuan	1.005	1.173	1.273	1.034	1.143	1.169	1.117	1.058	1.065	1.083	1.117	1.113	1.057	1.094
Guizhou	1.017	1.079	1.171	1.029	1.167	1.033	1.263	1.167	1.273	1.302	1.065	1.191	1.125	1.148
Yunnan	1.079	1.110	1.091	1.021	0.980	1.178	1.106	1.094	1.030	1.015	1.077	1.063	0.659	1.209
Tibet	1.020	0.943	1.105	0.984	1.185	1.072	1.076	0.904	1.310	0.896	1.070	1.165	1.201	1.159
Shaanxi	1.085	1.150	1.183	0.971	1.211	1.167	1.061	1.132	1.084	1.012	1.057	1.044	1.058	1.087
Gansu	1.077	1.115	1.085	1.067	1.112	1.073	1.350	0.841	1.019	1.060	1.091	1.200	1.100	1.140
Qinghai	1.070	1.103	3.120	1.025	1.138	1.280	0.441	2.464	1.084	0.937	1.209	0.484	1.086	1.440
Ningxia	1.317	1.210	1.219	1.162	1.070	1.111	1.057	1.116	1.047	1.144	1.002	1.070	1.131	0.963
Xinjiang	1.047	1.186	1.014	1.129	1.495	0.985	1.093	1.042	1.064	1.003	1.035	1.048	1.151	1.079
Average	1.017	1.110	1.148	1.097	1.293	1.100	1.054	1.027	1.088	1.044	1.029	1.015	0.988	1.108

Table 6.9 Technical efficiency of agricultural land in 31 Chinese provinces (cities)

Region	2006	2007	2008	2009	2010	2011	2012	2013	2014	2015	2016	2017	2018	2019
Beijing	1.168	1.010	1.034	1.025	2.091	1.000	0.991	0.486	0.808	1.154	0.995	1.154	1.543	0.870
Tianjin	1.016	1.012	0.914	1.063	1.830	0.994	1.291	0.637	0.950	1.142	0.836	1.004	1.832	1.006
Hebei	1.150	1.003	0.973	1.002	1.195	0.869	0.971	1.110	0.860	1.058	1.064	1.011	1.917	1.007
Shanxi	1.108	0.950	0.810	1.600	0.990	0.990	1.041	0.907	0.952	1.027	0.965	1.236	1.864	0.891
Inner Mongolia	1.095	0.915	0.955	0.964	1.014	1.026	1.101	1.360	0.936	1.108	1.045	0.934	1.534	0.960
Liaoning	1.166	0.952	0.940	0.880	1.110	0.972	1.133	0.903	0.902	1.250	0.931	1.108	1.770	1.032
Jilin	1.074	0.944	0.956	0.993	0.906	0.993	1.312	1.244	0.969	1.142	0.754	0.937	1.600	0.902
Heilongjiang	1.119	0.951	1.030	0.879	0.996	1.099	1.290	1.410	0.933	1.044	1.073	1.128	1.548	1.000
Shanghai	1.000	1.000	1.000	1.000	1.000	1.000	1.000	1.000	1.000	1.000	1.000	1.000	1.000	1.000
Jiangsu	1.086	0.996	1.004	1.072	1.142	1.040	1.084	1.044	0.972	1.249	1.115	1.039	1.464	1.000
Zhejiang	1.102	0.940	1.001	0.976	1.118	2.115	1.000	0.467	0.959	1.111	1.087	1.173	1.576	1.000
Anhui	1.137	0.967	1.023	0.931	1.131	0.933	0.968	1.050	0.968	1.120	1.116	1.022	1.813	1.047
Fujian	1.160	0.973	0.996	0.951	1.126	0.988	1.109	0.890	1.028	1.086	1.180	1.102	2.157	0.964
Jiangxi	1.117	0.964	1.012	0.937	2.373	0.430	0.985	2.397	0.950	1.092	1.048	1.209	0.848	0.994
Shandong	1.104	1.043	0.986	1.070	1.116	0.936	0.986	1.003	1.053	1.085	1.025	1.020	1.786	1.053
Henan	1.107	1.025	1.007	1.070	1.217	0.908	1.052	0.904	1.065	1.076	1.081	1.080	1.670	1.060
Hubei	1.104	0.969	1.000	0.947	1.195	1.020	0.996	1.006	0.961	1.151	1.180	1.065	1.894	1.062
Hunan	1.147	0.971	1.009	0.945	1.142	0.999	1.114	0.801	0.934	1.055	1.047	1.187	1.581	1.106
Guangdong	0.515	0.896	0.979	0.930	2.285	0.471	0.970	1.026	2.047	1.083	1.228	0.572	1.934	0.996
Guangxi	1.141	0.992	1.023	0.914	1.097	1.019	0.945	0.988	0.955	1.125	1.069	1.209	1.800	1.055
Hainan	0.451	0.838	1.011	2.347	1.130	1.019	0.436	2.171	1.069	1.157	1.073	1.000	1.000	0.998
Chongqing	0.935	1.035	0.992	0.984	1.891	1.040	0.992	0.568	0.953	1.109	1.162	1.180	1.855	1.044

(continued)

Table 6.9 (continued)

Region	2006	2007	2008	2009	2010	2011	2012	2013	2014	2015	2016	2017	2018	2019
Sichuan	1.060	1.014	1.153	0.943	1.101	1.027	1.008	0.992	0.964	1.137	1.091	1.262	1.523	0.991
Guizhou	1.073	0.932	1.060	0.939	1.084	0.937	1.142	1.094	1.152	1.366	1.041	1.349	1.369	1.000
Yunnan	1.139	0.959	0.988	0.931	0.941	1.035	0.996	1.026	0.932	1.065	1.052	1.204	1.130	1.055
Tibet	1.077	0.815	1.000	0.897	1.096	0.890	1.065	0.847	1.186	0.940	0.907	1.522	1.390	0.972
Shaanxi	1.144	0.994	1.072	0.886	1.179	1.013	1.074	1.207	0.955	1.090	1.137	1.013	2.006	1.038
Gansu	1.136	0.964	0.982	0.973	1.028	0.973	1.223	0.788	0.923	1.112	1.064	1.361	1.385	1.004
Qinghai	1.129	0.953	2.826	0.935	1.052	1.070	0.432	2.309	0.981	0.983	1.040	0.632	1.257	1.208
Ningxia	1.176	1.088	1.015	1.122	0.819	0.972	1.213	1.283	0.922	1.232	1.078	1.028	1.894	0.928
Xinjiang	0.946	1.066	0.844	1.090	1.145	0.880	1.268	1.164	0.937	1.080	1.114	1.006	1.671	1.000
Average	1.042	0.970	1.025	1.016	1.196	0.960	1.014	1.024	0.992	1.108	1.047	1.071	1.566	1.006

Table 6.10 Technical progress of agricultural land in 31 provinces (cities) in China

Regions	2006	2007	2008	2009	2010	2011	2012	2013	2014	2015	2016	2017	2018	2019
Beijing	0.998	1.157	1.104	1.096	1.080	1.103	1.109	0.964	1.120	0.953	1.022	0.884	0.651	1.119
Tianjin	1.120	1.113	1.201	1.036	1.262	1.121	0.854	0.934	1.135	0.928	0.929	1.042	0.616	1.058
Hebei	0.948	1.157	1.104	1.096	1.033	1.147	1.113	0.994	1.135	0.928	0.929	1.042	0.566	1.035
Shanxi	0.948	1.157	1.104	1.096	1.082	1.099	1.108	1.067	1.105	0.953	1.022	0.912	0.555	1.144
Inner Mongolia	0.948	1.157	1.137	1.049	1.185	1.150	0.939	0.850	1.135	0.928	0.929	1.042	0.697	1.117
Liaoning	0.948	1.157	1.104	1.096	1.059	1.119	1.109	1.055	1.127	0.939	0.946	0.923	0.612	1.055
Jilin	1.040	1.135	1.178	1.036	1.205	1.136	0.876	0.872	1.135	0.928	0.929	1.042	0.710	1.138
Heilongjiang	0.948	1.157	1.104	1.096	1.060	1.137	0.962	0.874	1.135	0.928	0.929	1.042	0.712	1.400
Shanghai	1.030	1.112	1.152	1.035	1.112	1.118	0.971	1.046	1.045	0.942	0.957	0.961	0.559	1.089
Jiangsu	1.016	1.094	1.128	1.041	1.020	1.120	1.046	1.044	1.115	0.928	0.929	1.042	0.692	1.044
Zhejiang	0.948	1.157	1.104	1.096	1.082	1.505	1.082	0.643	1.105	0.953	1.022	0.884	0.657	1.118
Anhui	0.948	1.157	1.104	1.096	1.026	1.154	1.113	1.036	1.135	0.928	0.929	1.042	0.563	1.025
Fujian	0.948	1.157	1.104	1.096	1.025	1.153	1.112	1.067	1.042	0.945	0.984	0.975	0.507	1.115
Jiangxi	0.948	1.157	1.104	1.096	1.082	1.099	1.108	1.067	1.105	0.953	1.022	0.884	0.644	1.099
Shandong	1.016	1.094	1.128	1.041	1.020	1.120	1.046	1.137	1.001	0.955	0.920	0.975	0.575	1.023
Henan	1.016	1.094	1.128	1.041	1.020	1.120	1.046	1.137	1.001	0.955	0.920	0.975	0.619	1.027
Hubei	0.948	1.157	1.104	1.096	1.022	1.154	1.103	1.054	1.124	0.928	0.929	1.042	0.563	1.043
Hunan	0.948	1.157	1.104	1.096	1.076	1.105	1.109	1.067	1.105	0.953	1.021	0.885	0.658	1.086
Guangdong	0.948	1.157	1.104	1.096	1.024	1.154	1.097	1.066	0.985	0.945	0.999	0.963	0.490	1.135
Guangxi	0.948	1.157	1.104	1.096	1.034	1.142	1.111	1.067	1.105	0.953	1.018	0.887	0.602	1.082
Hainan	0.948	1.157	1.104	1.096	1.020	1.158	1.058	1.067	0.989	0.945	1.008	0.989	0.499	1.130
Chongqing	0.948	1.157	1.104	1.096	1.032	1.144	1.112	1.067	1.105	0.953	1.014	0.892	0.607	1.055

(continued)

Table 6.10 (continued)

Regions	2006	2007	2008	2009	2010	2011	2012	2013	2014	2015	2016	2017	2018	2019
Sichuan	0.948	1.157	1.104	1.096	1.039	1.138	1.109	1.067	1.105	0.953	1.024	0.882	0.694	1.104
Guizhou	0.948	1.157	1.104	1.096	1.076	1.103	1.107	1.067	1.105	0.953	1.023	0.883	0.822	1.148
Yunnan	0.948	1.157	1.104	1.096	1.042	1.138	1.110	1.067	1.105	0.953	1.023	0.883	0.584	1.146
Tibet	0.948	1.157	1.104	1.096	1.082	1.205	1.011	1.067	1.105	0.953	1.180	0.766	0.863	1.192
Shaanxi	0.948	1.157	1.104	1.096	1.027	1.152	0.988	0.938	1.135	0.928	0.929	1.031	0.527	1.048
Gansu	0.948	1.157	1.104	1.096	1.082	1.103	1.104	1.067	1.105	0.953	1.025	0.881	0.794	1.135
Qinghai	0.948	1.157	1.104	1.096	1.082	1.196	1.022	1.067	1.105	0.953	1.162	0.766	0.863	1.192
Ningxia	1.120	1.113	1.201	1.036	1.305	1.142	0.871	0.870	1.135	0.928	0.929	1.042	0.597	1.038
Xinjiang	1.107	1.113	1.201	1.036	1.305	1.119	0.861	0.895	1.135	0.928	0.929	1.042	0.689	1.079
Average	0.976	1.144	1.120	1.079	1.081	1.145	1.040	1.003	1.097	0.943	0.983	0.948	0.631	1.102

Table 6.11 Pure technical efficiency of agricultural land in 31 Chinese provinces (cities)

Region	2006	2007	2008	2009	2010	2011	2012	2013	2014	2015	2016	2017	2018	2019
Beijing	2.585	0.451	2.216	0.490	2.039	1.000	1.000	0.486	0.839	1.115	0.995	2.207	1.000	1.000
Tianjin	1.150	0.991	0.966	1.000	2.121	0.977	1.098	0.556	0.997	1.079	0.786	1.005	1.852	1.023
Hebei	1.579	1.042	0.956	0.964	0.826	1.220	1.002	0.948	0.732	0.924	0.918	1.389	1.067	0.972
Shanxi	1.784	0.984	0.804	1.422	0.698	1.339	1.117	0.743	0.857	0.954	0.905	1.313	1.352	0.931
Inner Mongolia	1.564	0.967	1.074	0.817	0.860	1.134	0.939	1.040	1.003	0.947	0.906	0.938	1.107	0.990
Liaoning	1.809	1.014	0.936	0.803	0.775	1.315	1.172	0.757	0.855	1.105	0.790	1.147	1.424	0.989
Jilin	1.579	0.952	1.009	0.835	0.813	1.078	1.079	0.960	1.003	0.977	0.644	0.934	1.206	0.897
Heilongjiang	1.716	1.010	1.060	0.847	0.761	1.386	1.112	1.027	1.000	0.894	0.925	1.164	1.038	1.000
Shanghai	1.000	1.000	1.000	1.000	1.000	1.000	1.000	1.000	1.000	1.000	1.000	1.000	1.000	1.000
Jiangsu	1.000	1.000	1.000	1.000	1.000	1.000	1.000	1.000	1.000	1.000	1.000	1.000	1.000	1.000
Zhejiang	1.773	0.988	0.977	0.879	0.813	1.450	1.000	0.740	0.933	1.030	1.008	1.041	1.341	1.000
Anhui	1.446	1.033	1.009	0.931	0.782	1.273	0.982	0.960	0.825	0.968	0.987	1.288	1.002	1.017
Fujian	1.874	1.025	0.964	0.853	0.698	1.395	1.200	0.728	0.841	0.980	0.986	1.430	1.387	1.000
Jiangxi	1.906	0.980	0.988	0.888	1.546	0.625	1.017	1.825	0.904	1.016	0.986	1.104	0.726	0.964
Shandong	1.000	1.000	1.000	1.000	1.000	1.000	1.000	1.000	1.000	1.000	0.939	1.065	1.000	1.000
Henan	1.127	0.981	0.999	1.018	1.015	1.038	1.001	0.987	0.936	0.987	0.958	1.144	1.000	1.000
Hubei	1.986	1.036	1.063	0.980	0.615	1.764	0.958	0.958	0.636	0.961	0.970	1.609	1.043	1.012
Hunan	2.109	1.016	0.982	0.951	0.646	1.721	1.160	0.705	0.659	0.980	0.992	1.557	1.049	1.096
Guangdong	1.000	1.000	1.000	1.000	1.000	1.000	1.000	0.895	1.117	1.000	1.000	1.000	1.000	1.000
Guangxi	2.221	1.074	1.011	0.956	0.530	1.919	0.945	0.948	0.633	1.026	0.997	1.726	1.098	1.049
Hainan	0.618	0.941	1.069	1.608	1.000	1.000	0.665	1.503	1.000	1.000	1.000	1.000	1.000	1.000
Chongqing	1.503	1.131	1.017	0.866	1.251	1.403	1.051	0.498	0.848	1.033	1.084	1.155	1.627	1.011

(continued)

Table 6.11 (continued)

Region	2006	2007	2008	2009	2010	2011	2012	2013	2014	2015	2016	2017	2018	2019
Sichuan	1.942	1.058	1.128	0.999	0.571	1.783	0.984	0.927	0.669	1.056	1.036	1.500	1.000	1.000
Guizhou	1.450	0.968	1.068	0.931	0.844	1.145	1.292	0.874	1.126	1.276	0.978	1.196	1.124	1.000
Yunnan	1.618	1.000	1.000	1.000	0.622	1.609	1.000	1.000	0.707	0.990	1.000	1.429	0.712	1.082
Tibet	1.000	1.000	1.000	1.000	1.000	1.000	1.000	1.000	1.000	1.000	1.000	1.000	1.000	1.000
Shaanxi	1.969	1.073	1.087	0.911	0.769	1.512	1.025	0.887	0.908	0.947	0.969	1.253	1.313	0.998
Gansu	1.519	1.007	0.989	0.954	0.813	1.153	1.411	0.634	0.902	1.035	0.992	1.187	1.184	1.005
Qinghai	1.000	1.000	1.000	1.000	1.000	1.000	0.478	2.092	0.995	0.970	1.036	0.698	1.434	1.000
Ningxia	1.401	1.092	1.062	0.918	0.991	0.797	1.328	0.947	0.985	1.097	0.923	1.029	1.857	0.902
Xinjiang	1.277	1.000	1.000	0.914	1.094	0.939	1.065	0.930	0.968	0.932	0.966	1.029	1.198	1.000
Average	1.465	0.985	1.032	0.944	0.900	1.188	1.019	0.908	0.889	1.007	0.953	1.181	1.138	0.997

Table 6.12 Scale efficiency of agricultural land in 31 Chinese provinces (cities)

Region	2006	2007	2008	2009	2010	2011	2012	2013	2014	2015	2016	2017	2018	2019
Beijing	0.452	2.238	0.467	2.090	1.025	1.000	0.991	0.999	0.963	1.035	0.999	0.523	1.543	0.870
Tianjin	0.884	1.021	0.947	1.062	0.863	1.017	1.176	1.146	0.952	1.058	1.064	0.998	0.989	0.984
Hebei	0.729	0.963	1.017	1.039	1.446	0.712	0.969	1.171	1.175	1.145	1.159	0.728	1.797	1.035
Shanxi	0.621	0.965	1.008	1.125	1.417	0.739	0.932	1.221	1.112	1.076	1.066	0.942	1.379	0.958
Inner Mongolia	0.700	0.946	0.890	1.180	1.179	0.905	1.172	1.308	0.933	1.170	1.153	0.996	1.387	0.969
Liaoning	0.644	0.939	1.004	1.097	1.432	0.739	0.967	1.193	1.055	1.131	1.179	0.966	1.243	1.044
Jilin	0.680	0.991	0.948	1.190	1.114	0.921	1.216	1.295	0.966	1.168	1.171	1.003	1.326	1.006
Heilongjiang	0.652	0.942	0.971	1.038	1.309	0.793	1.161	1.374	0.933	1.167	1.160	0.969	1.490	1.000
Shanghai	1.000	1.000	1.000	1.000	1.000	1.000	1.000	1.000	1.000	1.000	1.000	1.000	1.000	1.000
Jiangsu	1.086	0.996	1.004	1.072	1.142	1.040	1.084	1.044	0.972	1.249	1.115	1.039	1.464	1.000
Zhejiang	0.622	0.951	1.024	1.111	1.375	1.459	1.000	0.631	1.029	1.078	1.078	1.127	1.176	1.000
Anhui	0.786	0.937	1.014	1.000	1.446	0.733	0.986	1.094	1.173	1.157	1.131	0.794	1.809	1.030
Fujian	0.619	0.949	1.033	1.116	1.613	0.709	0.924	1.223	1.223	1.108	1.198	0.771	1.555	0.964
Jiangxi	0.586	0.983	1.025	1.056	1.535	0.688	0.968	1.313	1.052	1.075	1.062	1.095	1.168	1.030
Shandong	1.104	1.043	0.986	1.070	1.116	0.936	0.986	1.003	1.053	1.085	1.092	0.958	1.786	1.053
Henan	0.982	1.044	1.008	1.051	1.200	0.875	1.051	0.916	1.138	1.090	1.128	0.944	1.670	1.060
Hubei	0.556	0.936	0.941	0.967	1.942	0.578	1.039	1.050	1.512	1.197	1.216	0.662	1.815	1.050
Hunan	0.544	0.955	1.027	0.993	1.766	0.581	0.961	1.137	1.418	1.077	1.055	0.762	1.507	1.009
Guangdong	0.515	0.896	0.979	0.930	2.285	0.471	0.970	1.147	1.832	1.083	1.228	0.572	1.934	0.996
Guangxi	0.514	0.923	1.012	0.957	2.070	0.531	1.000	1.042	1.507	1.097	1.072	0.700	1.640	1.006
Hainan	0.730	0.890	0.946	1.459	1.130	1.019	0.655	1.444	1.069	1.157	1.073	1.000	1.000	0.998
Chongqing	0.622	0.916	0.976	1.136	1.512	0.741	0.944	1.140	1.124	1.074	1.072	1.022	1.141	1.033

(continued)

Table 6.12 (continued)

Region	2006	2007	2008	2009	2010	2011	2012	2013	2014	2015	2016	2017	2018	2019
Sichuan	0.546	0.958	1.022	0.944	1.929	0.576	1.024	1.070	1.440	1.077	1.052	0.841	1.523	0.991
Guizhou	0.740	0.964	0.993	1.009	1.285	0.818	0.883	1.252	1.024	1.071	1.065	1.128	1.217	1.000
Yunnan	0.704	0.959	0.988	0.931	1.513	0.643	0.996	1.026	1.319	1.075	1.052	0.842	1.587	0.975
Tibet	1.077	0.815	1.000	0.897	1.096	0.890	1.065	0.847	1.186	0.940	0.907	1.522	1.390	0.972
Shaanxi	0.581	0.926	0.986	0.972	1.534	0.670	1.047	1.360	1.052	1.150	1.173	0.809	1.528	1.040
Gansu	0.748	0.957	0.993	1.020	1.265	0.844	0.867	1.242	1.023	1.075	1.073	1.147	1.170	0.999
Qinghai	1.129	0.953	2.826	0.935	1.052	1.070	0.903	1.104	0.986	1.013	1.004	0.907	0.877	1.208
Ningxia	0.839	0.996	0.956	1.222	0.827	1.220	0.913	1.355	0.936	1.124	1.168	0.999	1.020	1.029
Xinjiang	0.741	1.066	0.844	1.193	1.047	0.937	1.191	1.252	0.968	1.159	1.153	0.977	1.394	1.000
Average	0.711	0.985	0.993	1.077	1.329	0.808	0.995	1.128	1.116	1.100	1.098	0.907	1.376	1.009

References

Asamoah, E. F., Beaumont, L. J., & Maina, J. M. (2021). Climate and land-use changes reduce the benefits of terrestrial protected areas. *Nature Climate Change, 11*(12), 1105–1110. https://doi.org/10.1038/s41558-021-01223-2

Bai, Y., Deng, X., Cheng, Y., Hu, Y., & Zhang, L. (2021). Exploring regional land use dynamics under shared socioeconomic pathways: A case study in Inner Mongolia, China. *Technological Forecasting and Social Change, 166*, 120606. https://doi.org/10.1016/j.techfore.2021.120606

Deng, C., Liu, J., Nie, X., Li, Z., Liu, Y., Xiao, H., Hu, X., Wang, L., Zhang, Y., Zhang, G., Zhu, D., & Xiao, L. (2021). How trade-offs between ecological construction and urbanization expansion affect ecosystem services. *Ecological Indicators, 122*, 107253. https://doi.org/10.1016/j.ecolind.2020.107253

Dong, N., Liu, Z., Luo, M., Fang, C., & Lin, H. (2019). The effects of anthropogenic land use changes on climate in china driven by global socioeconomic and emission scenarios. *Earth's Future, 7*(7), 784–804. https://doi.org/10.1029/2018EF000932

Göpel, J., Schüngel, J., Schaldach, R., Meurer, K. H. E., Jungkunst, H. F., Franko, U., Boy, J., Strey, R., Strey, S., Guggenberger, G., Hampf, A., & Parker, P. (2018). Future land use and land cover in Southern Amazonia and resulting greenhouse gas emissions from agricultural soils. *Regional Environmental Change, 18*(1), 129–142. https://doi.org/10.1007/s10113-017-1235-0

Jain, P., & Jain, P. (2020). Are the sustainable development goals really sustainable? A policy perspective. *Sustainable Development, 28*(6), 1642–1651. https://doi.org/10.1002/sd.2112

Li, M., Klimont, Z., Zhang, Q., Martin, R. V., Zheng, B., Heyes, C., Cofala, J., Zhang, Y., & He, K. (2018). Comparison and evaluation of anthropogenic emissions of SO2 and NOx over China. *Atmospheric Chemistry and Physics, 18*(5), 3433–3456. https://doi.org/10.5194/acp-18-3433-2018

Liao, M. L., Chen, Y., Wang, Y. J., & Lin, M. S. (2019). Study on the coupling and coordination degree of high-quality economic development and ecological environment in Beijing-Tianjin-Hebei region. *Applied Ecology and Environmental Research, 17*(5), 11069–11083. https://doi.org/10.15666/aeer/1705_1106911083

Meinshausen, M., Jeffery, L., Guetschow, J., Robiou du Pont, Y., Rogelj, J., Schaeffer, M., Höhne, N., den Elzen, M., Oberthür, S., & Meinshausen, N. (2015). National post-2020 greenhouse gas targets and diversity-aware leadership. *Nature Climate Change, 5*(12), 1098–1106. https://doi.org/10.1038/nclimate2826

Peng, C., Li, B., & Nan, B. (2021). An analysis framework for the ecological security of urban agglomeration: A case study of the Beijing-Tianjin-Hebei urban agglomeration. *Journal of Cleaner Production, 315*, 128111. https://doi.org/10.1016/j.jclepro.2021.128111

Shen, L., & Sun, Y. (2016). Review on carbon emissions, energy consumption and low-carbon economy in China from a perspective of global climate change. *Journal of Geographical Sciences, 26*(7), 855–870. https://doi.org/10.1007/s11442-016-1302-3

Si, R., Aziz, N., & Raza, A. (2021). Short and long-run causal effects of agriculture, forestry, and other land use on greenhouse gas emissions: Evidence from China using VECM approach. *Environmental Science and Pollution Research, 28*(45), 64419–64430. https://doi.org/10.1007/s11356-021-15474-1

Strauss, B. H., Orton, P. M., Bittermann, K., Buchanan, M. K., Gilford, D. M., Kopp, R. E., Kulp, S., Massey, C., de Moel, H., & Vinogradov, S. (2021). Economic damages from Hurricane Sandy attributable to sea level rise caused by anthropogenic climate change. *Nature Communications, 12*(1), 2720. https://doi.org/10.1038/s41467-021-22838-1

Tao, Z., Guanghui, J., Ruijuan, Z., Qiuyue, Z., Wenqiu, M., Qinglei, Z., & Yuling, L. (2018). Addressing the rural in situ urbanization (RISU) in the Beijing-Tianjin-Hebei region: Spatio-temporal pattern and driving mechanism. *Cities, 75*, 59–71. https://doi.org/10.1016/j.cities.2018.01.001

Xiong, L., Yu, C., de Jong, M., Wang, F., & Cheng, B. (2017). Economic transformation in the Beijing-Tianjin-Hebei region: Is it undergoing the environmental Kuznets curve? *Sustainability, 9*(5), 869. https://doi.org/10.3390/su9050869

Yang, S., Sheng, D., Adamowski, J., Gong, Y., Zhang, J., & Cao, J. (2018). Effect of land use change on soil carbon storage over the last 40 years in the Shi Yang River Basin, China. *Land, 7*(1), 11. https://doi.org/10.3390/land7010011

Chapter 7
Renewable and Globally Scarce Resources: Agricultural and Food Resources

7.1 Introduction

Food has always been a strategic product of the national economy and people's livelihoods. Food, energy, and financial security are the three major types of economic security. Food security is an important aspect of global security. Food security is a global issue and the responsibility of the international community and the governments of all countries. The World Food Summit in 1996 held by the Food and Agriculture Organization (FAO) showed that 840 million people were struggling with starvation worldwide. The conference set the goal of halving the number of hungry people by 2015. However, by 2012, the number of hungry people worldwide had increased to 870 million. In the first half of 2020, nearly 690 million people worldwide were starving.[1] Today, the world is facing "big changes unseen in a century," and global food security is facing major challenges. The State of Food Security and Nutrition in the World 2021, released by the United Nations (UN) in July 2021, shows that global food security governance is still plagued by many factors, including economic recession, frequent extreme weather, continuous wars and conflicts, and the continued COVID-19 pandemic (FAO et al., 2021). Food security risks in the four dimensions of food availability, access, utilization, and stability were not eliminated. Global food security situation will be more severe, and the difficulty to achieving the sustainable development goal of "zero hunger" by 2030 will increase. The 2021 Global Food Crisis Report issued a dire warning that conflict, extreme weather, and the economic shock of the COVID-19 pandemic will continue to push millions of people into sudden food insecurity. Therefore, ensuring food security has become an arduous task, particularly during the outbreak of COVID-19. Increasing attention is being paid to ensuring global food security. In this context, ensuring the security of agricultural and food resources has become key to urgent global solutions.

[1] https://m.thepaper.cn/baijiahao_12693194.

X. Deng et al., *Environmental and Natural Resources Economics*,
https://doi.org/10.1007/978-981-99-9923-1_7

In response to the current global food shortage, all countries should stabilize food production and order, maintain a balance between food supply and demand, and ensure the safety of all links during the food production process. International organizations such as the FAO should play a positive role in strengthening international cooperation to maintain global food security. China has fully demonstrated its responsibilities as a major agricultural country. China feeds nearly 20% of the world's population, using 9% of arable land and 6% of freshwater resources worldwide to produce food, which makes great contributions to global food security. In the process of food production, China is actively exploring the development patterns of international food cooperation, such as promoting South-South cooperation in the field of food and deepening the model of food economic and trade cooperation along the "Belt and Road." Therefore, this chapter analyzes the scarcity of food resources, the mitigation effect of agricultural policies on food security, and the impact of climate change on food security based on the status of global food resources. This study has important theoretical significance and practical value for ensuring global food security.

7.2 Global Agriculture and Food Resources

7.2.1 Global Status

7.2.1.1 Global Population Growth and Urbanization Drive Food Demand

According to an estimate from the FAO, the current land area under cultivation is 1.4 billion hectares, and the global population is about 7.64 billion. The FAO predicts that the population will reach 10 billion in 2050, increasing by 25% over the next 25 years. The global urbanization rate also increased, reaching 55% in 2018. This will change the consumption structure of residents, such as by increasing the demand for livestock products, aquatic products, and food. However, owing to the uneven distribution of food production and population worldwide, food shortages in some developing countries, especially in sub-Saharan Africa, are becoming increasingly serious.

7.2.1.2 Frequent Natural Disasters Threaten Stability of Global Food Production and Supply

Food production is extremely sensitive to climate factors, particularly the frequent occurrence of extreme weather conditions, severe droughts, floods, and other natural disasters that threaten global food production. For example, owing to the frequent occurrence of natural disasters, wheat production in Australia fell continuously from

2006 to 2007, and corn production in the United States fell continuously from 2010 to 2012, which seriously threatened the stability of food production and supply. In addition, the FAO has lowered its food production forecast for four consecutive months since September 2020, indicating that the growth potential of the food supply has weakened due to the COVID-19 pandemic and unfavorable weather.

7.2.1.3 Total Global Food Supply is Sufficient While Soybean Supply Has Changed

In December 2020, the FAO estimated that global cereal production in 2020–2021 is expected to reach 2.74 billion tons, which is sufficient to meet the global consumption demand. Global soybean production in 2020–2021 is expected to reach 366 million tons, whereas soybean consumption is expected to reach 373 million tons. Its supply has not yet met market demand.

7.2.2 *China's Status*

7.2.2.1 Total Amount of Agricultural and Food Resources in China

Grain production in China showed a steady growth trend from 2016 to 2020 (Fig. 7.1). The grain planting area in China was 116.77 million hectares in 2020. Of this total area, the sown areas of main crops, such as paddy, wheat, and corn, were 30.08 million hectares, 23.38 million hectares, and 41.26 million hectares, respectively. In addition, the sown areas of cotton, oilseeds, and sugar crops were 3.17 million hectares, 13.13 million hectares, and 1.57 million hectares, respectively. Annual grain yield was 669.49 million tons. The yields of summer grain, early rice, and autumn grain were 142.86 million tons, 27.29 million tons, and 499.34 million tons, respectively. The yields of rice, wheat, and corn were 211.86 million tons, 134.25 million tons, and 260.67 million tons, respectively.

7.2.2.2 Temporal Evolution of Agricultural and Food Resources in China

Agricultural crops and grain sown area

From 2006 to 2019, the sown area of agricultural crops and grains in China first increased and then slowly declined (Fig. 7.2). China has long continued to increase its arable land protection. After the complete abolition of the agricultural tax, a binding indicator was put forward to keep the red line at 1.8 billion mu of arable land protection in 2006. In 2017, 1.55 billion mu of permanent basic farmland were demarcated. In recent years, the red line for arable land has been strictly guaranteed.

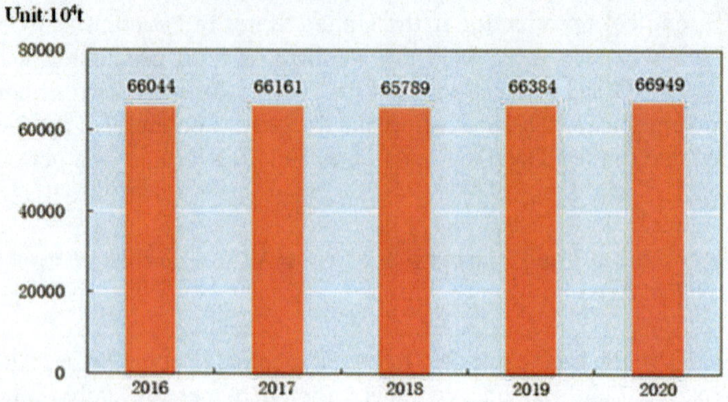

Fig. 7.1 Grain yield in China, 2016–2020

The main reason for the decrease in grain sown area in 2016 is that in response to the contradiction between the supply and demand of grain varieties, some measures were taken to optimize the agricultural planting structure and regional layout, including appropriately reducing the corn planting area in non-advantage areas, and adopting "corn-to-soybean," "grain-to-oil," and "grain-to-feed" measures.

Yield per unit area of agricultural crops and grain

From 2006 to 2019, the yield per unit area of crops and grains in China exhibited an overall upward trend (Fig. 7.3). This benefited from comprehensive factors, such as the improvement of arable land quality, scientific breeding, climatic conditions, continuous progress in production technology, and rational allocation of production factors. However, the yield per unit of crops and grains declined in 2016 and 2015, respectively. The main reasons for this decline are as follows: first, the reduction in the planting area of high-yield crops, such as corn. Second, the national agrometeorological disasters were more severe than in the previous year, and some areas were

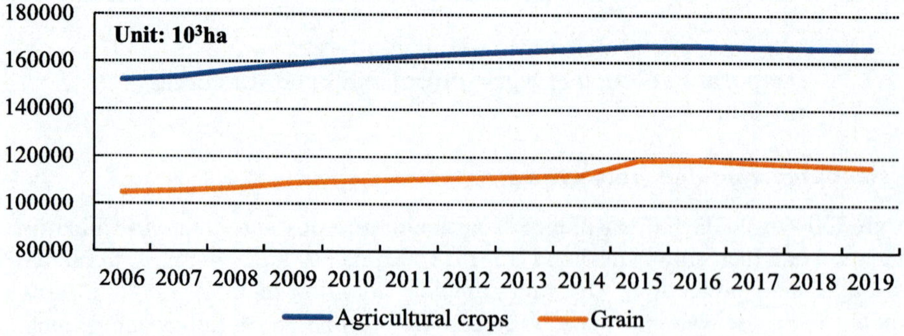

Fig. 7.2 Agricultural crops and grain sown area in China during 2006–2019

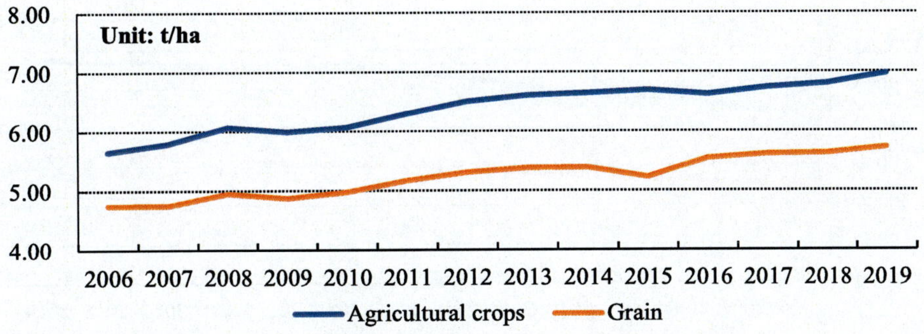

Fig. 7.3 Yields per unit area of agricultural crops and grain in China, 2006–2019

severely affected. The production of summer grains and early rice decreased during the disaster. During the early stages of autumn grain growth, many places in South China, especially Hubei and Anhui, suffered heavy rainfall. Some farmlands were repeatedly flooded, and the crops fell significantly. From late July to mid-August, some areas in South China experience continuous high-temperature weather, leading to an increase in the rate of empty rice husks. Different degrees of drought occurred in some areas of Northeast and Northwest China, which was unfavorable for corn production and grain filling.

Agricultural crops and grain yields

From 2006 to 2019, except in 2016, agricultural crop and grain yields showed an upward trend (Fig. 7.4). This is mainly due to improvements in variety and the continuous popularization of high-yield and high-quality seeds. The increased degree of mechanization in agriculture has greatly reduced labor intensity and improved labor efficiency. The largest decline in 2016 was caused by lower planting and lower yields. In 2016, the grain-sown area decreased by 4.72 million mu compared with the previous year, causing a reduction of 1.7 billion kilograms, accounting for 33.2% of the total grain yield reduction. The total grain yield fell by 3.5 billion kilograms owing to the decline in per-unit yield, accounting for 66.8% of the total grain yield reduction. In general, increasing the yields per unit area of agriculture and grain under conditions of scarcity of existing land resources is beneficial for increasing production.

7.2.2.3 Spatial Pattern of Agricultural and Food Resources in China

Agricultural crops and grain sown area

Mainland China is divided into four regions: eastern, central, western, and north-eastern China. Eastern China includes Beijing, Tianjin, Hebei, Shanghai, Jiangsu, Zhejiang, Fujian, Shandong, Guangdong, and Hainan. Central China includes Shanxi, Anhui, Jiangxi, Henan, Hubei, and Hunan. Western China includes Inner

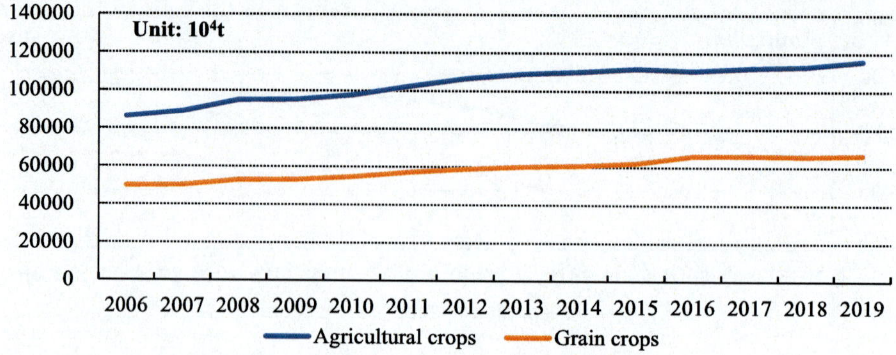

Fig. 7.4 Agricultural crops and grain yields in China during 2006–2019

Mongolia, Guangxi, Chongqing, Sichuan, Guizhou, Yunnan, Tibet, Shaanxi, Gansu, Qinghai, Ningxia, and Xinjiang. Northeast China includes Liaoning, Jilin, and Heilongjiang.

The sown area of agricultural crops is highest from the western to the central, eastern, and northeastern regions, in order of size. Meanwhile, the grain-sown area is highest from the central to the western, eastern, and northeastern regions, in order of size (Figs. 7.5 and 7.6). In general, the sown areas of crops and grain are larger in the central and western regions because of their larger land area. Comprehensively determined by strategic adjustment and regional planning, the eastern region has a higher level of urbanization and a smaller crop sown area.

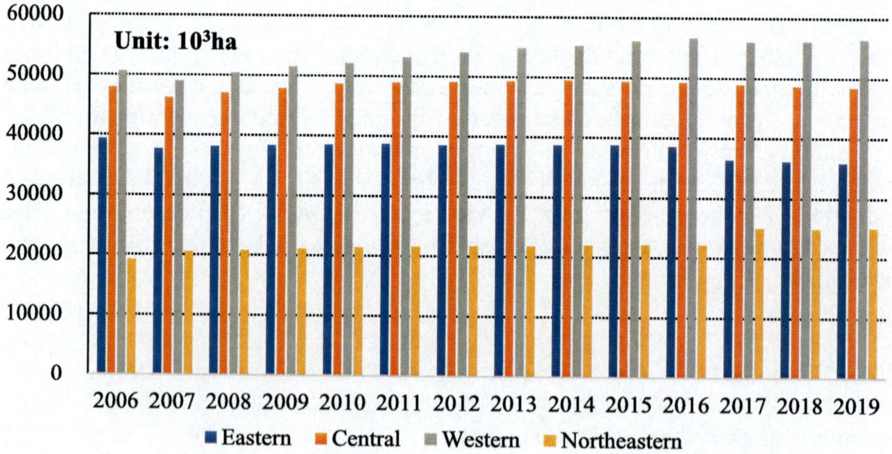

Fig. 7.5 The spatial distribution of agricultural crops sown area in China, 2006–2019

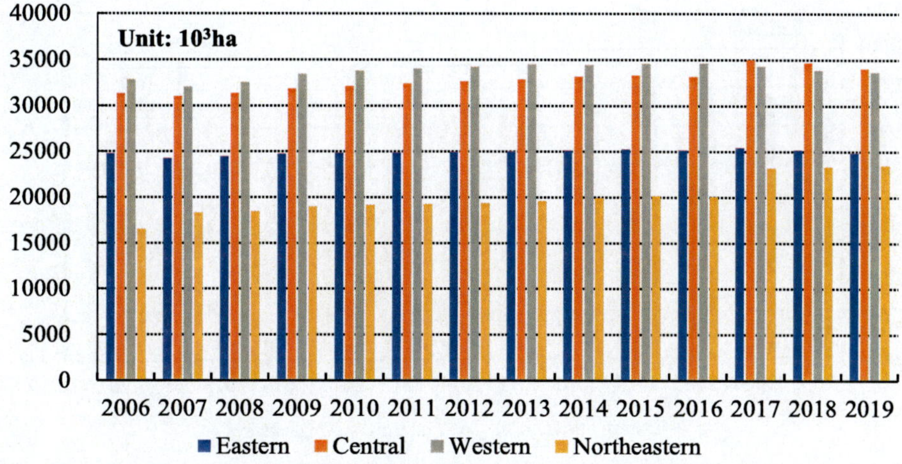

Fig. 7.6 The spatial distribution of grain sown area in China during 2006–2019

Yield per unit area of agricultural crops and grain

The yield per unit area of crops is highest from the eastern to the northeastern, western, and central regions, in order of size. Meanwhile, the yield per unit area of grain in the eastern region is higher than that in the northeastern region first and then overtaken, followed by the central and western regions (Figs. 7.7 and 7.8). In general, the yields per unit area of crops and grain in the eastern and northeastern regions are highest, the central region has the lowest crop yield, while the western region has the lowest grain yield. This may be related to climate change and soil structure in the western region, which reduce the proportion of grain production in agriculture.

Agricultural crops and grain yields

The agricultural crop yield is highest from the western to the northeastern, central, and northeastern regions, in order of size, whereas the grain yield is highest from the central to the western, eastern, and northeastern regions, in order of size. The grain-sown area in the western region is much larger than that in the northeast region, making the total grain yield in the western region rank ahead of that in the northeast region (Figs. 7.9 and 7.10).

7.3 Food Security

Food security is an important aspect of global security. It is not only a priority issue of national governance, but also a core concern from the Millennium Development Goals (2000–2015) to the Sustainable Development Goals (2015–2030). The State of Food Security and Nutrition in the World 2021, released by the UN

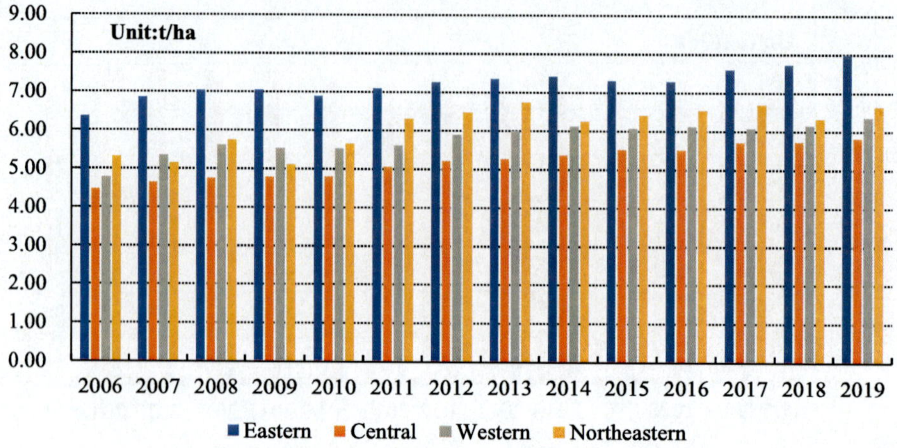

Fig. 7.7 The spatial distribution of yield per unit area of agricultural crops in China during 2006–2019

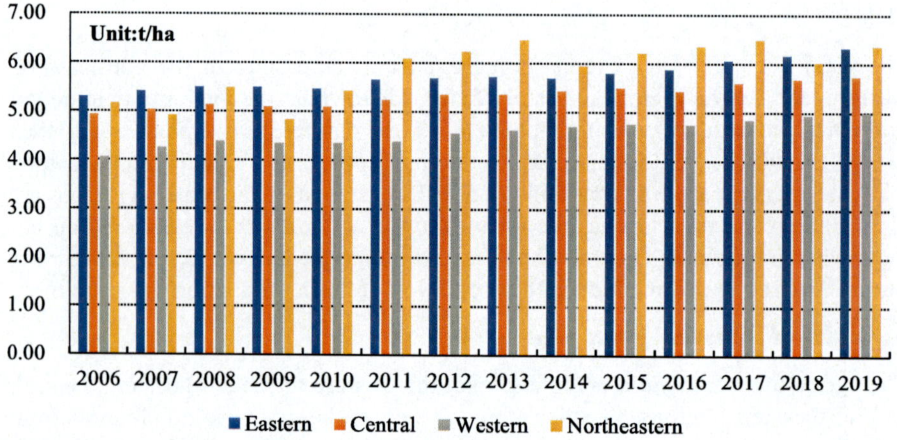

Fig. 7.8 The spatial distribution of yield per unit area of grain in China during 2006–2019

in July 2021, showed that global food security governance is still plagued by many factors, including economic recession, frequent extreme weather, continuous wars and conflicts, and the continued COVID-19 pandemic. Food security risks in the four dimensions of food availability, access, utilization, and stability were not eliminated. The global food security situation will be more severe, and the difficulty by achieving the sustainable development goal of "zero hunger" by 2030 will increase. Therefore, global food security governance remains critical, requiring the cooperation of state and non-state actors to realize food system transformation and build a more resilient, inclusive, and sustainable food system. Based on this, this section demonstrates the

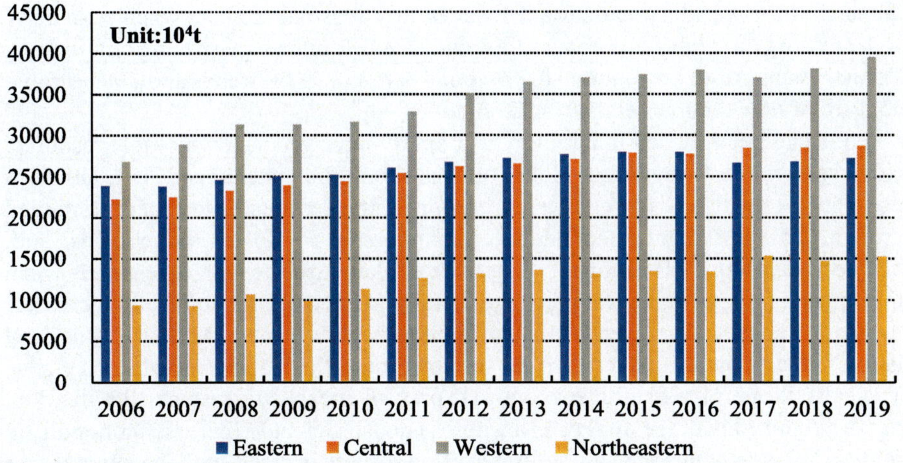

Fig. 7.9 The spatial distribution of agricultural crops yields in China during 2006–2019

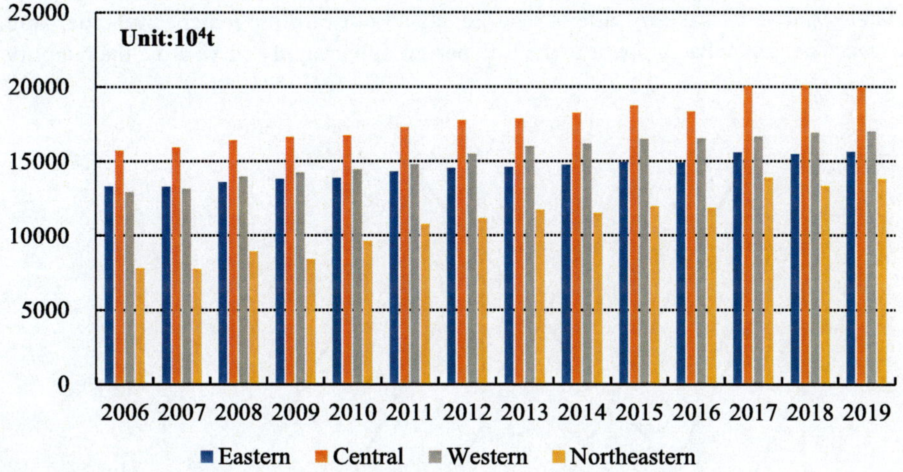

Fig. 7.10 The spatial distribution of grain yield in China during 2006–2019

scarcity of food resources, mitigation effects of agricultural policies on food security, and prospects for future food security situations.

7.3.1 Scarcity of Food Resources

Most authorities agree that food production is adequate. A typical example is a study by the FAO, which reported that the available food supply exceeds the nutritional demand of the world's population. However, this does not indicate whether scarcity is decreasing or increasing because this evidence is limited to a certain point in time. Further assumptions are required to prove and extrapolate the trends and a measurable concept of how the market distributes food.

Food can be produced indefinitely if properly managed as a renewable resource. However, two facts regarding global hunger should be considered: First, although population growth has slowed, it has not stopped. It is reasonable to expect continued growth in food demand. Second, the primary input for food production is land; however, land supply is ultimately fixed. We explain how the market responds when the demand for renewable resources produced by a fixed production factor increases.

Most arable land in Western countries is privately owned. Access to this land is limited, and owners have the right to crowd out others and reap what they sow. Typically, farmers have sufficient control to prevent unexpected resource devaluation, but their control over the market as a whole is not sufficient to increase monopolistic profits. What conclusions can be drawn from a market faced with ever-increasing demand and a fixed supply of land? What does scarcity mean? How can scarcity be detected? The key to this answer lies in the supply and-demand curves (Fig. 7.11).

Assuming an equilibrium state at the beginning, the quantity supplied at price P_O at Q_O. The passage of time is recorded as an outward movement in the demand curve. Consider what happens in the fifth period. If the supply curve is S_a, the quantity

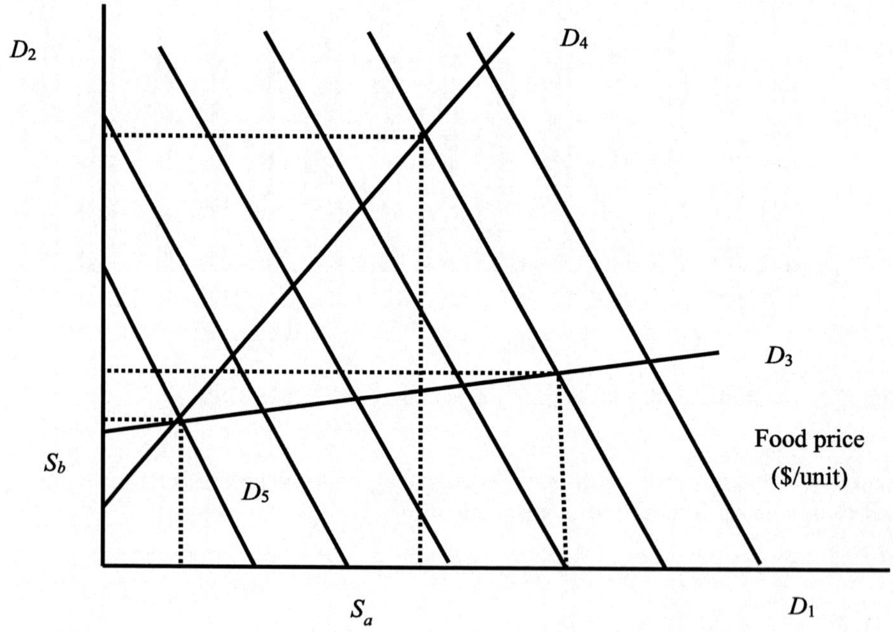

Fig. 7.11 Food market

supplied will rise to Q_{5b}; however, if the supply curve is S_b, the quantity supplied will rise to Q_{3b} and the price will rise to P_{3a}.

This analysis inspired us to consider the scarcity of global food markets. In other words, such a shortage does not necessarily indicate a scarcity. Even in a relatively unfavorable supply environment described by supply curve S_b, the quantity of food supply and demand will be equal. When prices rise, the underlying demand is hindered and increased supply emerges. Some critics argue that food demand is not price sensitive. They argue that food is an essential commodity for survival and that its demand is inelastic and invariant to price changes. Not all foods are necessary. When we observe ordinary vending machines in developing countries for some time, we find that some foods are unnecessary.

There are many examples of price-responsive foods. One occurred in the 1960s, when meat prices skyrocketed in a relatively short period. It was not long before hamburgers in supermarkets were mostly replaced by soy food. As a result, meat consumption has greatly reduced. This is an important example because livestock rearing in Western countries consumes the largest number of grains. This indicates that price affects the balance between the direct consumption of grains and indirect consumption through meat.

We already know enough about the demand side, but not about the supply side. What factors determine whether S_a or S_b describes past and future conditions more fully? As prices necessarily stimulate supply to respond accordingly, the question is how large is this response? When food demand increases, the food supply can be increased in several ways: by increasing the cultivation area, increasing the yield of arable land, or a combination of the two. Historically, each of these avenues has been important.

Of particular note is that the most fertile land is cultivated first. These lands were then cultivated intensively until they became barren and new, less fertile land was cultivated. As newly cultivated land is relatively infertile, it is only brought into production when prices rise to a point where farming is profitable. Therefore, the arable land supply curve (as long as land is an important factor in food production, this is the supply curve for food) can be assumed to be upward sloping.

Evidence can be used to test both forms of the global scarcity hypothesis. The assumption of absolute shortage suggests that food production per unit of capital is declining. This hypothetical absolute scarcity implies that the slope of the supply curve is so steep that the production increase cannot keep pace with the demand increase caused by population growth. If the absolute scarcity assumption holds, food production per unit of capital decreases. If established, this form would provide theoretical support for "lifeboat ethics."

The relative scarcity assumption of the global scarcity hypothesis holds, even when production per unit of capital increases over time. The theory holds that the slope of the supply curve is very steep; therefore, in general, the food price increases faster than the price of other commodities, and the relative food price increases gradually over time. If the relative scarcity assumption holds, welfare per unit of capital decreases even as production increases. This problem is more closely related to food costs than to food availability; when food supply increases, food costs increase faster than other commodities.

7.3.2 *Mitigation Effect of Agricultural Policies on Food Security*

Many countries around the world have issued relevant food policies to ensure national food security. China has also issued a series of food policies, which have played a role in ensuring food security. Facing a series of severe situations such as accelerated industrialization, frequent natural disasters, and labor loss, it is particularly important to choose appropriate food policies to ensure food security.

From the perspective of food development and evolution in China, in addition to an increase in food production over several years, this has also been accompanied by years of food production reduction. The main reasons include some natural factors, but also the situation of "low grains hurting farmers," resulting in a relative surplus of food, which inhibits the production enthusiasm of farmers and hinders the increase in food production. To this end, the Chinese government has adopted a series of macro-control measures for agricultural production since 2004, including farmland and arable land, food production, market prices, imports and exports, and inventories. Protection policies involving farmland and arable land form the basis for ensuring food security. Food production can be stabilized by formulating a minimum purchase price policy. Regulating food market prices and increasing farmers' incomes can effectively promote enthusiasm for cultivation. Food subsidy policies can be classified according to the perspective of "social reproduction," and the national minimum food purchase price level for the current year in the sales process should be formulated. This can effectively regulate and stabilize the food market at the macro level and ensure food security.

7.3.3 *Food Security Prospects*

The current COVID-19 pandemic has led to instability in the global food supply chain (FSC). The turbulent international situation has increased the uncertainty in global food security. Thus, the global food export market is highly concentrated. This has caused problems such as increased uncertainty in global food trade and the global food crisis, which threaten food security. Thus, ensuring food security is a primary issue faced by all countries.

7.3.3.1 Global Prospects

Increasing capital investment in global arable land. Statistics show that some major food-producing countries in the world still have the potential to increase their yields. For example, Ukraine and Russia have low grain yields, and there is still much room for further development. Brazil has more than 400 million hectares of arable land that has not yet been developed. Increasing the proportion of agricultural capital investment can effectively increase food production and meet human consumption demands.

Increasing aid for agriculture in developing countries. Statistics show that the current agricultural productivity of the least developed countries is only approximately 50% of that of developed countries. Therefore, developed countries should actively disseminate agricultural technology and knowledge to developing countries to exploit their food production potential fully.

Shaping a fair and reasonable international agricultural trade order. For human survival, each country must coordinate its interests and make relevant concessions. Specifically, the export of agricultural products by the United States should significantly reduce export subsidies. Developed regions, such as the European Union, Japan, and South Korea, should reduce tariffs and quota restrictions. Developing countries should also lower their market-entry thresholds as much as possible. The global food supply and demand should be flexibly allocated and adjusted through an orderly international food market.

7.3.3.2 China's Prospects

According to a survey by the National Bureau of Statistics (NBS), the per capita food consumption in the urban and rural areas of China in 2018 was 116 kg, while the per capita rice and wheat production totaled 246 kg, which is more than twice the per capita food consumption. From 2013 to 2018, China's per capita food consumption decreased from 139 to 116 kg. In the future, per-capita ration consumption will continue to decline, which can offset the increase in ration consumption caused by population growth. Overall, food consumption in China is generally stable or even declining. Food production capacity will continue to grow to meet the needs of food consumption.

The per capita consumption of animal-sourced food in China increased from 63 to 72 kg between 2013 and 2018. These developmental trends will continue in the future. Therefore, the demand for corn continues to increase. Between 2009 and 2019, corn production increased by 87.52 million tons, at a rate of 51%. The corn-sown area increased by 25%, and the yield per unit area increased by 20%. There is currently a shortage of corn. The key to achieving basic grain self-sufficiency depends on the development of corn production. Whether expanding the area or increasing the yield, certain constraints and challenges still exist. Therefore, greater efforts should be made to increase corn production.

Three decisive factors are related to ensuring production capacity: arable land, technology, and innovation in the production organization. The quantity and quality of arable land must be guaranteed and improved. Therefore, it is necessary to take hard measures to implement the strictest arable land protection system. At the same time, it is important to reduce the occupation of arable land in plains areas through animal husbandry by moving pigs and chickens into buildings and forests into hills. In addition, financial investment should be increased for high-standard farmland construction to meet the needs of agricultural mechanization and the application of advanced technology.

In terms of agricultural science and technology, more importance should be attached to basic and applied research, especially modern biotechnology, to carry out a variety of innovations, accelerate research on forage crops, and strengthen innovation in modern animal husbandry technology. Scientific and technological innovations in agriculture are fundamental for increasing production capacity.

In terms of production organization, the contradiction between small production and large markets can be solved through organizational innovation, which is conducive to the adoption of advanced agricultural technology. Through the enterprising of agricultural production and cooperation between companies and farmers, high-quality and efficient agriculture can be promoted for agricultural modernization.

In terms of scientific and technological support, in addition to strengthening agricultural scientific and technological innovation, it is necessary to pay more attention to the innovation achievements to support the development of modern agricultural formats and realize modern agriculture such as mechanization of field planting, mechanization of field planting, production facilities of horticulture, and intensification of aquaculture.

In terms of moderate imports, an open vision is necessary to alleviate the constraints of insufficient arable land resources in China through moderate imports. Generally, the current structural characteristics of agricultural trade should be maintained by exporting labor-intensive products and importing land-intensive products.

Ensuring food security requires the market to play a decisive role in resource allocation and the government to play a better role. The market's role is mainly to deal with short-term production structure adjustments, while the government's role is to deal with long-term production capacity improvements. As for the government, it is necessary to give full play to the leading role of the central government in the reform of macro-mechanisms and policy support for agriculture as well as to the positive role of local governments in county-level industrial development planning, technology promotion, marketing management, financial investment, and production organization construction.

The global food system is currently undergoing an important transformation. International and regional mechanisms should give full play to their respective advantages, making up for the shortcomings of food security governance and enhancing the resilience of the food system. As the largest developing country in the world, China should take measures to ensure food security according to the developmental demands of different countries and regions within the "Belt and Road" framework.

By combining hard power, such as capital and technology, with soft power, such as experience and knowledge, the Silk Road can be built into a "Road to Harvest" and "Road to No Hunger," and then build a food community between China and the world.

7.4 Climate Change and Food Security

7.4.1 The Impact of Climate Change on Agriculture

Climate change is a key factor affecting agricultural production. Agricultural production has become diversified owing to the periodicity of crop growth and the temporal and spatial characteristics affected by the instability of climate change. Therefore, the impacts of climate change are inseparable from those of seed selection, growth, development, and crop harvesting. This further shows that climate change has an all-round and multilayered impact on agricultural production, with both positive and negative impacts.

Specifically, climate change has had a positive impact on food production. The external environment, which directly affects crop growth, has undergone subtle changes, including rising temperature, increasing concentrations of greenhouse gases, and changing regional heat conditions. Regarding the growing environment for crops, climate warming reduces disasters, advances the spring phenological period, prolongs the planting period, and supplements heat during the growing period, which is conducive to food production. Global climate change has caused changes in the regional heat conditions in China, which has also changed the pattern of agricultural production. The boundary suitable for crop planting has moved significantly northward. Rice varieties in South China have gradually expanded to North China, and the boundary of winter wheat planting has moved northward and expanded westward, greatly increasing the rice and wheat planting areas. The national multiple cropping index showed an upward trend.

Climate change has also negatively affected food production. Rising surface temperature lead to higher respiration rates in crops, accelerates the process of photosynthesis and the growth and development of crops, hinders the normal development of crops, and affects crop yield. In addition, crop diseases, insects, and weeds increase as the temperature increases, which also directly affects the benign growth of crops. Climate change affects the crop growth environment. On the one hand, the increase in temperature can improve the activity of soil microorganisms and accelerate the loss of organic soil matter and nitrogen, causing soil degradation, erosion, and salinization and weakening the resistance of agricultural ecosystems to natural disasters. On the other hand, regional precipitation and its distribution patterns have also changed with regional climate change, increasing the occurrence of extreme precipitation events such as floods and droughts. Climate change shortens the growth period of crops and reduces their ability to withstand abnormal climatic conditions. For example, crops such as wheat and rapeseed in East China mostly mature early. With the warming of the winter climate, the growth period of crops is correspondingly shortened, which weakens the cold resistance of plants and seriously damages crop yield. This poses new challenges for the adjustment of planting systems in China.

7.4.2 The Cycle of Harvest and Famine

The rest of the global food issues stem from occasional fluctuations in the food supply caused by different weather and planting decisions. Even if the average food supply is appropriate, it comprises a series of years of abundance and famine, which means that society as a whole can be balanced by such fluctuations.

The volatility of food supply appears to be very high, and the price is even more volatile. Why? One feature of agriculture is that farmers' production decisions increase, or at least prolong, this volatility. The cobweb model explored this trend (Fig. 7.12).

Let us suppose that owing to weather-induced shortages, the quantity supplied is Q0, causing the price to rise to P0. In the following planting season, the farmers had to plant well before harvesting. The decision on how much to plant depends on the

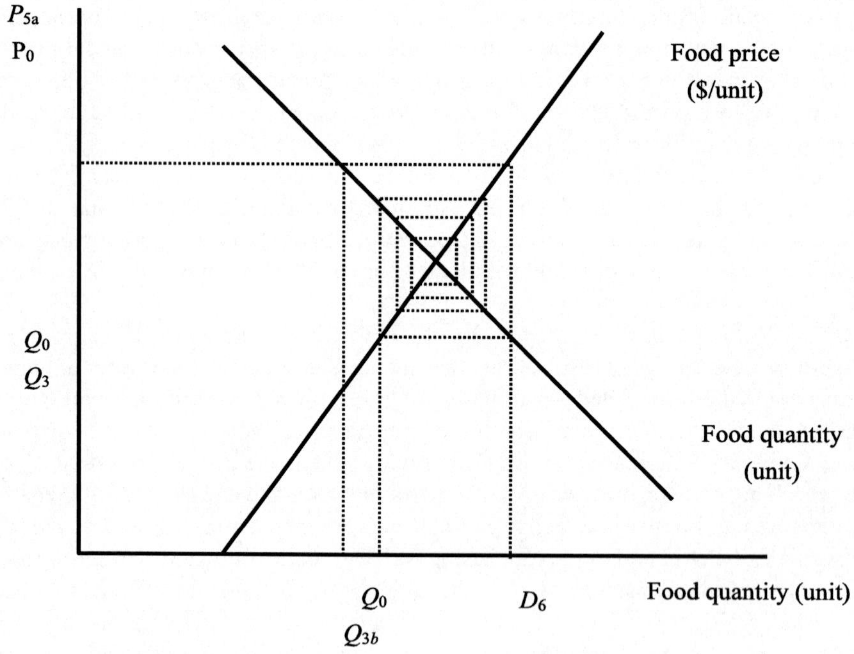

Fig. 7.12 The cobweb model

expected price. Let us assume that they use the price this year as their estimate for the next year.

They would plan to supply Qi (and are able to supply it in the absence of bad weather). At price P0, the market cannot absorb as much of the commodity; therefore, the price falls to P1. If farmers planned food production at this price in the following year, they would produce Q3. This yield causes the price to increase to P2.

Here, damping vibrations occur. In the absence of further supply shocks, the amplitudes of price and quantity fluctuations decrease over time until equilibrium price and quantity are reached.

Food demand tends to be inelastic, especially in developing countries. This has some important implications. The more price inelastic the demand curve is in the event of a weather-induced shortage, the higher the price that will bring demand and supply into balance (Fig. 7.13).

Figure 7.13 records an initial equilibrium state in which demand D1 and supply S0 are equal at point A. A decrease in supply is described on the supply curve as a change from S0 to S1. Of the two demand curves presented in the figure, D1 is more price elastic. Note that the more inelastic the demand curve, such as D2, the higher the market-clearing price.

Figure 7.13 also shows the impact of price volatility on producer income. How do supply changes affect farmers' income? At first glance, the results seem ambiguous because in times of shortage, they receive higher prices (increase) but have fewer products to sell (decrease). What influence is dominant?

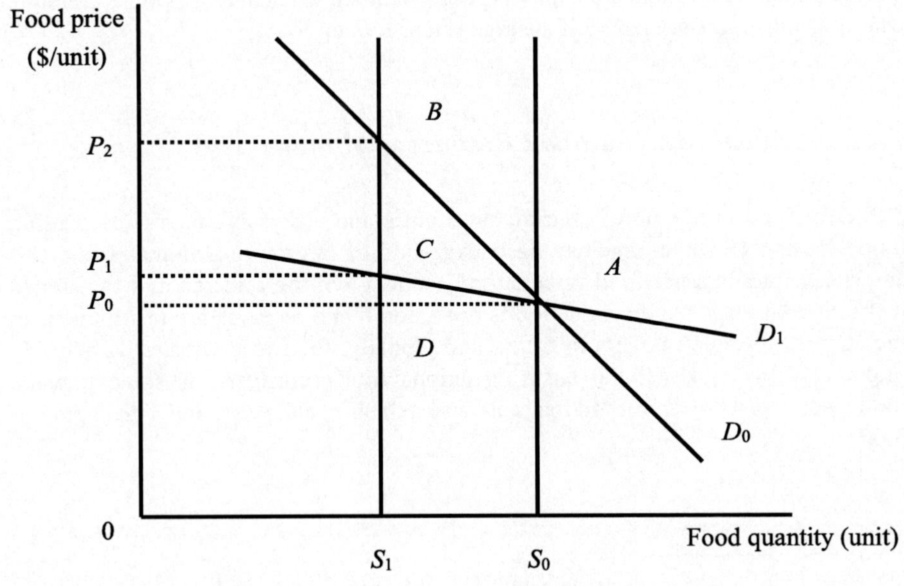

Fig. 7.13 Price elasticity of demand and price fluctuation range

Because producer income is equal to the price multiplied by the quantity, it is represented by a rectangle. The size of the rectangle depends on its perimeter. Before the supply change, the income received is represented by rectangle OP0AS0. After the supply change, the income corresponding to the elastic demand curve (D1) is OP1CS1 and the income corresponding to the inelastic demand curve is OP2BS1.

Does income increase when supply changes? To answer this question, we divide the area of P0P1CD (net income increase under an elastic demand curve) or P0P2BD (net income increase under an inelastic demand curve) by DAS0S1 (revenue loss due to reduced production).

It is clear that the more inelastic the demand curve, the more farmers as a whole benefit from production reduction. Since the demand curve is price-inelastic in the relative range (where food production can generally be satisfied in the short run), farmers can improve their income by reducing supply.[2]

This is a very different image for consumers. Undoubtedly, consumers experience curtailment and benefit from overproduction. The more inelastic the demand curve, the greater the reduction in consumer surplus caused by production curtailment and the greater the increase in consumer surplus caused by overproduction.

This produces an interesting (but difficult stimulus mechanism from a policy perspective) stimulus mechanism. The producer group does not have any interest in controlling supply declines, but has a potential interest in controlling oversupply. Meanwhile, consumers have no objection to excess supply but want to prevent supply from shrinking. Although society as a whole would benefit from price and yield stabilization, different social classes have highly divergent views on how to achieve this stabilization. Farmers are happy if prices stabilize when average prices are high, whereas consumers are happy if average prices remain low.

7.4.3 Agriculture and Food Resources Management

Sufficient food production is the material basis and a decisive factor in ensuring food security. However, against the background of large global demand for food and rising international food prices, food security will be affected and threatened if the circulation of food production is not smooth and reserves are insufficient. In the current unstable markets in China and globally, food reserves are particularly important. This is a key link in ensuring national food security. At the same time, the food reserve is a "reservoir" to maintain a sustainable, balanced, and safe supply of food.

[2] Of course, this does not necessarily hold for every farmer. If supply reduction is concentrated only on a small number of farmers, their situation would undoubtedly be worse, and the situation of other farmers would be better. The reason is that the income of the latter would outweigh the losses of the former.

During periods of unbalanced food supply and demand, and large fluctuations in market prices, appropriate adjustment of food reserves can stabilize food production and prices and configure efficient monitoring, early warning, and regulation mechanisms. For example, food demand elasticity changes little, but supply elasticity tends to be large, which may be caused by fluctuations in food output and poor circulation between production and marketing areas. Thus, food prices fluctuated significantly. In this regard, relevant departments will implement government interventions to carry out macro-control and adjust food reserves. Food reserves act as reservoirs during the bumper harvest period by absorbing inventory, reducing imports, and expanding transformation. During a period of poor harvest, the market supply should be increased by selling inventories, increasing imports, and restricting transformation to promote the basic balance between supply and demand and the basic stability of prices.

Promoting the transformation of food systems can ensure global food security and address climate change challenges. Food systems emit more than one-third of the global GHG emissions, and agriculture-forestry-land use is an economic sector with huge carbon sink potential. 2022 Global Food Report notes that transforming food systems requires major policy reforms, substantial investments, and an enabling environment for innovation (IFPRI, 2022). These include adopting an integrated landscape governance approach to coordinate the sustainable use of land, water, forest, and energy resources and adopting national food dietary guidelines for consumers to make healthy, sustainable food choices. It would also be beneficial to conduct trade and investment based on climate-smart practices, invest in green innovation technologies based on agricultural subsidies, and create new financial resources through green bonds.

7.5 Summary

This chapter analyzes food security from various aspects, including the current situation of agriculture and food resource development, scarcity of food resources, agricultural policy, climate change, future food development prospects, cycles of food harvests and famine, and food reserves. The results showed that food demand increased with population growth and urbanization levels. However, the frequent occurrence of natural disasters threatens the stability of global food production and supply, resulting in an insufficient global food supply. In recent years, China's agricultural and food resources have steadily increased. Temporally, the sown area, yield per unit area, and total yield of crops and food have generally increased. Spatially, the sown area of crops is highest from the western to the central, eastern, and northeastern regions, in order of size, whereas the grain-sown area is highest from the central to the western, eastern, and northeastern regions, in order of size. The yield per unit area of crops is highest from the eastern to the northeastern, western, and central regions, in order of size. Meanwhile, the yield per unit area of grain in the eastern region is higher than that in the northeastern region first and then overtaken, followed by the central and western regions. The crop yield is highest from the western to the northeastern, central, and northeastern regions, in order of size, whereas the grain yield is highest from the central to the western, eastern, and northeastern regions, in order of size.

Based on food resource scarcity, the mitigation effect of agricultural policies on food security, and prospects of the future food security situation, we analyzed the scarcity of food resources through the balance of supply and demand and explained that the characteristics of the food industry, including weaknesses, resource dependence, and public welfare, determine that its development is inseparable from the support of national food policy. Food security is an important aspect of global security.

We discuss climate change and food security issues from three perspectives. First, climate change has two effects on food security: multiscale, multilevel, positive, and negative impacts on food production. Second, the remaining global food issues are due to occasional fluctuations in food supply caused by different weather and planting decisions. Even if the average food supply is appropriate, it is composed of a series of years of abundance and famine, which means that society can achieve balance through such fluctuations. Third, sufficient food production and supply are the bases for ensuring food security. In agriculture and food resource management, food reserves are an important link in ensuring national food security, especially when the food market is unstable. Food reserve systems are an important guarantee for realizing food security.

References

FAO, IFAD, UNICEF, WFP, and WHO. (2021). *The state of food security and nutrition in the world 2021. Transforming food systems for food security, improved nutrition and affordable healthy diets for all.* FAO.

International Food Policy Research Institute (IFPRI). (2022). *2022 global food policy report: Climate change and food systems.* IFPRI.

Chapter 8
Storable and Renewable Resources: Forest Resources

8.1 Introduction

Forest resources occupy an important position in the forest ecosystem, and changes in forest resources can cause changes in the forest ecosystem, improve or deteriorate the ecological environment, and affect national socioeconomic development (Liu et al., 2015; Wang, 2014). Forest ecosystems play a pivotal role in global ecosystems. Therefore, strengthening the construction of ecological civilization requires attaching great importance to the conservation and utilization of forest resources in order to maintain the sustainable utilization of forest resources. (Zhang et al., 2021). The construction of an ecological civilization is always an important issue in national development, and it is extremely important and urgent to strengthen the protection of forest resources and enhance the construction of an ecological civilization (Song et al., 2015). Forest ecosystems have complex structures and functions. They have a variety of ecological functions such as water conservation (Wu et al., 2014), clean air, wind and sand control, biodiversity conservation, and forest protection, which provide a variety of services for humans and maintain the safety and comfort of terrestrial life systems.

Although the contribution of forests to our daily lives can easily be overlooked, numerous studies have demonstrated the importance of forest resources. Forests provide a wide variety of products and services, and the raw materials for homes, wood products, and paper are derived from forests. Wood is used for a variety of fuels worldwide, and trees absorb carbon dioxide and increase oxygen to maintain the air balance. Forests provide habitats for wildlife and reproduction and play an important role in maintaining water bodies that provide large amounts of drinking water (Sgroi, 2020; Zhou et al., 2020). As the role of forests in human survival and development has been gradually emphasized, countries worldwide have begun to pay attention to the protection of forest resources and have defined forests as irreplaceable and important environmental resources that should be protected and utilized more, giving full play to their powerful ecological benefits (Zhao et al.,

X. Deng et al., *Environmental and Natural Resources Economics*,
https://doi.org/10.1007/978-981-99-9923-1_8

2004). In 1992, the United Nations adopted the Statement of Principles on Forests, which promoted global conservation, management, development, and utilization of forest resources. Forest resources are closely related to each other, and they provide important material resources for human survival and development. They are also an important part of the global ecological environment, with ecological functions and values, and are the most important terrestrial ecosystems (Grinkevich et al., 2019). Forests are renewable resources that can be used everywhere through proper management of forest resources (Tong, 2021). Domestic and foreign scholars have varying understandings of the connotations of forest resources. As early as 1978, in the Economics of Forestry in the USSR, Soviet scholars defined forest resources as the territory that has been occupied by forests and allocated to forestry operations in accordance with prescribed procedures, which brings about multiple benefits for forests. Obviously, this was a relatively narrow definition, and in 1979, in the *Soviet Handbook of Forestry Economics*, forest resources were broadly defined as "the complete national forest resource, which is the entire forest and the uncovered land needed for the reproduction of forest products and which plays an auxiliary role in the production process." Most scholars define forest resources as a broader concept that includes not only forest resources but also various plants, animals, microorganisms, and the environments on which they depend.

In 1986, China defined forest resources, arguing that they included forestland, wild animals, and plants within the forest area. Some scholars have pointed out that forest resources are a systemic sum of woody plants as the main body in forest areas, including animals, plants, microorganisms, and the forest environment. They believe that forest resources are renewable, play an important role in and benefit human beings, and are an indispensable ecological subsystem and material basis for human survival (Wu et al., 1999). Forest resources are complex ecosystems, and their quantity and quality are important indicators of the ecological status of a country and region. Compared with other types of natural resources, forest resources are characterized as follows.

8.1.1 Renewability

The main part of forest resources, that is trees, can be regenerated and reused under certain conditions, and it is a renewable resource (Cheng et al., 2010). Forest resources can be naturally renewed and used sustainably as long as they are used reasonably. By contrast, if forest resources are destroyed, they will be exhausted. Forest regeneration is the natural reproduction of biology that has grown for decades or even longer. Understanding the regeneration of forest resources and their laws is crucial for the sustainable utilization of forest resources.

8.1.2 Complexity

Forest resources include those that are valuable to humans in forest ecosystems. Forest ecosystems have a wide variety of resources, both physical and non-physical, and forestry departments often cannot manage them comprehensively within a certain period of time (Nord-Larsen & Schumacher, 2012) At the same time, the forestry department has not been able to carry out detailed statistics on some forest by-product resources, which makes information on the various biological resources contained in forests relatively vague (Negi & Maikhuri, 2017). From the definition of forest resources, it can be observed that they have a wide range and are complex. However, resources such as trees and woodlands remain the main body and foundation of forest resources.

8.1.3 Externalities

Forest resources can not only directly provide people with various material products, but also provide non-physical forms of ecological and landscape services. Forest resources have irreplaceable functions such as regulating climate, conserving water sources, maintaining soil and water, preventing wind and sand fixation, reducing natural disasters, and improving soil fertility (Kleinn et al., 2005). Simultaneously, forest resources have powerful social functions such as cultural and tourism benefits. Forest resources can maintain the balance of the ecological environment, play a key role in controlling natural disasters, such as floods, droughts, and sandstorms, and maintain a good living environment for humans. The ecological environment function of forest resources is non-competitive and non-exclusive in terms of consumption, and has negative externalities. Forestry development is subject to external conditions to a certain extent and is related to the stock of wood or the rapidity of harvesting from forests. These external conditions make it difficult to efficiently allocate forestry, induce forest loss, and reduce the number of forestry agencies.

8.2 Forest Resources

8.2.1 The Total Amount of Forest Resources

In the 1990s, the world's forest resources were severely damaged and the total amount of global forest resources continued to decrease. The main reasons for the continuous reduction in forest resources are the frequent occurrence of various natural disasters, deforestation leading to wasteland, and the increasing demand for forest products owing to social development. However, total forest resources in some countries, including China, continue to increase. The world's forest resources will continue to decrease in the twenty-first century; however, the prospects for forest resources are still positive.

Forest resources not only provide a variety of material resources for human social life, but also lay the foundation for the sustainable development of social ecology, which is closely related to human daily life, reproduction, and development (Feng et al., 2016). Therefore, it is very important to understand the current situation of forest resources worldwide and realize the comprehensive and coordinated development of forest resources, the environment, and the economy. The five continents of the world have a land area of 13.236 billion square hectometers, forestry land area of 4.172 billion square hectometers, forest land area of 3.526 billion square hectometers, an average forest coverage rate of 27%, and a total forest volume of 336.9 billion cubic meters. Forest resources on Earth are relatively abundant and distributed across all continents (except Antarctica). Forest resources can be divided into forest land, canopy forest land, sparse forest land, scattered forest land, and shrub land according to the degree of forest density. According to the *2020 Global Forest Resources Assessment Report* released by the FAO, the world's total forest area is 4.06 billion hectares, accounting for 31% of the total land area, which is equivalent to 0.52 hectares per person. However, the global distribution of forests among the world population is uneven. The forest area of Russia, Brazil, Canada, the United States, and China accounts for more than half of the total forest area in the world. China ranks fifth in forest area after Russia, Brazil, Canada, and the United States, and its forest volume ranks seventh after Russia, Brazil, Canada, the United States, Zaire, and India (Fig. 8.1).

The total area of global forest resources has decreased since 2010 (Fig. 8.2). Many scholars have pointed out that rapid global industrialization will affect climate and desert areas as well as indirectly damage forest resources, which requires the rapid development of industries related to forest governance and protection.

In 2020, the world's total forest-growing stock was estimated to be 557 billion cubic meters. The highest growing stock per unit area is found in tropical regions, mainly South America, Central America, West Africa, and Central Africa; however, in some regions, such as Europe, temperate and boreal forests are also high. Brazil is estimated to have the largest growing stock of all countries at 120 billion cubic

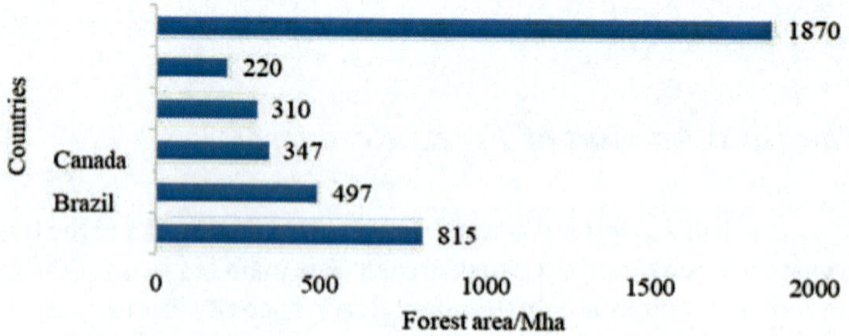

Fig. 8.1 Top five countries by forest area in 2020

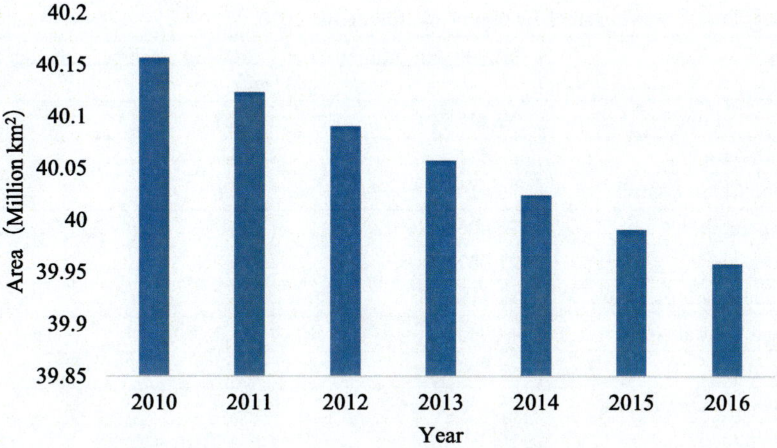

Fig. 8.2 The area of forest resources in the world from 2010 to 2016

meters, which is equivalent to nearly 22% of the total global growing stock. Russia, Canada, and the United States also have a considerable growing stock (Table 8.1).

Forest ecosystems are the largest terrestrial carbon sinks. Therefore, understanding the status and trends of various forest carbon storage pools is important for clarifying the role of forests in the global carbon cycle. Global forest carbon stocks declined from 668 to 662 Gt between 1990 and 2020, which was attributable to an overall reduction in forest area. However, this trend varies considerably between regions. For example, forest biomass carbon stocks increased substantially in East, West, and Central Asia; Europe; and North America, whereas carbon stocks in South America, West, and Central Africa decreased significantly. Although global forest carbon stocks declined between 1990 and 2020, the carbon stocks per hectare increased for all storage pools (Table 8.2).

China is extremely rich in species and has a wide variety of ecosystems. China is one of the five major forestry countries in the world, with a large forest area and complete ecosystem. China's forest resources are rich in species, with various vegetation resources and an ever-changing topography that can attract many tourists and provide good tourism benefits. The distribution of forest areas in China's forest resources is uneven, and there is a large gap in the volume per hectare and the proportion of planted forests. The main indicators reflecting the quantity of forest resources are forest area and forest stock volume. According to the *China Forestry and Grassland Statistical Yearbook (2019)*, the forest area of China in 2019 was 218.22 million hectares, the forest coverage rate was 22.96%, and the forest stock volume was 17,058.20 million cubic meters. As shown in Table 8.3, the total forest resources in China grew from 2003 to 2019, and the forest area increased from 158.94 million hectares in 2003 to 218.22 million hectares in 2019. However, forest resources in China still have problems such as an insufficient total amount and low quality. Therefore, management of forest resources in China faces several challenges.

Table 8.1 Forest growing stock by region and subregion, 2020

Region	Total growth (million m^3)	Growing stock per unit area (m^3/ha)
Eastern and Southern Africa	19,146	64.7
Northern Africa	841	23.9
West and Central Africa	56,419	184.6
Africa in total	76,406	120
East Asia	27,049	99.7
South and South East Asia	31,518	106.5
West and Central Asia	3935	71.2
Asia in total	62,502	100.4
Europe, excluding Russia	35,158	173.9
Europe in total	116,230	114.2
Caribbean Region	725	91.9
Central America	4233	188.9
North America	90,108	124.7
North and Central America in total	95,067	126.3
Oceania in total	18,867	101.8
South America in total	187,455	222.1
World in total	556,526	137.1

Data source Global Forest Resources Assessment 2020

Despite its large land area, the forest coverage rate in China is generally low. Nationally, 54% of forest land is of poor quality, and only 10% is of good quality. The overall quality of forest resources is not high, and the available resources Due to the small amount, the contradiction between the supply and demand of timber is prominent, and the supply of forest resources cannot meet the needs of social and economic development. However, in recent years, the quantity, quality, and ecological functions of forest resources in China have increased continuously. This initially formed a rational pattern in which state-owned forests are mainly public welfare forests, which are mainly commercial forests (Zhang, 2019), and wood supply is mainly plantation forests.

8.2.2 *Spatial Distribution of Forest Resources*

The global distribution of forest resources is currently uneven. The forest coverage rate in Europe is 49%, ranking first worldwide, whereas that in Africa is only 2%, which is the lowest worldwide. Table 8.4 shows that Russia has the largest forest area in the world, up to 8.153 million square kilometers, and its forest coverage rate is

Table 8.2 The total forest carbon stock by region and subregion, 1990–2020

Region	Forest carbon stocks (Mt)			
	1990	2000	2010	2020
Eastern and Southern Africa	20,932	29,642	27,978	26,250
Northern Africa	2338	2242	2190	2090
West and Central Africa	61,005	58,253	55,745	52,546
Africa in total	94,274	90,137	85,913	80,886
East Asia	27,110	30,261	33,908	37,907
South and South East Asia	45,804	43,792	43,071	41,468
West and Central Asia	4180	4511	4959	5358
Asia in total	77,093	78,564	81,938	84,733
Europe, excluding Russia	31,625	34,260	36,833	39,192
Europe in total	158,744	162,457	168,069	172,442
Caribbean Region	1552	1783	1977	2098
Central America	4988	4617	4270	4069
North America	136,644	137,730	139,324	139,951
North and Central America in total	143,184	144,131	145,572	146,118
Oceania in total	33,338	33,111	33,077	33,063
South America in total	161,765	154,917	147,917	144,846
World in total	668,399	663,316	662,485	662,088

approximately 49.8%, which is at the leading level in the world. Brazil has the second-largest forest area in the world, at approximately 4,965,562 square kilometers, and with a forest coverage rate of approximately 58.9%. However, in recent years, Brazil's hotspot rainforests have suffered huge damage, such as fires and deforestation, which has reduced the area of Brazil's rainforests by 15–20%. Canada has the third-largest forest area in the world, approximately 3.47 million square kilometers, and with a forest coverage rate of approximately 38.2%. Fourth, the United States has a forest area of approximately 3.1037 million square kilometers, and a forest coverage rate of approximately 33.9%. The fifth-largest area was China. Australia's forest area is approximately 1.25 million square kilometers, with a coverage rate of 16.3%. The forest coverage rate of Japan is approximately 68.5%. South Korea's forest coverage rate was approximately 63.4%, Indonesia's forest coverage rate was approximately 49.9%, and Germany's rate was approximately 32.7%. Thailand is about 32.2%, France is about 31.2%, and India is about 23.8%. Compared to other countries, China's forest quality needs to be improved.

As shown in Table 8.5, the average annual net loss of forest area in North and Central America between 1990 and 2020 was 293,000 hectares, and the annual net increase was 184,000 hectares between 2000 and 2010. This fluctuation mainly reflects changes in the data collection of its National Forest Survey but not true forest area dynamics. Between 2010 and 2020, Asia had the highest net growth in forest

Table 8.3 Statistics on China's forest resources

Year	Forest coverage rate (%)	Forest area (10^4 ha)	Plantation area (10^4 ha)	Forest stock ($10^4 m^3$)
2003	16.55	15,894.09	4708.95	1,126,659.15
2004	18.21	17,490.92	5364.99	1,245,584.58
2005	18.21	17,490.92	5364.99	1,245,584.58
2006	18.21	17,490.92	5364.99	1,245,584.58
2007	18.21	17,490.92	5364.99	1,245,584.58
2008	18.21	17,490.92	5364.99	1,245,584.58
2009	20.36	19,545.22	6168.84	1,372,080.36
2010	20.36	19,545.22	6168.84	1,372,080.36
2011	20.36	19,545.22	6168.84	1,372,080.28
2012	20.36	19,545.22	6168.84	1,491,268.19
2013	21.63	20,768.73	6933.38	1,513,729.72
2014	21.63	20,768.73	6933.38	1,513,729.72
2015	21.63	20,768.73	6933.38	1,513,729.72
2016	21.63	20,768.73	6933.38	1,513,729.72
2017	21.63	20,768.73	6933.38	1,513,729.72
2018	22.96	22,044.62	8003.10	1,756,022.99
2019	22.96	21,822.05	8003.10	1,705,819.59

Data source China Forestry and Grassland Statistical Yearbook

Table 8.4 Top 10 countries by forest area

Rank	Countries	Forest area (10^3 ha)	Proportion in the world (%)
1	Russia	815,312	20
2	Brazil	496,620	12
3	Canada	346,928	9
4	United States	309,795	8
5	China	219,978	5
6	Australia	134,005	3
7	Congo	126,155	3
8	Indonesia	92,133	2
9	Peru	72,330	2
10	India	72,160	2

Data source Global Forest Resources Assessment 2020

Table 8.5 Average annual net change in forest area by region and subregion, 1990–2020

Region	Change of forest area (%)					
	1990–2000		2000–2010		2010–2020	
	10^3 ha/yr	%	10^3 ha/yr	%	10^3 ha/yr	%
Eastern and southern Africa	−1,345	−0.40	−1,773	−0.55	−1,907	−0.62
Northern Africa	−182	−0.47	−127	−0.34	−168	−0.47
West and Central Africa	−1,748	−0.50	−1,503	−0.45	−1,862	−0.59
Africa in total	−3,275	−0.45	−3,403	−0.49	−3,938	−0.60
East Asia	1,917	0.88	2,332	0.97	1,901	0.73
South and South East Asia	−1,843	−0.58	−262	−0.09	−941	−0.31
West and Central Asia	129	0.26	285	0.55	213	0.39
Asia in total	202	0.03	2,355	0.39	1,173	0.19
Europe, excluding Russia	763	0.40	585	0.30	330	0.16
Europe in total	795	0.08	1,171	0.12	348	0.03
Caribbean Region	85	1.34	69	0.97	39	0.51
Central America	−218	−0.81	−211	−0.85	−130	−0.56
North America	−160	−0.02	327	0.05	−57	−0.01
North and Central America in total	−293	−0.04	184	0.02	−148	−0.02
Oceania in total	−165	−0.09	−231	−0.13	423	0.23
South America in total	−5,102	−0.54	−5,249	−0.58	−2,597	−0.30
World in total	−7,838	−0.19	−5,173	−0.13	−4,739	−0.12

Data source Global Forest Resources Assessment 2020

area, with East Asia accounting for most of it; with a net annual increase of 1.94 million hectares. Since 1990, there has been an overall net increase in forest area in Asia, increasing by 202,000 hectares per year between 1990 and 2000, 2.35 million hectares per year between 2000 and 2010, and 1.17 million hectares per year between 2010 and 2020.

Net forest losses occurred in South and Southeast Asia; however, these losses were partially offset by increases in forest areas in India and Vietnam. During this period, the rate of net forest loss in South and South-East Asia decreased significantly, from 1.84 million hectares per year between 1990 and 2000 to 941 000 hectares per year between 2010 and 2020. This decline was mainly due to a dramatic reduction in the rate of forest loss in Indonesia, which fell from 1.73 million hectares per year between 1990 and 2000 to 753,000 hectares per year between 2010 and 2020. Between 2010 and 2020, Oceania had the second-highest net annual increase in the face value of forests worldwide (after Asia), reaching 423,000 hectares, reversing the negative growth trend of previous decades in the region. This reversal largely reflects changes in Australian reporting from an average annual net loss of 207,000 hectares between 1990 and 2000 and 227,000 hectares between 2000 and 2010 to an average annual net gain of 446,000 hectares between 2010 and 2020. In the three decades to 2020, there was a net increase in forest area in Europe. The average annual net increase increased from 795,000 hectares between 1990 and 2000 to 1.17 million hectares between 2000 and 2010, before falling to 348,000 hectares between 2010 and 2020 (Table 8.6).

Between 2010 and 2020, the area of naturally regenerated forests decreased in all regions, except Europe and Oceania, with the greatest losses occurring in sub-Saharan Africa. Overall, the average annual loss of naturally regenerated forest in Africa was 3.32 million hectares between 1990 and 2000, 3.57 million hectares between 2000 and 2010, and 4.01 million hectares between 2010 and 2020. The increase in the loss rate in the last decade is mainly due to eastern and southern Africa, which increased from 1.83 million hectares per year in 2000–2010 to 1.96 million hectares per year in 2010–2020. Between 2010 and 2020, losses in the subregion were dominated by Angola, with 548,000 hectares per year. North and Central America reported a net

Table 8.6 Naturally regenerated forest area by region and subregion, 1990–2020

Region	Natural regeneration forest (10^3 ha)			
	1990	2000	2010	2020
Eastern and southern Africa	33,987	326,366	308,090	288,639
Northern Africa	38,542	36,627	34,984	33,168
West and Central Africa	355,885	338,136	322,316	303,441
Africa in total	734,301	701,128	665,390	625,248
East Asia	152,423	160,773	165,507	173,264
South and South East Asia	313,562	286,574	277,679	264,578
West and Central Asia	44,965	45,738	47,806	49,288
Asia in total	510,950	493,085	490,992	487,130
Europe, excluding Russia	116,352	118,921	118,854	118,819
Europe in total	912,651	912,829	914,376	915,250
Caribbean Region	5451	6277	6735	7008
Central America	27,928	25,687	23,438	22,014
North America	698,721	687,735	683,341	676,632
North and Central America in total	732,099	719,699	713,514	705,654
Oceania in total	181,705	179,067	176,037	179,949
South America in total	966,621	913,239	855,289	823,941
World in total	4,038,327	3,919,046	3,815,598	3,737,172

Data source Global Forest Resources Assessment 2020

annual loss of 786,000 hectares of naturally regenerated forest between 2010 and 2020, largely attributable to North America, with a slight increase in the associated area of the Caribbean. The region's average annual loss rate was much lower than that between 1990 and 2000, mainly because of a decline in annual loss rates in North America. Next, the United States' average annual loss fell from 354,000 hectares in 1990–2000 to 88,200 hectares in 2010–2020. The total annual loss of naturally regenerated forests in Asia between 2010 and 2020 was 386,000 hectares, mainly attributable to losses in South and Southeast Asia. During the same period, the areas increased in East, West, and Central Asia. The average annual loss rate in Asia between 2010 and 2020 was higher than that between 2000 and 2010; however, this figure was still much lower (80%) than the 1.79 million hectares in 1990–2000. The decline in loss rates is largely attributable to lower losses in South and Southeast Asia, particularly Indonesia, where the rate of loss of naturally regenerated forest was 2.1 million hectares per year between 1990 and 2000, and 787,000 hectares per year between 2010 and 2020. In Europe, the area of naturally regenerated forests has increased every year for the past 30 years, but the rate of growth slowed by almost half between 2010 and 2020, from 155,000 hectares per year to 87,400 hectares per year compared to 2000–2010. This is mainly attributable to the Russian Federation, where the average annual increase in area decreased from 161,000 hectares between 2000 and 2010 to 90,900 hectares between 2010 and 2020. Excluding the Russian Federation, the area of naturally regenerated forests in Europe decreased slightly between 2000 and 2010 (6690 hectares per year) and between 2010 and 2020 (3490 hectares per year).

With the rapid development of human society and the concept of sustainable development, the quantity and quality of China's forest resources can no longer meet the needs of existing social and economic development. The main manifestations are that the area and stock volume of forest resources are low, the distribution and structure of forest resources are seriously unbalanced, and the quality of plantations is low. The geographical distribution of forest resources in China is uneven. For example, the northeastern and southwestern regions of China have the most concentrated areas of forest resources, and the forest coverage in the central region is also relatively extensive. However, in addition to this, the forest coverage in western, northern, and central China is very low. In addition, China has a large population base and, therefore, a low per-capita forest share. China is a vast territory and ranks fourth in the world in terms of land area; however, its per capita forest share is only one-fourth of the world average, lagging behind most countries. Nearly half of China's forests are concentrated in six provinces (Inner Mongolia, Heilongjiang, Yunnan, Sichuan, Tibet, and Guangxi), of which the former three provinces of Inner Mongolia, Yunnan, and Sichuan account for 30.64% of China's forest area. However, Hainan, Jiangsu, Ningxia, and other provinces had smaller forest areas, accounting for less than 2% of the national forest area. Owing to historical problems and natural climatic conditions, the distribution of forest resources in North and Northwest China is low, whereas there are relatively more forest resources in the northeast, southwest, and southeast regions, and the distribution of forest resources in China is extremely uneven (Table 8.7).

Table 8.7 Forest resources in provinces of China, 2019

Province	Forest coverage rate (%)	Forest area (10^4 ha)	Forest stock (10^4 m^3)
Beijing	43.77	71.82	2437.36
Tianjin	12.07	13.64	460.27
Hebei	26.78	502.69	13,737.98
Shanxi	20.50	321.09	12,923.37
Inner Mongolia	22.10	2614.85	152,704.12
Liaoning	39.24	571.83	29,749.18
Jilin	41.49	784.87	101,295.77
Heilongjiang	43.78	1990.46	184,704.09
Shanghai	14.04	8.9	449.59
Jiangsu	15.20	155.99	7044.48
Zhejiang	59.43	604.99	28,114.67
Anhui	28.65	395.85	22,186.55
Fujian	66.80	811.58	72,937.63
Jiangxi	61.16	1021.02	50,665.83
Shandong	17.51	266.51	9161.49
Henan	24.14	403.18	20,719.12
Hubei	39.61	736.27	36,507.91
Hunan	49.69	1052.58	40,715.73
Guangdong	53.52	945.98	46,755.09
Guangxi	60.17	1429.65	67,752.45
Hainan	57.36	194.49	15,340.15
Chongqing	43.11	354.97	20,678.18
Sichuan	38.03	1839.77	186,099.00
Guizhou	43.77	771.03	39,182.90
Yunnan	55.04	2106.16	197,265.84
Xizang	12.14	1490.99	228,254.42
Shaanxi	43.06	886.84	47,866.70
Gansu	11.33	509.73	25,188.89
Qinghai	5.82	419.75	4864.15
Ningxia	12.63	65.6	835.18
Xinjiang	4.87	802.23	39,221.50

Data source China Forestry and Grassland Statistical Yearbook

Owing to the differences in natural climatic conditions, historical development, and economic development among the provinces in China, the distribution of forest resources in China is unbalanced, showing a distribution pattern of more in the southeast and less in the northwest. The density of roads in China is high, the level of mechanization is high, and the management of forested areas is relatively good. Forest resources in the southwest region are mainly distributed in the mountains and valleys of western Sichuan, northwestern Yunnan, and southeastern Tibet, where timber production is relatively difficult, the logging rate is low, and investment in capital construction is ineffective.

China has numerous forest resources. From north to south, there are cold temperate coniferous, temperate coniferous and deciduous broad-leaved mixed, warm temperate deciduous, subtropical evergreen broad-leaved, and tropical monsoon forests. There are more than 20,000 species of woody plants worldwide, of which there are more than 8,000 species in China and more than 2,000 species of trees, which form a colorful forest landscape. Forest resources in China have a long history of change and natural evolution, and have gradually formed five major forest areas with obvious regional characteristics closely related to natural geographical conditions and social and economic development: northeast Inner Mongolia forest area, southwest high mountain forest area, southeast low mountain hill forest area, northwest high mountain forest area, and tropical forest area. Many forest types in China have high economic or ornamental landscape value, such as ginkgo, yew, sumac, rubber trees, redwood, Eucommia, mulberry, and tea. The distribution of the age structure of forests in China is unreasonable. Young and middle-aged forests account for 71.1% of the national forest area and 74.4% of timber forests. The trend of younger forest resources in the hills and mountains of the southeast is more pronounced.

In general, the geographical distribution of forest resources in China is extremely uneven, mainly in the northeastern and southwestern forest areas. Mature forest resources have been exhausted in many areas, and most existing mature forests are located in undeveloped remote mountainous areas. The lack of mature forest resources will lead to the comprehensive logging of near-mature forest resources, making the forest age structure unreasonable and leading to declining forest quality. However, the relative shortage of total forest resources, low quality, and uneven distribution have not been fundamentally resolved, and the development of forest resources remains a challenge.

8.3 Analysis of the Utilization Efficiency of Forest Resources

8.3.1 Problems in the Utilization of Forest Resources

As the top priority of the global ecosystem, forest resources are not only a key factor in maintaining a good ecological environment but also an important source of materials for human survival and development, providing a variety of wood for human production services and the development of human society. key material basis (Okumu & Muchapondwa, 2020). In China, the quantity and area of forest resources are not high, the supply of wood is insufficient, and deforestation is a serious phenomenon that seriously damages the structure of forests and has a negative impact on the ecological environment (Sun et al., 2017). Since the reform and opening up, China has implemented strict forest protection policies and increased afforestation. Plantations have continued to grow, and both the area and stock volume have increased. However, there are still many deficiencies in the development, utilization, and management of forest resources, which restrict the development of forest resources. Forest resources are key components of terrestrial ecosystems that provide humans with diverse products and production materials that contribute to economic, social, and environmental development (Jannat et al., 2018). Forest resources have enormous environmental and economic value. However, if people improperly use forest resources, they will suffer significant resource losses.

8.3.1.1 Inadequate Management of Forest Resources

Generally, China's forest resources are affected by extensive management, low productivity of forestland, and a decline in forest quality. With the development of the national economy and an increase in the population, the consumption of forest resources has increased sharply (Liu, 2021). While the vast majority of the world's forests is threatened by overexploitation, many others are at risk of abandonment because of unmanaged management (Starr et al., 2019). Sustainable management of forests is essential for maintaining their potential and ecosystem services. Ensuring the versatility of forests is one of the most important challenges in the coming years. With the development of the social economy, people are gradually paying more attention to the supervision of resources to use them more efficiently and obtain greater benefits (Wang, 2021). However, although the management of forest resources has been strengthened, there are still many deficiencies, such as local forestry failing to properly handle accountability and compensation for damage to forest resources, and failing to perform timely and effective supervision.

Macro-management of forest resources in China is mainly based on afforestation and greening, and the coordinated development and allocation of trees, shrubs, bamboos, and grasses are managed as a whole. The key to forest resource management is to solve the problems of land greening planning, location planning, greening modes, and greening space management at the national level (Lu, 2021). Owing to the vast land area of China, the distribution of forest resources is relatively wide, which means that the management of forest resources must increase the inputs of labor and financial resources. However, a lack of funds and management technicians for forestry projects in many regions makes it difficult to manage and protect forest resources.

8.3.1.2 Serious Deforestation

Deforestation is considered one of the factors contributing to global climate change, and the impact of deforestation on the exchange of water vapor and carbon dioxide between the atmosphere and land surface is the most important issue for the climate system (He et al., 2016). Deforestation can cause soil erosion, and soil erosion can lead to sediment entering lakes, streams, and other water sources, which can reduce local water quality and lead to poor health of the population in the region. However, stopping deforestation means giving up jobs and income, either from logging or the land that comes from deforestation. Factors relevant to deforestation investigated in the literature include geophysical and demographic factors, economic variables, government policies, and other actions. Among the main causes of deforestation, half of the forest is illegally derived from trees used as fuel, some are used to provide more land for housing and urbanization, and some are used to obtain wood for commercial projects, such as paper and furniture (Table 8.8).

An estimated 420 million hectare of forest was lost to deforestation between 1990 and 2020, but this rate slowed over this period. Deforestation rates were 15.8 million hectares per year between 1990 and 2000, 15.1 million hectares per year between 2000 and 2010, 11.8 million hectares per year between 2010 and 2015, and 10.2 million hectares per year between 2015 and 2020. Between 2015 and 2020, Africa had the highest annual deforestation rate (4.41 million hectares), followed by South America (2.96 million hectares) and Asia (2.24 million hectares). In Africa, most deforestation occurs in Eastern and Southern Africa (2.2 million hectares per year), and West and Central Africa (1.9 million hectares per year). In Asia, deforestation occurs mainly in South and Southeast Asia (1.96 million hectares per year). Deforestation rates in Africa have been rising since 1990, but decreased slightly between 2015 and 2020 compared to 2010–2015. However, deforestation rates in Asia and South America are now almost half of those in the 1990s.

Although China stipulates the quantity of forest logging, there is still a large amount of excessive logging, which has become a major problem in forest management worldwide. With the rapid progress of human economy and society, the social demand for forest products is also increasing. In China, the rising trend in the average consumption of forest resources in recent years is not clear in Mingxi County; however, it continues to grow. The indiscriminate deforestation of forests not only destroys their ecological structure but also accelerates global warming and reduces biological species diversity, thereby causing various ecological and environmental problems. Cutting is an important method for forest resource utilization, which

Table 8.8 Deforestation rates by region and subregion for four periods, 1990–2020

Region	Deforestation rate (10^3 ha/yr)			
	1990–2000	2000–2010	2010–2015	2015–2020
Eastern and southern Africa	1781	2240	2116	2199
Northern Africa	461	442	330	316
West and Central Africa	1854	1631	1998	1899
Africa in total	4096	4314	4444	4414
East Asia	399	353	369	170
South and South East Asia	3689	2232	2460	1958
West and Central Asia	82	99	96	107
Asia in total	4170	2684	2925	2235
Europe in total	88	92	201	69
Caribbean Region	3	2	23	5
Central America	228	222	142	168
North America	740	475	253	263
North and Central America in total	972	699	418	436
Oceania in total	655	662	458	42
South America in total	5837	6667	3354	2953
World in total	15,818	15,117	11,801	10,150

Data source Global Forest Resources Assessment 2020

directly affects stock growth, forest structure, and forest regeneration. Therefore, a key path for forest resource management is the formulation of a scientific and reasonable logging plan to promote the sustainable use and development of forest resources.

8.3.1.3 Frequent Natural and Human-Caused Disasters

In the prevention and control of forest resources, forest fires and damage caused by various pests and diseases cause serious harm to forest resources, and forest pests and diseases cause serious losses to forest resources, sometimes even devastating blows. Forest fires are a source of forest damage and a top priority in forest resource protection (Wang et al., 2021). Based on whether trees cause losses and the size of the fire area, forest fires can be classified into general forest fires, larger forest fires, major forest fires, and major forest fires (Li et al., 2021). The following Table 8.9 shows the statistics for the number and severity of forest fires in China from 2008 to 2019.

There are many fires in China, and forest fire prevention work is of great importance for strengthening protection (Table 8.9). At the same time, most of the fires are caused by human factors. Careless production, life, and slash-and-burn cultivation habits can cause serious forest fires (Shang, 2021). In addition, in recent years, many

Table 8.9 The situation of forest fires in China from 2008 to 2019

Year	Forest fires	General fires	Large fires	Major fires	Catastrophic fires	Total fire area (ha)
2008	14,144	5673	0	13	0	184,494.5
2009	8859	4945	3738	35	1	213,635.9
2010	7723	4795	2902	22	4	116,243.4
2011	5550	2993	2548	9	0	63,416.21
2012	3966	2397	1568	1	0	43,171.17
2013	3929	2347	1582	0	0	42,890.42
2014	3703	2080	1620	2	1	55,340
2015	2936	1676	1254	6	0	33,077.62
2016	2034	1340	693	1	0	18,161.46
2017	3223	2258	958	4	3	44,428.4
2018	2478	1579	894	3	2	28,595.2
2019	2345	1534	802	8	1	39,704.7

young people in rural areas have gone out to work, and those left behind in the villages are the elderly and children. When a fire occurs, it is impossible to put it out and call the police in time, resulting in greater losses in the event of a fire (Jia, 2021). Although China's investment in forest fire prevention has increased annually in recent years, the construction of forest fire prevention infrastructure is still lagging, resulting in insufficient forest fire monitoring, isolation, communication equipment, and material reserves (Liang et al., 2021). To effectively deal with and respond to fires, it is necessary to establish a complete firefighting plan and firefighting measures in advance according to the characteristics of each terrain and forest resource to prevent serious losses caused by fires (Wang & Wang, 2021). Owing to the complex terrain in many forested areas, traffic and communication are inconvenient, and fires cannot be put out in time. At the same time, forest areas with complex terrain often lack maintenance and a relative lack of forest rangers, which further increases their potential safety hazards.

8.3.1.4 Contradiction Between Protection and Utilization

The effective protection of forest resources and ensuring that forest resources are not damaged have always been the focus of continuous research and the attention of resource management departments (Yang, 2021). In particular, system and mechanism construction, information monitoring, forest land protection, forest fires, and rodent pest control research in forest resource management has never been interrupted. This is related to the security of the national ecological environment and sustainable development of forests. Therefore, intensifying research on forest resource protection and management strategies is necessary to improve forest resource protection and ecological restoration (Jing & Jing, 2019). With the accelerated pace of industrialization in various countries worldwide, the continuous development of the social economy, and the continuous increase in the total population, the demand for forest trees is increasing, and the area of occupied forest land will increase annually (Xu, 2021). However, forestry plays a pivotal role in coping with global climate change and the sustainable development of the ecological environment, and the protection of forestry should be continuously strengthened (Xia, 2020). The demand for forestry is high, and its protection should be strengthened. The contradiction between forestry ecological construction land and forestry industry construction land and the contradiction between supply and demand of forest land are increasingly intensifying. At the same time, the management of forest land is insufficient in some areas, which further increases the conflict between the protection and utilization of forest resources (Li, 2020). From the perspective of China, the unreasonable development of forest resources in some areas has led to the destruction of forests in important ecological function areas and continuous deterioration of the ecological environment.

During periods of rapid economic development, people tend to overexploit forest resources and ignore sustainable development of forest resources (Van Hoang et al., 2020). At the same time, the problem of insect pests in China's forest resources is serious. Therefore, it should be strictly quarantined to determine the presence of pests and diseases on the surface of the forest and ensure timely treatment of diseased wood. The contradiction between the effective supply of forests in China and growing demand is still prominent. Owing to the high degree of dependence on external timber, few available resources, and few large-diameter timber and precious timber species, the structural contradiction between timber supply and demand is very prominent.

8.3.2 Efficiency Calculation and Analysis

Tourism in forest parks is an important channel for utilizing forest resources. Forest resources have not only ecological benefits but also greater economic benefits. Forest parks play an important role in developing the economic and tourism benefits of forest resources, which can not only exert economic benefits but also consider the important task of protecting the ecological environment (Huang & Lin, 2015). Forest tourism provides a good ecological environment that brings people closer to nature and is loved by many people (Liu & Zhou, 2020). Simultaneously, forest tourism has been widely favored and rapidly developed in recent years. Research on forest parks has focused on the efficiency of forest parks (Tan, 2021; Zhu et al., 2017). We take 31 provinces in China as research objects to analyze the tourism efficiency of national forest parks. Considering the differences in forest parks in various provinces, data envelopment analysis (DEA) is used to analyze the tourism utilization efficiency of forest parks. This study uses the super-efficiency SBM model in DEA to measure efficiency under variable returns to scale for each year.

Forest park operations offer several benefits. This section focuses on the tourism efficiency of forest parks. The output indicators included the number of tourists received and tourism income, and the input indicators included forest park area, annual investment capital, and number of employees. Data are obtained from the *2011–2019 China Forestry Statistical Yearbook, China Tourism Statistical Yearbook*, and provincial statistical bulletins of national economic and social development.

The overall tourism efficiency of China's forest parks is not high; the utilization rate of forest parks still has much room for improvement, and the development of forest parks is still in a state of extensive development (Table 8.10). Second, there are differences in the efficiency of forest park tourism development in different provinces of China, and each province should develop its tourism efficiency of forest parks based on its own actual situation. In general, the allocation of tourism resources in China's forest parks needs to be further improved, and tourism resources should be fully utilized to maximize benefits. Owing to the constraints of the selection of indicators and methods, the provincial forest park efficiency calculated in this study was a relative value and did not fully reflect the current management status of forest parks. In future, more accurate methods and indicators should be selected to measure the tourism efficiency of forest parks.

8.3.3 Reasons for Inefficient Utilization of Forest Resources

The utilization efficiency of forest resources is affected by several factors. Future studies should focus on the efficient use of forest resources (Liu et al., 2021). The status and development of forest resources are related to the survival and development of species. People can use forest resources rationally and efficiently to promote sustainable development, and quickly destroy them (Barnes et al., 2010). Therefore, the rational use of forest resources is crucial to the virtuous circle of human society and the ecological environment. Ensuring the efficient use of forest resources can bring great value to society and the environment. Forest resources have powerful functions, provide various materials and services to human beings, and have economic, ecological, and social value. Although forest resources in China have made some progress, overall forest resources are still poor and insufficient to meet the needs of human society for development and utilization. At the same time, owing to the constraints of economy, technology, and other factors in the process of development and utilization, the overall utilization efficiency of forest resources is not high, and the phenomenon of resource waste is relatively serious. Presently, the reasons for the low utilization efficiency of forest resources mainly include the following.

8.3.3.1 Excessive Deforestation by People

Scholars believe that excessive deforestation owing to profit-maximizing decisions causes forest inefficiency. Private forests are an important component of this resource.

Table 8.10 Forest park tourism efficiency value, 2011–2019

Province	2011	2012	2013	2014	2015	2016	2017	2018	2019
Anhui	0.803	1.072	0.462	0.236	0.238	0.285	0.273	0.276	0.282
Beijing	1.093	0.220	0.273	0.130	0.130	0.127	0.146	0.129	0.218
Fujian	0.203	0.210	0.179	0.086	0.078	0.056	0.120	0.176	1.004
Gansu	0.259	0.098	0.051	0.037	0.050	0.052	0.063	0.054	0.069
Guangdong	0.187	0.186	0.197	0.109	0.101	0.120	0.129	0.167	0.293
Guangxi	0.008	0.100	0.053	0.027	0.038	0.040	0.050	0.064	0.119
Guizhou	0.044	0.231	0.114	0.132	0.140	0.216	0.240	0.230	0.162
Hainan	1.221	0.174	0.103	0.084	0.075	0.083	0.104	0.110	0.097
Hebei	0.197	0.135	0.096	0.051	0.057	0.059	0.054	0.055	0.067
Henan	0.151	0.469	0.176	0.130	0.119	0.115	0.141	0.155	0.133
Heilongjiang	0.013	0.127	0.028	0.026	0.027	0.026	0.023	0.027	0.049
Hubei	0.265	0.228	0.177	0.083	0.090	0.105	0.114	0.136	0.142
Hunan	0.114	0.184	0.116	0.052	0.066	0.064	0.111	0.122	0.139
Jilin	0.087	0.120	0.144	0.024	0.022	0.029	0.029	0.043	0.057
Jiangsu	1.720	1.446	1.200	0.538	0.647	0.667	0.776	1.079	1.115
Jiangxi	0.307	0.555	0.360	0.215	0.390	0.238	0.238	0.256	0.402
Liaoning	2.411	0.140	0.120	0.050	0.074	0.094	0.112	0.104	0.087
Inner Mongolia	0.013	0.030	0.024	0.013	0.017	0.021	0.024	0.022	0.014
Ningxia	0.056	0.101	0.079	0.072	0.086	0.064	0.070	0.074	0.057
Qinghai	1.081	1.617	0.456	0.097	0.064	0.076	0.076	0.076	0.015
Shandong	1.048	0.654	0.243	0.148	0.135	0.165	0.239	0.292	0.429
Shanxi	0.105	0.124	0.058	0.073	0.068	0.076	0.080	0.095	0.075
Shaanxi	0.073	0.179	0.128	0.077	0.093	0.113	0.145	0.174	0.130
Shanghai	1.421	1.297	0.708	1.027	0.533	0.552	0.792	0.660	1.459
Sichuan	0.102	0.242	0.132	0.117	0.127	0.114	0.208	0.360	1.121
Tianjin	1.289	1.573	0.575	0.224	0.278	0.306	0.331	0.394	0.348
Xizang	0.069	0.099	0.033	0.022	0.000	0.012	0.018	0.012	0.008
Xinjiang	0.152	0.193	0.073	0.030	0.036	0.028	0.033	0.052	0.098
Yunnan	0.133	0.043	0.161	0.074	0.080	0.080	0.065	0.074	0.088
Zhejiang	1.152	0.249	0.323	0.201	0.191	0.234	0.350	0.368	1.159
Chongqing	0.085	0.183	0.147	0.088	0.116	0.151	0.161	0.178	0.170

However, private forest policymakers are plagued by various external costs, and providing sustainable flows of wood fiber is not the only social purpose of forests. When the act of harvesting wood imposes costs on other valuable aspects of the forest (e.g., water bodies maintenance, soil erosion control, and biodiversity conservation), these costs are not borne by decision-makers; these facility costs are usually not borne by decision-makers in profit maximization. Regarding the value of existing forests as a wildlife habitat or a key element of a local ecosystem as an external cost can lead to inefficient decision-making that threatens biodiversity. Failure to recognize the full social value of existing forests incentivizes not only the inefficient harvesting of large volumes of wood in deforestation but also harvesting, even when conservation of forests is an effective alternative.

8.3.3.2 Inappropriate Government Incentives

Profit maximization does not produce effective results when policymakers are confronted with counterintuitive incentives. Unfortunately, forestry provides a multitude of situations in which perverse incentives produce inefficient and unsustainable results. Private forests are a very important force worldwide, but in some countries such as the United States, they are the dominant force. Private forest decisions are plagued by various external costs, and providing sustainable flows of wood fiber is not the only social purpose of forests. When the act of harvesting wood imposes costs on other valuable aspects of the forest, such as water bodies maintenance, prevention of soil erosion, and protection of biodiversity, these costs are not borne by decision makers. These facility costs are often not adequately considered in profit-maximization decisions. The fact that the value of existing forests as wildlife habitats or key elements of local ecosystems is an external cost can lead to inefficient decision-making that threatens biodiversity. Failure to recognize the full social value of existing forests not only incentivizes inefficient logging of large volumes of wood in deforestation, but also incentivizes logging, even when protecting forests is an effective alternative.

Government policies can also create negative incentives for landowners. For example, when the Brazilian government reduced taxes on agricultural income, this substantive treatment of agriculture was overvalued. This overvaluation leads to profitable deforestation and the conversion of land to agriculture, which would not be profitable even in the absence of discriminatory tax policies. This tax system encourages the conversion of land from forest to pasture at efficient rates, and subsidizes activities that would normally not be economically viable without tax discrimination. In essence, Brazilian taxpayers unknowingly subsidized deforestation, which devalued its natural capital stock, and Brazil's land title system played a role in the early history of deforestation. The Brazilian government has lowered agricultural taxes and attempted to cut forests and convert forest land into agricultural land for profit. Brazil's tax system encourages a relatively high conversion efficiency from forests to pasture or farmland, and funding this activity is not economically feasible in areas where there are no tax differentials. Another source of inefficiency in the use of forest resources is concession agreements, which specify the conditions under which public forests can be harvested. For loggers, cutting down existing forests has a significant advantage over planting new ones, and old trees can be cut down immediately for profit. Considerable economic rent is associated with forests because of the commercial value of larger, older trees.

8.3.3.3 Poverty and Debt

Poverty and debt are major sources of pressure on forests. Farmers perceive unclaimed woodlands as opportunities for landowners. Countries with large numbers of farmers see territorial or communal forests as a more politically viable source of land for the landless, rather than forcibly taking land from the rich. In Eastern and Southern Africa, positive feedback loops have created a downward cycle in which poverty and deforestation reinforce each other. Most natural forests have long been logged for timber and fuelwood as well as for growing crops on reclaimed land. As forests disappear, the rural poor are forced to devote more time to finding new sources of fuel. Once firewood is no longer available, dry animal manure is burned, eliminating it as a source of fertilizer to nourish depleted soil. The loss of trees leads to increased soil erosion, and soil depletion leads to reduced nutrient levels. Nutrient deficiencies exacerbate the threat to human health because of the inability to find or afford sufficient fuel, wood, or animal waste for cooking and boiling unclean water. Deteriorating health drains energy, increases susceptibility to diseases, and reduces productivity. Survival strategies may sacrifice long-term goals to avoid starvation or death, and forests are typically early victims.

Forest resources are essential for the provision of ecosystem and environmental services such as biodiversity conservation, fresh air, carbon sequestration, water flow maintenance, and soil fertility renewal. Consequently, rural communities worldwide rely on forests, which contribute significantly to their livelihood. Over the years, forest cover in many developing countries has steadily declined. This is usually due to advances in technology and population growth, as well as poverty and other social difficulties, resulting in an overreliance on forest resources and increased demand for forest ecosystem services. New strategies are required to organize the reduction of forests and to place the remaining forests under safe and effective management.

8.3.3.4 Sustainable Forestry

Sustainable forestry can only be achieved when forests are adequately protected and are capable of perpetuating harvests. In addition, sustainable forestry requires that harvesting be limited by the growth of the forest, leaving the amount of wood unaffected. However, efficiency does not necessarily satisfy the definitions of sustainable forestry. Maximizing the present value involves making an implicit comparison between the value added by delaying harvesting (mainly because of an increase in volume) and the value added by harvesting wood and investing in the benefits. The volume growth rate was lower in the slow-growing species. Selecting a harvest age that maximizes the present value of the net benefit of slow-growing forests will likely involve harvests that are higher than the net forest growth. The search for economically sustainable forestry practices has led to the development of new practices. One of the priorities is growing fast-growing tree species in plantations. Fast-growing species increase the economic attractiveness of replanting because investment capital takes less time. Species raised in plantations can be harvested and replanted at a low cost. Forest plantations have been established for various purposes, such as supplying fuelwood in developing countries and pulp for paper mills in industrialized and developing countries. However, plantations are controversial. Not only are plantations often planted with a single tree species, leading to poor wildlife habitats, but they also require large amounts of fertilizers and pesticides. In some parts of the world, the natural resilience of forest ecosystems is sufficiently high to be sustainable, despite unsustainable levels of harvest decades ago. In the United States, for example, at some point in the 1940s, the net growth of forestland exceeded timber deforestation. Subsequent surveys confirmed that despite the considerable and growing demand for wood, net growth continued to outpace harvests. Total forest biomass in the United States has been growing since World War II. Harvests during this period were sustainable for the whole country, although this was not the case for certain species in specific regions.

8.4 Public Policy

Public policies on forest resources aim to promote the coordinated development of the ecology, economy, and society of forest resources. Since the late 1980s, China has implemented three main forest resource management policies: forest resource logging restriction policy, national key public welfare forest protection policy, and the natural forest protection project. Judging from the implementation history of forest resource policies, these policies were not proposed under the condition of stable and sustainable forest construction, but under the condition that forest resources were destroyed and natural disasters continued.

8.4.1 Forest Resource Logging Restriction Policy

The contradiction between deforestation and forest resource management is mainly a contradiction between the protection, development, and utilization of forest resources (Li et al., 2021). Although the protection and utilization of forest resources can be coordinated and unified, excessive deforestation of forest resources is an extremely serious issue. When the over-mature and mature forests in the forest area are cut, the contradiction between deforestation and forest resource management becomes particularly prominent. The premise of forest resource protection is to obtain the monitoring data of forest resources, so as to ensure that forest resources are not excessively cut down and maintain their sustainable development and utilization.

Surveying forest resource planning and design is one way to obtain data on forest cutting quotas. Through survey data, a forest resource file can be established to formulate a reasonable forest-cutting quota and provide a data basis for planning by the forestry department. However, during the course of the investigation, there may be problems such as relatively backward equipment and technical means, resulting in incomplete investigation data and failure to investigate the latest data in a timely manner. Incomplete data and data lags cause data, such as the quantity, quality, and structure of forest resources to deviate from the actual situation. This deviation also makes local government departments unable to effectively prevent excessive deforestation of forest resources.

8.4.2 National Key Public Welfare Forest Protection Policy

The key national public welfare forest protection policy originated in the reform of the forestry classification management system. In 1995, China promulgated the *General Outline of the Reform of the Forestry Economic System*, which included shelter and special-purpose forests in the category of public welfare forests. Simultaneously, it stipulates that public welfare and commercial forests should be divided according to the purposes of forest use, production, and management. However, in public welfare forest policy, there is a phenomenon of focusing on demarcation and neglecting management and control. Therefore, the China Forestry Administration promulgated the Interim Measures for the *Administration of National Public Welfare Forests (Draft for Comment)*, which clearly stipulates that first-level national public welfare forests should not carry out production and operation activities in principle, and second- and third-level national public welfare forests should not damage the ecological environment. Under the premise of functional functions, forestland resources can be used reasonably according to law. It can be seen from this that protection of public welfare forests in China is gradually improving, and the relevant regulations for the management of national public welfare forests are becoming increasingly mature.

8.4.3 Natural Forest Protection Project

Natural forests are an important part of forest ecosystems and are the main source of forest resources. The natural forest protection project began to take strict control when the ecosystems of various river basins in China were continuously destroyed, and completely stopped the logging of natural forests in the upper reaches of the Yangtze River and the Yellow River. It is evident that the protection of natural forests is crucial to the construction of the ecological environment, which provides a protective barrier for our survival and stable development. Natural forest protection projects can fundamentally improve the ecological environment, protect biodiversity, and promote the coordinated and sustainable development of the human society. Simultaneously, the natural forest protection project implemented strict protection measures for natural forests, greatly reduced the amount of deforestation, strictly controlled forestry resources, and adjusted and optimized the structure of forestry. Natural forest protection is a great project that aims to optimize and strengthen the protective barrier function of forest resources with new methods and ideas and ensure the sustainable development of human resources and the environment.

8.5 Summary

Forest resources are typical storable and renewable resources. Sustainable forestry can only be achieved when the forests are adequately protected. In addition, sustainable forestry requires that harvesting be limited by the growth of the forest, leaving the amount of wood unaffected. Forestry has three distinct characteristics: slow growth in the early stage, rapid growth in the middle stage, and slow growth in the mature stage. The owner of the harvested forest can earn income from the sale of wood, but the owner of the delayed-harvest forest receives additional income. The extent of forest growth depends on the growth cycle of forests. From an economic point of view, the effective harvesting time of general wood is when the present value of the net benefit is the largest; that is, the marginal benefit of delaying harvesting for one year is equal to the marginal cost of delayed harvesting. China's forest resources have shown good momentum in general, but the management of forest resources should also be strengthened to consolidate the achievements of forest resource development and achieve sustainable management of forest resources. During the process of forest resource protection and management, forestry departments and managers should strictly abide by the relevant laws and regulations, implement relevant policies and measures, and strengthen the management of forest resources. By rationally developing and utilizing forest resources, the sustainable development of forest resources is realized. In the future, forest resource management should be strengthened in the following aspects.

8.5.1 Strengthen the Research on Forest Disaster Census and Prevention

The frequent occurrence of forest disasters is an important factor that hinders the sustainable development of forest resources. For China, forest disasters are far higher than the global level, and the damage is at a high face value; therefore, the prevention and control of forest disasters should be strengthened. The control of forest fires can be strengthened in terms of publicity, management, and emergency responses, and the occurrence of forest fires can be reduced. As for the problem of pests in forestry, biological research on forestry pests should be strengthened, the source of the disaster should be investigated, and pest control work should be conducted to further optimize the development of forestry.

8.5.2 Strengthen the Management of Forest Resources

Forest resource management is extremely important and requires significant human, material, and financial resources. However, the efficient management of forest resources is a question worthy of consideration. At this stage, technical means can provide convenience for the management of forest resources, such as statistical sampling, databases, model simulation, and 3S technologies. These technical means can provide convenience for the management of forest resources but have not yet reached a mature level. It is necessary to continue promoting the research and development of forest resource management technology to ensure the efficient management of forest resources. How to accurately, quickly, efficiently, scientifically, and rationally develop and utilize forest resources and the ecological environment structure of swimming lakes have become urgent problems to be solved in current greening and forestry management. Relevant departments should strengthen the level of information protection and management of forest resources, combine advanced information technology to measure the status of forest resources in real time, and focus on forest risk issues. Simultaneously, in the event of a forest fire, modern media and communication methods should be used to help forest protection personnel understand the current situation and address it on time to minimize the loss of forest resources. In summary, the management of forest resources is a long-term process that requires the continuous investment of certain human and material resources to realize the protection of forest resources and constantly rectify the problems existing in the process of forest management to maintain the ecological and economic benefits of forest resources.

8.5.3 Strengthen the Protection of Forest Resources

The total amount of forest resources is increasing worldwide. However, the results of forest protection studies are mixed, and little is known about the factors that influence the success of collective action in forest protection. For forestry operators to achieve sustainable development of forest resources, it is necessary to strengthen their management and protection of forest resources. The unilateral pursuit of economic interests, while ignoring the protection of forest resources, will make the forest ecosystem unbalanced, the ecological environment continues to deteriorate, and long-term sustainable development and utilization cannot be achieved. Strengthening the protection of forest resources is key to their sustainable development of forest resources and is a necessary measure to maintain the stability of the ecosystem. In the future, a scientific and reasonable forest development plan should be formulated in light of the current situation of forest resources, and scientific methods should be used to effectively reduce forest stealing and logging and avoid the aggravation of forest resource loss. Therefore, each region should combine local resource endowment characteristics, select native tree species, and rationally design, plan, and implement forest-protection plans. At the same time, managers should scientifically design a development system based on the needs of forest protection to promote a balance between forest protection and forest development.

8.5.4 Implement a Forest Resources Talent Revitalization Strategy

Forest resources not only have ecological benefits but also have public welfare and social properties that cannot be ignored. Therefore, we should respond to the national strategy of "rural revitalization," cultivate talents for the efficient utilization and protection of resources, and realize the revitalization of forest resources. The first was to sign contracts with students majoring in agriculture and forestry to train all types of professionals urgently needed in forest areas. The second is to strengthen the construction of technical forestry colleges, cooperate with colleges and universities, provide financial and technical support, and train professionals in the development and protection of forestry. Overall, the rapid development of modern society should not be at the expense of resources but should contribute to the sustainable development of resources, cultivate talent, and realize digital forestry through the innovative application of modern information technology to build a forest resource protection management system.

References

Barnes, J. I., MacGregor, J. J., Nhuleipo, O., & Muteyauli, P. I. (2010). The value of Namibia's forest resources: Preliminary economic asset and flow accounts. *Development Southern Africa, 27*(2), 159–176.

Cheng, S., Xu, Z., Su, Y., & Zhen, L. (2010). Spatial and temporal flows of China's forest resources: Development of a framework for evaluating resource efficiency. *Ecological Economics, 69*(7), 1405–1415.

Feng, J., Wang, J., Yao, S., & Ding, L. (2016). Dynamic assessment of forest resources quality at the provincial level using AHP and cluster analysis. *Computers and Electronics in Agriculture, 124*, 184–193.

Grinkevich, A., Grinkevich, L. S., Bobek, S., Tominc, P. (2019). Taxation of the use of forest resources: The case of the Russian Federation. *Drewno: Prace Naukowe, Doniesienia, Komunikaty, 62*.

He, W., Zhang, H., & Xu, J. (2016). Can the forest harvesting quota management system play a role in protecting forest resources: A literature review. *China Rural Insights, 2*, 84–93.

Huang, X., & Lin, X. (2015). The tourism efficiency of forest parks in my country and its influencing factors. *Forestry Science, 02*, 137–146.

Jannat, M., Hossain, M. K., Uddin, M. M., Hossain, M. A., & Kamruzzaman, M. (2018). People's dependency on forest resources and contributions of forests to the livelihoods: A case study in Chittagong Hill Tracts (CHT) of Bangladesh. *International Journal of Sustainable Development & World Ecology, 25*, 554–561

Jia, B. (2021). Current situation and countermeasures of forest fire prevention in nature reserves in Liaoning Province—Taking 6 national forest ecological type reserves in Liaoning Province as examples. *Agriculture and Technology, 10*, 80–82.

Jing, L., & Jing, L. (2019). Analysis of forest resources protection and ecological environment construction countermeasures. *Agriculture and Technology, 20*, 79–80.

Kleinn, C., Ramírez, C., Holmgren, P., Valverde, S. L., & Chavez, G. (2005). A national forest resources assessment for Costa Rica based on low intensity sampling. *Forest Ecology and Management, 210*(1–3), 9–23.

Li, H. (2020). Analysis of the impact of climate change and deforestation on hydropower planning in the Brazilian Amazon River Basin. *Water Resources and Hydropower Express, 05*, 4.

Li, C., Bai, Z., Tian, M., & Peng, C. (2021). The positive role of forest logging quotas in forest resource protection. *Modern Horticulture*, (13), 164–165+167.

Liang, S., Gao, D., & Ding, S. (2021). Research on modern forest fire prevention industry under the background of "Internet+" based on SWOT-PEST model—Taking Zhenjiang City as an example. *Economic Research Guide*, (24), 29–31+57.

Liu, Z. (2021). The status and role of forestry survey planning and design in the protection, management and utilization of forest resources. *Southern Agriculture, 17*, 99–100.

Liu, L., & Zhou, X. (2020). Forest ecotourism and forest protection. *Heilongjiang Science, 16*, 134–135.

Liu, W., Wang, X., Lu, F., & Ouyang, Z. (2015). Estimation of carbon storage and carbon sequestration capacity of global forest ecosystems and their regional characteristics. *Chinese Journal of Applied Ecology, 09*, 2881–2890.

Liu, Y., Zhao, M., & Zhang, Y. (2021). Forest certification policy and management strategy for forest resources in Heilongjiang forest area. *China Forest Products, 04*, 81–83.

Lu, R. (2021). Discussion on forest resource protection and forest fire prevention management measures. *New Agriculture, 19*, 30–31.

Negi, V. S., & Maikhuri, R. K. (2017). Forest resources consumption pattern in Govind wildlife sanctuary, western Himalaya, India. *Journal of Environmental Planning and Management, 60*(7), 1235–1252.

Nord-Larsen, T., & Schumacher, J. (2012). Estimation of forest resources from a country wide laser scanning survey and national forest inventory data. *Remote Sensing of Environment, 119*, 148–157.

Okumu, B., & Muchapondwa, E. (2020). Determinants of successful collective management of forest resources: Evidence from Kenyan community forest associations. *Forest Policy and Economics, 113*, 102122.

Sgroi, F. (2020). Forest resources and sustainable tourism, a combination for the resilience of the landscape and development of mountain areas. *Science of the Total Environment, 736*, 139539.

Shang, D. (2021). Discussion on the application of "one picture" management of forest fire prevention: Taking Qixian County, Jinzhong City as an example to explore a new model of forest fire prevention. *Modern Rural Science and Technology, 09*, 95–96.

Song, C., Sun, X., & Wang, G. (2015). Research progress on the relationship between carbon and water in forest ecosystems and its influencing factors. *Chinese Journal of Applied Ecology, 09*, 2891–2902.

Starr, M., Joshi, O., Will, R. E., Zou, C. B., & Parajuli, R. (2019). Understanding market opportunities utilizing the forest resources of the cross-timbers ecoregion. *Journal of Forestry, 117*(3), 234–243.

Sun, Q., Han, L., & Xiao, Y. (2017). The impact of land use change on carbon storage in forest ecosystems. *Modern Agriculture, 10*, 42–43.

Tan, Z. (2021). Research on the development of ecotourism resources in forest parks: Taking forest parks in Northwest Sichuan as an example. *Economic Research Guide, 13*, 122–126.

Tong, H. (2021). Research on the reform of forest management and forest resources management system in key state-owned forest areas in my country. *Modern Agriculture Research, 06*, 100–101.

Van Hoang, C., Tran, T. Q., Nguyen, Y. H. T., & Nguyen, L. T. (2020). Forest resources and household welfare: Empirical evidence from North Central Vietnam. In *Natural resources forum* (Vol. 44, No. 4, pp. 311–333).
Wang, G. (2021). Problems and countermeasures in forest resource management and ecological forestry construction. *Modern Agricultural Science and Technology, 11*, 145–146.
Wang, J., & Wang, M. (2021). Research on forest resource protection and forest fire prevention management measures. *Modern Agriculture, 06*, 43–44.
Wang, K., Cao, Q., & Luo, Q. (2021). The application of forest resource management technology in fire prevention and extinguishing engineering: A review of "Introduction to Forest Resource Management." *Journal of Tropical Crops, 07*, 2198.
Wang, H. (2014). A study on influencing factors and efficiency of state-owned forest resources allocation in Heilongjiang Province. Northeast Forestry University.
Wu, Y., Zhou, G., & Guo, R. (1999). Early warning analysis of sustainable development of regional forest resources. *Journal of Zhejiang Forestry University, 01*, 57–62.
Wu, S., Yan, X., & Zhang, L. (2014). The relationship between energy value and service function value of forest ecosystems in China. *Acta Geographica Sinica, 03*, 334–342.
Xia, L. (2020). The origin and ecological impact of deforestation in the eastern United States before the 19th century. *Journal of Chaohu University, 04*, 82–88.
Xu, Z. (2021). Research on the balance between forest protection and forest resource development. *Special Economic Animals and Plants, 08*, 103–104.
Yang, L. (2021). Discussion on measures to strengthen the management and protection of forest resources. *Farm Staff, 16*, 159–160.
Zhang, P. (2019). Analysis of the development status and trend of forest resources in my country. *Green Science and Technology, 17*, 204–206.
Zhang, H., Li, X., Tan, X., Ma, L., & Li, L. (2021). A brief discussion on the classification system of natural resources. *Resources, Environment and Engineering, 04*, 547–550.
Zhao, T., Ouyang, Z., Zheng, H., Wang, X., & Miao, H. (2004). China forest ecosystem service function and its value evaluation. *Journal of Natural Resources, 04*, 480–491.
Zhou, Y., Feng, Z., & Chen, S. (2020). Research on innovation of monitoring system for forest resource management in my country. *World Forestry Research, 02*, 83–89.
Zhu, L., Hu, J., Zhou, B., Xu, Y., Chen, G., & Li, Y. (2017). China Province: The measurement of tourism development efficiency in forest parks and the evolution of spatial and temporal patterns. *Resources and Environment in the Yangtze River Basin, 12.*

Chapter 9
Depletable and Non-renewable Energy Resources: Coal and Oil Resources

9.1 Introduction

This chapter, from the perspective of the exhaustible and non-renewable energy resources in China, focuses on the evaluation and calculation of the production, consumption, and utilization efficiency of coal and oil resources and the total and regional distribution of coal and oil resources in China. Based on this understanding, the DEA model and Malmquist index analysis method are used to evaluate China's provinces and cities' coal and oil utilization efficiency. Finally, public policies are analyzed and divided into energy resource substitution and efficiency improvement policies. The specific analysis ideas are as follows.

The first subsection studies the status quo of coal and oil resources and explains the total amount and spatial distribution. After analysis, the rich and deficient areas of raw coal resources in China are studied, changes in coal and oil reserves over time are explored, and the regional economy is analyzed to determine the impact between the regional economy and energy consumption.

The second section analyzes the utilization efficiency of coal and oil resources. First, we analyzed the problems that exist in the utilization of these. Second, the DEA model is used to evaluate the utilization efficiency of coal and oil, and the meaning of the utilization efficiency of coal and oil resources is explained. According to the construction principles of the evaluation index system, including rationality, representativeness, feasibility, and scientific nature, we established an evaluation index system for the utilization efficiency of coal and petroleum resources from the three aspects of "financial resources," "human resources" and "material resources," and obtained the DEA analysis results of coal utilization efficiency in 30 provinces and the DEA analysis results of crude oil utilization efficiency in 27 provinces. Based on the DEA analysis, the Malmquist index analysis method is selected to measure the time-series changes in coal and oil utilization efficiency in each province, and the time-series changes in total factor productivity are analyzed based on the Malmquist

X. Deng et al., *Environmental and Natural Resources Economics*,
https://doi.org/10.1007/978-981-99-9923-1_9

index value to accurately evaluate the overall efficiency and determine the reasons for low utilization efficiency.

The third section analyzes the public policies in China, which are divided into energy and resource substitution, and efficiency improvement policies. They are analyzed from three aspects: the number and form of policies, the subject and strength of policy documents, and policy tools.

9.2 Status of Coal and Oil Resources

9.2.1 *Total Coal and Oil Resources*

By the end of 2020, proven coal reserves in China accounted for 13.3% of the world's coal reserves, and crude oil energy reserves were low at only 25 billion barrels (Wang et al., 2021). Since its reform and opening up, China's economy has developed rapidly, creating a miracle of economic development that is rarely observed at the time. The level of industrialization and urbanization has continued to deepen, and large consumption of energy resources accompanies these developments.

Figure 9.1 shows China's economic development and energy consumption from 2000 to 2020. China's gross domestic product (GDP) has been growing since 2000. As early as 2010, the GDP exceeded 40 trillion yuan, four times the 2000 GDP. In 2020, it exceeded 100 trillion yuan, and the growth rate of the secondary industry's output value accelerated from 4 trillion yuan in 2000 to 38 trillion yuan in 2020. Energy consumption has also increased with economic development. The energy consumption in 2020 is 3.41 times that in 2000. These data indicate that large investments in energy resources accompany the miracle of China's economic development. Total energy production rose until 2014 and declined in 2015 and 2016. Due to the consumption of many resources in the early development process, development is gradually becoming limited. The sustainable development strategy proposed by China is constantly adjusting its energy structure, which has also gradually flattened the consumption of crude oil and coal since 2014. The consumption of crude oil has been below 300,000 tons of standard coal from 2015 to 2020, and the consumption of crude oil has been growing slowly and has not yet exceeded 100,000 tons of standard coal in 2020.

Comparing the proven coal reserves in 2020 (which refers to the proven reserves at the end of 2020) and the reserve-production ratio, the coal energy reserves in China are mid-level at 143,197 million tons (Fig. 9.2). Coal combustion is a relatively abundant energy source in China, and its proven coal reserves will account for only 13.3% of global coal reserves by the end of 2020. In terms of reserve-production ratio, compared with 513.58 years in the United States, 407.14 years in Russia, and 146.79 years in India, economies of a similar scale, the reserve-production ratio in China is much lower than that of the other six countries, only 36.69 years, and the level of production guarantee is also 36.69 years. This is because raw coal consumption

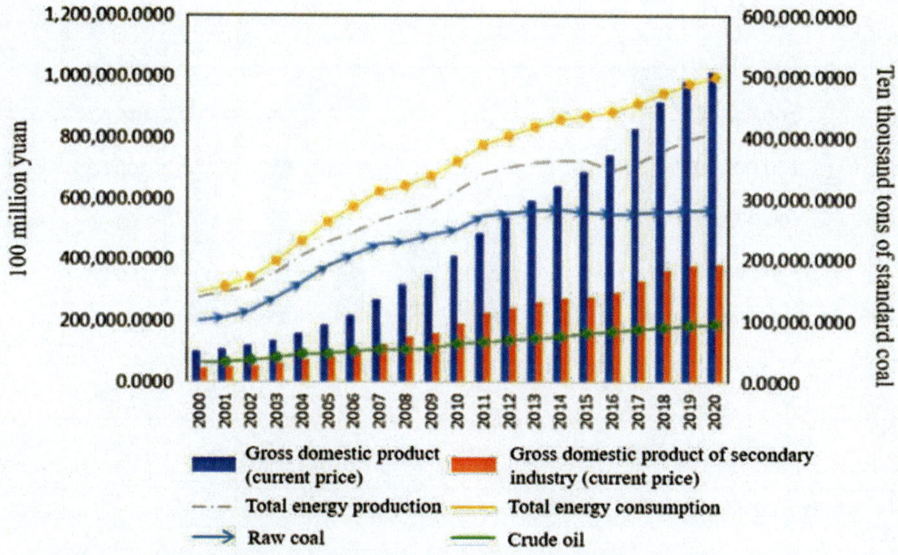

Fig. 9.1 The economic development and energy consumption of China from 2000 to 2020. *Note* The data comes from the Wind Financial Terminal

in China is expected to account for 56.8% of the world's total raw coal consumption by 2020. As the reserve-production ratio depends on the reserve and consumption ratios, the structural form of the low and high consumption ratios inevitably leads to relatively high and low storage-to-production ratios. Although China's coal reserves are not the lowest among the seven countries listed, second only to the United States, Russia, and Kazakhstan, China's coal reserve-to-production ratio is the highest among the countries shown in the figure because of its low reserves and high consumption ratios.

According to Fig. 9.3, comparing the proven crude oil reserves and reserve-production ratios of various countries in the world in 2020, the crude oil energy reserves of China are relatively low: only 25 billion barrels, less than 25% of neighboring Russia, accounting for only 36% of proven crude oil reserves in the United States. In terms of reserve-production ratio, China is relatively high, at 18.18 years, which is not only higher than the 11.40 years of the world's largest economy but also higher than that of Australia (13.89 years), India (16.10 years) and the EU (16.79 years), but lower than that in Russia (27.61 years) and Kazakhstan (45.26 years).

Figures 9.4 and 9.5 show China's main fossil energy production and consumption structures, from which the most abundant resource is raw coal. From the perspective of production structure, raw coal accounted for more than 70% of total energy production in 2000 and 2010. In 2020, raw coal production was 67.6%. Although the proportion of coal in the overall energy production structure has decreased, the proportion of production structure still accounts for more than 65% of the overall energy structure,

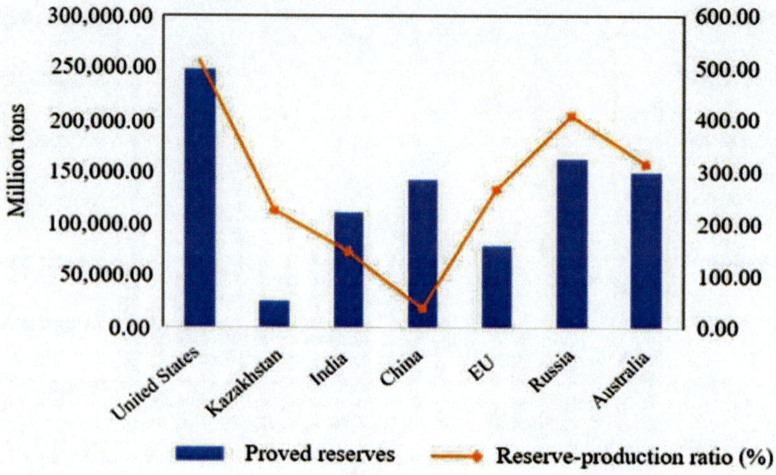

Fig. 9.2 Proven coal reserves and reserves-to-production ratio of countries around the world in 2020

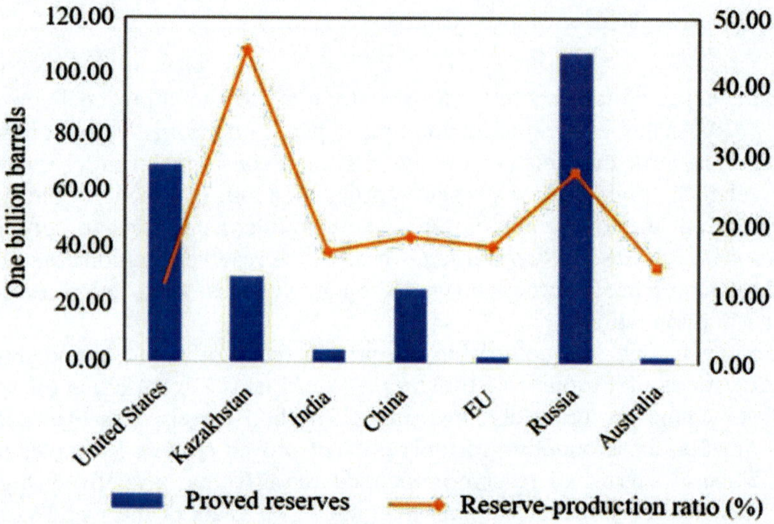

Fig. 9.3 Proven crude oil reserves and reserves-production ratios of countries around the world in 2020. *Note* The data comes from the Wind Financial Terminal

and coal is still the main resource supply for the production of various industries in China. From the perspective of consumption structure, raw coal accounted for approximately 70% of total energy consumption between 2000 and 2010. By 2020, the proportion of raw coal consumed will be 56.8%. The coal-dominated energy consumption structure makes the cleanliness of the energy production structure in China still relatively low.

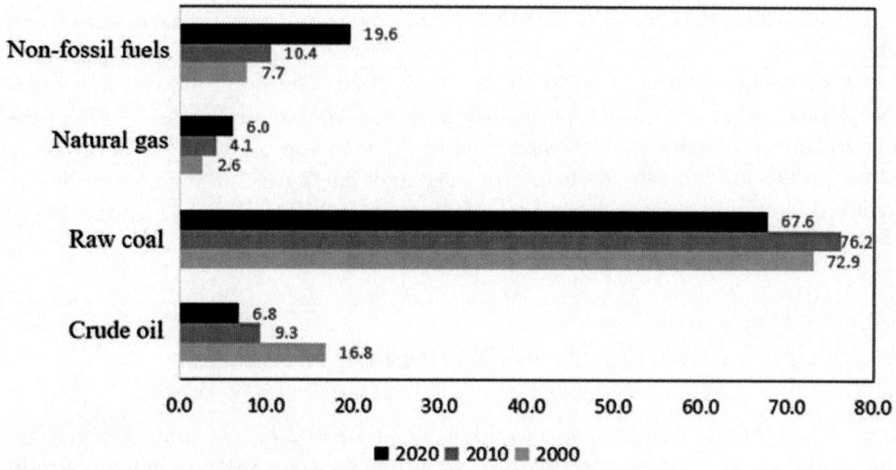

Fig. 9.4 Production structure of major fossil energy in 2000, 2010 and 2020

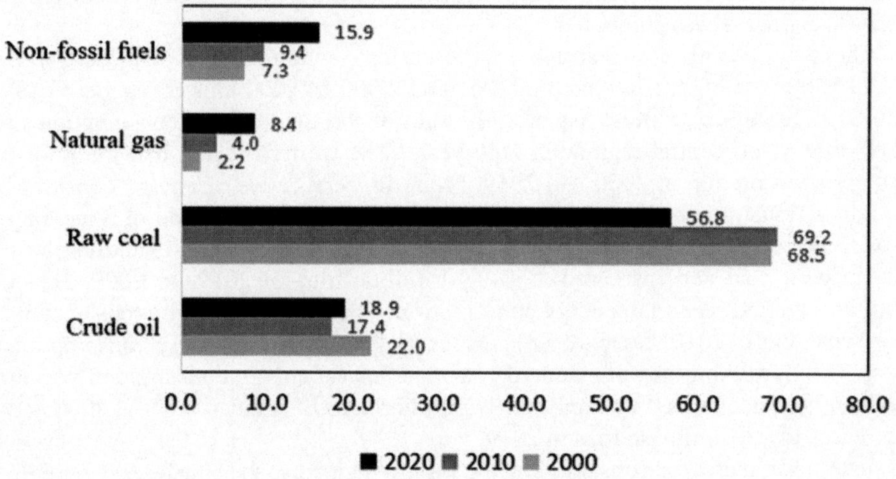

Fig. 9.5 Consumption structure of main fossil energy in 2000, 2010 and 2020

Comparing the data from 2000, 2010, and 2020, from the perspective of oil production and consumption structure, annual oil production is not sufficient to supply consumption. In 2020, crude oil accounted for the lowest proportion of total energy production in China (6.8%), whereas the consumption proportion reached 18.9%, much larger than production. In the energy resource consumption structure, raw coal and crude oil still occupy a high proportion, whereas natural gas and other non-fossil energy sources (such as hydropower and solar energy), although their proportions are gradually increasing, are still relatively low. Most of China's crude oil resources are imported with high external dependence. However, comparing the data in 2000 and 2020, it can be seen that the proportion of natural gas production and consumption is constantly rising, from about 6.5% of China's total energy production and consumption in 2000 to about 7% in 2020. The proportion of non-fossil energy production and consumption also increased from approximately 7.5% of the total energy production and consumption in 2000 to approximately 20% in 2020. Although the growth rate is slow, replacing new energy technologies will not be achieved overnight, and economic transformation and development will inevitably undergo a long-term transition.

9.2.2 *Spatial Distribution of Coal and Oil Resources*

Regarding regional competitive development and resource constraints, there are differences in the energy consumption structure between regions, which mainly depend on the unbalanced and insufficient economic development among regions and the quality of energy resources. Table 9.1 shows that the total energy consumption of 29 provinces (Data in Hong Kong, Macau, Taiwan, and Tibet were missing) in 2000 and 2019 was compared.

The total energy consumption in the eastern region was the highest, with 74,214,700 tons of standard coal in 2000 and 2,300,240,000 tons of standard coal in 2019 (Table 9.1), almost reaching the sum of the total energy consumption in the western and central regions in both years. The western region had the lowest energy consumption in 2000 and 2019. From the perspective of energy consumption in various provinces, in 2000 and 2019, the energy consumption of Shandong was the largest among the 29 provinces. In 2000, it reached 113.619 million tons of standard coal and increased to 413.90 million tons in 2019. In 2000, Hebei Province ranked second in energy consumption with 111.9571 million tons of standard coal, but in 2019, Guangdong Province ranked second in energy consumption with 341.42 million tons of standard coal. The lowest energy consumption was in Hainan Province, which was only 4.7995 million tons of standard coal in 2000 and increased to 22.64 million tons in 2019.

In terms of increased consumption, the eastern region has seen the largest increase in energy consumption in the past 19 years, with an increase of 1,558,095,300 tons of standard coal, followed by the central region, with an increase of 982,506,500 tons of standard coal, and finally the western region, with an increase of 846,907,900 tons of standard coal. Among the provinces in the eastern region, the largest increase in energy consumption was in Shandong Province, which had as much as 300.281 million tons of standard coal, followed by Guangdong Province, which had 246.943 million tons of standard coal. The lowest increase in energy consumption was observed in Hainan Province, which only increased by 17.8405 million tons of standard coal. Among the provinces in the central region, the largest increase in energy consumption was in Inner Mongolia, with as much as 217,966,700 tons of standard coal, followed by Henan Province, with an increase of 143,811,600 tons of standard coal, and Jilin Province, with the least increase in energy consumption,

Table 9.1 Total energy consumption by province in 2000 and 2019

Region	2000	2019	Region	2000	2019	Region	2000	2019
East	74,214.47	230,024.00	West	30,509.21	115,200.00	Central	45,852.35	144,103.00
Beijing	4144.00	7360.00	Guangxi	2669.34	11,270.00	Shanxi	6728.00	20,859.00
Tianjin	2793.71	8241.00	Chongqing	2428.00	8889.00	Inner Mongolia	3549.33	25,346.00
Hebei	11,195.71	32,545.00	Sichuan	6517.82	20,791.00	Jilin	3766.41	7132.00
Liaoning	10,655.82	23,749.00	Guizhou	4278.61	10,423.00	Heilongjiang	6166.15	11,614.00
Shanghai	5499.48	11,696.00	Yunnan	3468.33	12,158.00	Anhui	4878.82	13,870.00
Jiangsu	8612.43	32,526.00	Shaanxi	2731.23	13,478.00	Jiangxi	2505.00	9665.00
Zhejiang	6560.37	22,393.00	Gansu	3011.62	7818.00	Henan	7918.84	22,300.00
Fujian	3463.40	13,718.00	Qinghai	897.20	4235.00	Hubei	6269.09	17,316.00
Shandong	11,361.90	41,390.00	Ningxia	1179.43	7648.00	Hunan	4070.71	16,001.00
Guangdong	9447.70	34,142.00	Xinjiang	3327.63	18,490.00			
Hainan	479.95	2264.00	Tibet					

Note Data were obtained from Wind Financial Terminal. The unit was 10,000 tons of standard coal. Owing to the lack of data for Hong Kong, Macao, Taiwan, and Tibet, a data analysis of the total energy consumption of Hong Kong, Macao, Taiwan, and Tibet is not within the scope of this study with an increase of only 33,655,900 tons of standard coal. Among the provinces in the western region, Xinjiang had the largest increase in energy consumption, with 151.6237 million tons of standard coal, followed by Sichuan Province, which had increased by 142.7318 million tons of standard coal, and Qinghai, which had the least increase in energy consumption, with an increase of only 33.378 million tons. Standard coal, combined with the GDP data of various provinces in Table 9.2, shows that most of the total energy consumption is in provinces with high GDP, such as Guangdong, Shandong, and Jiangsu; however, the provinces with faster GDP growth and energy consumption are not necessarily high. The economy has a certain impact on energy consumption; however, its impact is not decisive.

According to Table 9.3, the basic coal reserves in the central region were the highest, with 218.096 billion tons in 2007 and 167.786 billion tons in 2017, nearly twice the coal base reserves of the eastern and western regions combined. The western region was the second, with 84.221 billion tons in 2007 and 65.89 billion tons in 2017. The eastern region was the least, accounting for 23.809 billion tons in 2007 and 16.514 billion tons in 2017.

Shanxi is an area rich in coal resources in China. The basic reserves of coal resources are the largest, with 105.613 billion tons in 2007 and 91.73 billion tons in 2017, surpassing all coal base reserves in the eastern and western regions in both years. Inner Mongolia is also rich in coal resources, with 80.84 billion tons in 2007 and 50.763 billion tons in 2017, which is close to all coal resources in the western region and greatly exceeds all coal resources in the eastern region. Jilin, Jiangxi, and Hubei have the least coal resources in the central region. In 2007, the basic coal reserves were 1.253 million tons, 792 million tons, and 332 million tons, respectively. In 2017, the basic coal reserves were 9.64 million tons, 267 million tons, and 319 million tons, respectively. Guizhou, Shaanxi, and Xinjiang have relatively abundant coal reserves in the western region. In 2007, the basic coal reserves were 14.676, 27.624 billion tons, and 12.324 billion tons, respectively. In 2017, the basic coal reserves were 12.162 billion tons, 15.933 billion tons, and 17.41 billion tons, respectively. Provinces with relatively scarce coal reserves are Guangxi, Qinghai, and Tibet. In 2007, the basic coal reserves were 851, 2,053, and 12 million tons, respectively. In 2017, basic coal reserves were 0.89 million tons, 1,233 million tons, and 12 million tons, respectively. The basic coal reserves in Tibet remain unchanged. Coal resources were the scarcest in the eastern region. Shandong, Hebei, and Liaoning were the provinces with relatively abundant coal reserves. In 2007, the basic coal reserves were 9.625, 6.346, and 4.336 billion tons, respectively. In 2017, the basic coal reserves were 7.398 billion tons, 4.327 billion tons, and 2.659 billion tons, respectively. Provinces with relatively scarce coal reserves are Guangdong, Hainan, and Zhejiang. In 2007, the basic coal reserves were 1.89 million tons, 0.9 million tons, and 49 million tons. In 2017, the basic coal reserves were 1.19 million tons, 0.23 million tons, and 43 million tons. The basic coal reserves of Zhejiang decreased by only 0.06 billion tons.

In terms of changes in basic coal reserves, the central region saw the largest reduction in basic coal reserves, with a decrease of 50.31 billion tons from 2007 to 2017, followed by the western region with a decrease of 18.331 billion tons, and

Table 9.2 Provincial GDP and growth rate in 2000 and 2019

Region	2000	2019	Rate	Region	2000	2019	Rate	Region	2000	2019	Rate
East	57,413	536,071	8	West	15,737	187,973	11	Central	25,543	261,290	9
Beijing	3162	35,371	10	Guangxi	2080	21,237	9	Shanxi	1846	17,027	8
Tianjin	1702	14,104	7	Chongqing	1791	23,606	12	Inner Mongolia	1539	17,213	10
Hebei	5044	35,105	6	Sichuan	3928	46,616	11	Jilin	1952	11,727	5
Liaoning	4669	24,909	4	Guizhou	1030	16,769	15	Heilongjiang	3151	13,613	3
Shanghai	4771	38,155	7	Yunnan	2011	23,224	11	Anhui	2902	37,114	12
Jiangsu	8554	99,632	11	Shaanxi	1804	25,793	13	Jiangxi	2003	24,758	11
Zhejiang	6141	62,352	9	Gansu	1053	8718	7	Henan	5053	54,259	10
Fujian	3765	42,395	10	Qinghai	264	2966	10	Hubei	3545	45,828	12
Shandong	8337	71,068	8	Ningxia	295	3748	12	Hunan	3551	39,752	10
Guangdong	10,741	107,671	9	Xinjiang	1364	13,597	9		25,543	261,290	9
Hainan	527	5309	9	Tibet	118	1698	13		1846	17,027	8

Note Data were obtained from Wind Financial Terminal. The unit is 100 million yuan

Table 9.3 Basic coal reserves by province in 2007 and 2017

Region	2007	2017	Region	2007	2017	Region	2007	2017
East	238.09	165.14	West	842.21	658.9	Central	2180.96	1677.86
Beijing	6.75	2.63	Guangxi	8.51	0.89	Shanxi	1056.13	917.3
Tianjin	2.97	2.97	Chongqing	20.23	17.69	Inner Mongolia	808.4	507.63
Hebei	63.46	43.27	Sichuan	49.31	51.63	Jilin	12.53	9.64
Liaoning	43.36	26.59	Guizhou	146.76	121.62	Heilongjiang	74.15	62.47
Shanghai			Yunnan	79.68	57.52	Anhui	80.88	83.7
Jiangsu	17.58	10.31	Shaanxi	276.24	159.33	Jiangxi	7.92	2.67
Zhejiang	0.49	0.43	Gansu	58.42	25.45	Henan	117.8	84.77
Fujian	4.44	3.54	Qinghai	20.53	12.33	Hubei	3.32	3.19
Shandong	96.25	73.98	Ningxia	59.17	38.22	Hunan	19.83	6.49
Guangdong	1.89	0.23	Xinjiang	123.24	174.1			
Hainan	0.9	1.19	Tibet	0.12	0.12			

Note Data were obtained from Wind Financial Terminal. The units are 100 million tons (data on Shanghai's coal resources are missing)

the eastern region with a decrease of 7.295 billion tons. Among the provinces and cities, Inner Mongolia had the largest reduction in basic coal reserves, decreasing by 30.077 billion tons between 2007 and 2017, accounting for approximately 60% of the reduction in basic coal reserves in the eastern and central regions. Xinjiang increased the most, with an increase of 5.086 billion tons.

Owing to a lack of data, we selected nine provinces with relatively rich oil content in China for our research. The oil content in the eastern region was low, and most of the oil was distributed in the central and western regions. According to Table 9.4, Xinjiang is an oil-rich area in China, and its basic reserves are the largest among the sample provinces and cities. In 2004, the reserves amounted to 397.827 million tons, which increased to 595.763 million tons in 2016. This was followed by Shaanxi, whose basic oil reserves exceeded 300 million tons in 2012 and reached 383.756 million tons in 2016. The third largest oil reserve is in Gansu, with basic oil reserves of 83.286 million tons in 2004, nearly 100 million tons in 2008, and 282.617 million tons in 2016. These three provinces are all located in western China. In 2004, the combined base oil reserves of the three provinces accounted for approximately 90% of China's base oil reserves. In 2016, this proportion was as high as 95%. Oil resources are the most abundant in western China. Anhui, Guangxi, and Yunnan had the lowest basic petroleum reserves among the sample provinces. In 2016, the reserves were 2,385,000 t, 1,540,000, and 122,000 tons, respectively. Yunnan has the least reserves, with only 105,000 tons in 2004 and approximately 120,000 tons from 2008 to 2016. Regarding changes in basic oil reserves, Shaanxi Province has the largest increase in basic oil reserves, with an increase of 221.304 million tons from 2004 to 2016. Gansu, Xinjiang, and Yunnan increased by 199.331 million tons, 197.936 million tons, and

Table 9.4 Basic oil reserves in each province

Province	2004	2008	2012	2016
Anhui	103.30	161.00	260.06	238.50
Henan	5986.40	5183.00	5160.24	4427.00
Hubei	1062.40	1210.00	1328.70	1185.90
Guangxi	129.00	187.00	139.00	154.00
Sichuan	243.80	338.00	804.63	623.40
Yunnan	10.50	12.00	12.21	12.20
Shaanxi	16,245.20	23,047.00	31,397.94	38,375.60
Gansu	8328.60	9114.00	19,184.32	28,261.70
Xinjiang	39,782.70	43,643.00	56,464.74	59,576.30

Note Data were obtained from Wind Financial Terminal. The unit is 10,000 tons

17,000 t, respectively. Basic oil reserves in Henan Province decreased annually by 15.594 million tons from 2004 to 2016.

9.3 Coal and Oil Resource Utilization Efficiency

9.3.1 Problems Existing in the Utilization of Coal and Oil Resources

9.3.1.1 The Natural Environment in Coal-Rich Areas is Relatively Harsh, and the Conditions for Resource Extraction Are Poor

China is rich in coal resources, accounting for approximately 70% of its total energy production; however, the distribution of resources is uneven, and the environment in resource-rich areas is also relatively harsh, with a lack of water resources, dry climate, little rainfall, serious land desertification, and a fragile ecological environment (Liu et al., 2020). Compared with other countries' environments, China's coal resources are buried deeper, coupled with the impact of the harsh natural environment. The coal mining conditions in China are also poor, which hinders the mining and utilization of coal resources and increases the cost of workforce and material resources, which hinders the development of the regional economy.

9.3.1.2 The Oil Safety Risk Factor Is Increasing Day by Day, and the Utilization Efficiency Needs to Be Improved

China's oil reserves are not rich enough to be self-sufficient; therefore, they have never been able to keep up with the development of the national economy. Oil security involves political, diplomatic, economic, military, and other strategic issues and plays a vital role in a country's economic development (Wu, 2014). However, China has not yet established a diversified oil reserve system. There are no perfect oil reserve laws or regulations that clarify the obligations, reserve scale, or plan of the oil reserve theme. As a result, the stability of the oil supply cannot be guaranteed, making China increasingly dependent on oil from other countries and regions. Regarding oil utilization efficiency, China's industrialization started late and developed rapidly, which lowered oil development and utilization efficiency, causing problems such as the waste of oil resources and excessive consumption of human, material, and financial resources.

9.3.1.3 Negative Impact of Coal and Oil Resource Utilization on the Environment

The constraints of China's economic and resource status, along with the excessive exploitation and use of traditional fossil energy resources, especially coal and oil, have caused huge damage to ecological civilization and have seriously affected the construction of a resource-saving and environment-friendly society. From the perspective of the proportion of coal and oil resources in China's energy utilization, these resources will remain dominant in the energy structure for a long time. However, large-scale development and utilization of coal and oil have increased environmental pressure. In recent years, climate issues have become a serious global problem that the world needs to face. Although China is reforming and upgrading its energy consumption structure under carbon peaking and carbon neutrality goals, and the proportion of clean energy in total energy consumption is gradually increasing, traditional fossil energy still occupies a relatively high proportion of the overall energy consumption structure, causing irreversible damage to land during the extraction of coal and oil (Zhang et al., 2022). For example, the pollution caused by slag produced in the process of coal mining and equipment, such as the drilling floor for oil extraction, is also accompanied by various natural disasters, such as landslides, earthquakes, and the collapse of ground buildings.

The impact on water areas cannot be underestimated, such as lowering the groundwater level and polluting water sources, which has greatly impacted ecological and environmental protection. With respect to the global environment, the use of coal and oil resources has changed global climate conditions, inevitably leading to a continuous reduction in the amount of ice and snow and an increase in the frequency and intensity of extreme weather events. For example, environmental problems such as smog have erupted in large areas of cities, and acid rain has caused serious damage to water resources. In summary, the development and utilization of fossil fuels, especially coal and petroleum resources, have brought enormous pressure on the environment, and the impact on the environment is often irreversible.

9.3.2 Evaluation of Coal and Oil Utilization Efficiency Based on DEA

9.3.2.1 Implications of Resource Utilization Efficiency of Oil and Coal

This study measured the utilization efficiency of oil and coal at both the micro and macro levels. The micro level is concerned with achieving the maximum economic output and minimum environmental impact of coal and oil under the same input resource conditions. The macro-level focuses on the economic and environmentally effective coal and oil output during utilization. Specifically, coal and oil utilization efficiency can be measured from the following three aspects.

Input perspective: coal, oil, and other input elements such as human and financial resources.

Utilization efficiency: the relevant technical efficiency of coal or oil utilization.

Output perspective: Economic efficiency is measured by the output of the economy and environment and the economic output of the total factor input in coal or oil utilization. The environmental output of the total factor input is called environmental efficiency.

Based on relevant research, the energy consumption per unit of GDP was decomposed by analyzing the development mechanisms that can affect coal and oil utilization efficiency. At the same time, according to a comparison of the industrialization processes of developed and developing countries, the changing trend of unit GDP can be analyzed. The specific trends are as follows.

The overall energy consumption per unit GDP followed a U-shaped curve (Fig. 9.6). With the same energy consumption per unit of GDP, the per capita GDP of developed countries is much larger than that of developing countries. In terms of energy consumption per unit GDP, developed countries are in a stage of declining GDP energy consumption, whereas developing countries are in a stage of rising GDP energy consumption. From previous research and analysis, the main factors affecting energy consumption per unit of GDP in China are found to be economic development and industrial structure. Adjustments in energy consumption intensity and industrial structure often cause direct changes in GDP energy consumption, and energy consumption per unit of GDP mainly depends on economic output and energy input. Therefore, the efficiency of coal or oil utilization is determined by the input and output of coal or oil. Based on this, this study used a data envelopment analysis (DEA) model to evaluate the relative efficiency of coal and oil production and utilization in China and explored the internal and external factors that affect its efficiency. This study selected the industrial sector as the research object.

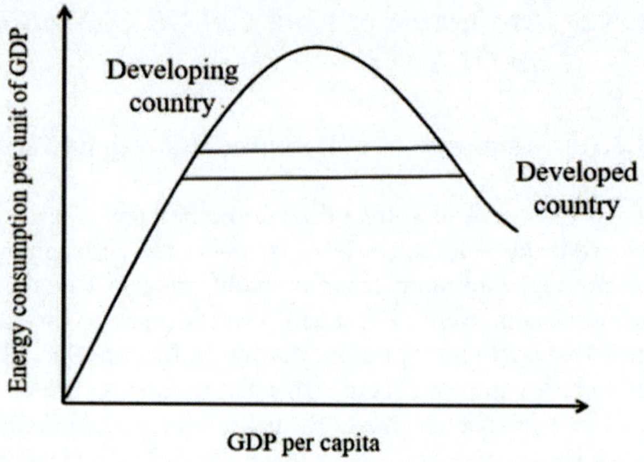

Fig. 9.6 Schematic diagram of energy consumption per unit of GDP

9.3.2.2 Principles for Constructing the Evaluation Index System for the Utilization Efficiency of Coal and Petroleum Resources

Considering the complexities of energy, economy, and environment, when constructing an evaluation index system for coal and oil utilization efficiency, it is necessary to select indicators that can be reasonably quantified based on the actual situation of industrial development. Specifically,

Principle of rationality: The meaning of the indicators can be understood by most people, and they have the characteristics of simplicity and reasonableness. Data indicators from the China Energy Statistical Yearbook can be used as references.

Principle of representativeness: The selected indicators should have sufficient professional representativeness and be able to describe research issues to draw relevant conclusions accurately.

Feasibility principle: The selected indicators can be quantified, and there must be certain methods to obtain the data. The data can be obtained directly from the energy website and the China Statistical Yearbook.

Scientific principles: To study the internal mechanisms of coal and oil utilization efficiency, a logical system of scientific planning is required to establish the overall process of evaluation indicators. Therefore, the selection of indicators was based on scientific principles.

9.3.2.3 Establishment of the Evaluation Index System for the Utilization Efficiency of Coal and Petroleum Resources

Before establishing the model, selecting indicators based on inputs and outputs was necessary. Based on previous research, the following index systems for evaluating coal and crude oil utilization efficiency were established:

From Tables 9.5 and 9.6, three variables in the input index are described from three perspectives: capital stock, financial resources, labor force, human resources, coal

Table 9.5 Input–output indicators of coal utilization efficiency

Category		Indicator	Calculation method
Input	x1	Capital stock	Completed amount of fixed asset investment in the whole society
	x2	Labor force	Employed person: total
	x3	Coal input	Coal consumption: total
Output	y1	Economy	GDP
	y2	Environmental	100,000/discharge of industrial wastewater

Table 9.6 Input–output indicators of crude oil utilization efficiency

Category		Indicator	Calculation method
Input	x1′	Capital stock	Completed amount of fixed asset investment in the whole society
	x2′	Labor force	Employed person: total
	x3′	Crude oil input	Crude oil consumption: total
Output	y1′	Economy	GDP
	y2′	Environmental	100,000/discharge of industrial wastewater

consumption, crude oil consumption, and material resources. The output indicator has two variables. Considering the availability of indicator data, the economic output was measured by the GDP of each province to reflect the economic effects of coal and crude oil utilization. Environmental output was measured by the discharge of industrial waste gas and wastewater, which represents the environmental impact caused by the utilization of coal and crude oil. Therefore, the input–output index system of coal and oil utilization efficiency, constructed in this study, is scientific and comprehensive, and meets the purpose of evaluating coal and crude oil utilization efficiency.

9.3.2.4 Evaluation Model of Utilization Efficiency of Coal and Oil Resources

Data Envelopment Analysis (DEA) is a systematic method of analyzing inputs and outputs objectively. This model does not require dimensionless processing of data and does not require assigning weights to indicators, which makes it more rigorous. Therefore, this study used the DEA model to evaluate coal and oil utilization efficiency.

The DEA model is divided into two categories: the CCR model, which does not consider scale benefits, and the BCC model, which does. The creation of the CCR model refers to the generation of a theoretical DEA method. Under the assumption of constant returns to scale, the technical efficiency measured by the CCR model is mixed technical efficiency, including scale efficiency. Traditional DEA methods are either input- or output-oriented. Therefore, the input-oriented CCR model is used as an example to explain the planning formula of the model and better understand the improvement methods adopted in this study.

Assuming that the model is selected by the decision-making unit (DMU) called k, each decision-making unit is recorded as DUM_j ($j = 1, 2, 3, \ldots n$). For each DMU, there are p kinds of input indicators and q kinds of output Output indicators, set as x_i ($i = 1, 2, \ldots p$) and y_r = ($r = 1, 2, \ldots q$). The weight of the input index is v_i ($i = 1, 2, \ldots p$), and the weight of the output index is u_r ($r = 1, 2, \ldots q$), then the efficiency evaluation DEA-BCC model of the province k is expressed as:

$$\max \frac{\sum_{r=1}^{q} u_r y_{rj}}{\sum_{i=1}^{p} v_i x_{ij}} = \frac{u_1 \cdot y_{1j} + u_2 \cdot y_{2j} + \cdots + u_q \cdot y_{qj}}{v_1 \cdot x_{1j} + v_2 \cdot x_{2j} + \cdots + v_p \cdot x_{pj}} \tag{9.1}$$

$$s.t. \begin{cases} \frac{\sum_{r=1}^{q} u_r y_{rj}}{\sum_{i=1}^{p} v_i x_{ij}} = \frac{u_1 \cdot y_{1j} + u_2 \cdot y_{2j} + \cdots + u_q \cdot y_{qj}}{v_1 \cdot x_{1j} + v_2 \cdot x_{2j} + \cdots + v_p \cdot x_{pj}} \le 1 \\ u_r \ge 0,\ v_i \ge 0 \\ i = 1, 2, \ldots p;\ r = 1, 2, \ldots q;\ j = 1, 2, \ldots n \end{cases} \tag{9.2}$$

The conditions for judging the validity of the CCR model are as follows:

$$s.t. \begin{cases} \sum_{j=1}^{n} X_j \cdot \lambda_j + S^- = \theta X_0 \\ \sum_{j=1}^{n} Y_j \cdot \lambda_j - S^+ = Y_0 \\ \lambda_j \ge 0,\ j = 1, 2 \ldots n;\ S^-, S^+ \ge 0 \end{cases} \tag{9.3}$$

Among them, θ is the effective value of the DUM, and S− and S+ are the slack variables for input and output, respectively. When $\theta < 1$, DUMj is valid for non-DEA efficient, when $\theta = 1$, and S− $\neq 0$ or S+ $\neq 0$, then DUMj is valid for weak DEA efficient, when $\theta = 1$, and each optimal solution satisfies S− $= 0$ and S+ $= 0$, which indicates that the decision-making unit DUMj is strong DEA efficient.

Based on the CCR model, the BCC model can be constructed by introducing the restrictive condition $\sum \lambda_J = 1$, so that the pure technical and scale efficiency can be further considered, and DUMj can be judged to be in the situation of return to scale.

The first-stage DEA model exists in an ideal data selection state; however, the data selected in reality contain many factors, such as policy factors, environmental factors, and random errors. To obtain more accurate results, it is necessary to eliminate errors caused by variables such as the external environment. The Stochastic Frontier Regression method (SFA) was used in this study. The first-stage model was constructed first, and the corresponding slack variables were further calculated according to the data measured in the first stage. Second, based on the input-dominated BCC model in the first stage, with environmental variables and random errors as the explained variables, a stochastic frontier method is constructed to decompose the output slack variables as follows:

$$S_{ni} = f(Z_i; \beta_n) + \varepsilon_{ni} + \mu_{ni} \tag{9.4}$$

where Zi is the environmental factor, βn is the coefficient of the environmental factor, ni is the management inefficiency error term, and ni is the random error term, which represents the influence of the slack variable Sni on the output n of the DMUi i. Both μ_{ni} and ε_{ni} obey the standard normal distribution, $\varepsilon_{ni}+\mu_{ni}$ is the mixed error term of the model.

If the outputs before and after the adjustment are X_{ni} and X_{ni}^{A}, respectively, then the conversion formulas before and after the model adjustment are:

$$X_{ni}^{A} = X_{ni} + \left[\max\left(f\left(Z_i; \widehat{\beta_n}\right)\right) - f\left(Z_i; \widehat{\beta_n}\right)\right] + \max(\varepsilon_{ni}) - \varepsilon_{ni} \tag{9.5}$$

We replace the adjusted input with the original input value before adjustment, re-substitute the output values of different time periods and provinces into the first-stage DEA model, and calculate each province's coal and oil utilization efficiency from 2010 to 2019. Compared with the results obtained in the first stage, this model eliminated factors such as random error terms, was more accurate and objective, and scientifically and comprehensively reflected the true level of coal utilization efficiency in each province.

9.3.2.5 Results of the Utilization Efficiency of Coal and Oil Resources

According to the relevant data on coal input and output in each province from 2010 to 2019, a three-stage DEA model was constructed, and the DEAP2.1 software was used to solve the problem, and the following results were obtained.

Before the model had random errors or other factors, the average coal utilization efficiency of the 30 provinces was 0.570 (Table 9.7). After the three-stage DEA eliminates environmental factors and random disturbances in the input data, technical efficiency increases to 0.582 (Table 9.7). This shows that factors such as random disturbances influence coal utilization efficiency, which is shown as an inhibitory effect. By replacing technical efficiency with pure technical efficiency and scale efficiency, pure technical efficiency increased by 0.001, and scale efficiency increased by 0.006. Therefore, the improvement in technical efficiency was primarily due to improvements in scale efficiency.

The comprehensive technical efficiency of Beijing, Shanghai, and Hainan is 1, which is at the effective frontier. The comprehensive technical efficiency levels of Anhui, Jiangxi, Chongqing, and Shaanxi were relatively low (less than 0.4), indicating that the comprehensive technical efficiency of coal utilization in these provinces has room for further improvement. Further analysis shows that the low value of the technical efficiency of coal utilization in a few provinces is mainly due to their own low pure technical efficiency, that is, due to less investment in research and development, including labor and fixed asset investment in various provinces, such as Anhui, Jiangxi, and Chongqing. From the results in Jiangsu and Ningxia, the

Table 9.7 DEA analysis results of coal utilization efficiency in 30 provinces

Province	crste	vrste	Scale	
Beijing	1.000	1.000	1.000	–
Tianjin	0.872	0.957	0.911	irs
Hebei	0.402	0.420	0.958	drs
Shanxi	0.455	0.503	0.906	irs
Inner Mongolia	0.634	0.691	0.917	irs
Liaoning	0.506	0.555	0.912	drs
Jilin	0.433	0.511	0.847	irs
Heilongjiang	0.459	0.492	0.933	irs
Shanghai	1.000	1.000	1.000	–
Jiangsu	0.568	1.000	0.568	drs
Zhejiang	0.672	0.774	0.868	drs
Anhui	0.319	0.335	0.952	irs
Fujian	0.593	0.598	0.992	irs
Jiangxi	0.376	0.400	0.939	irs
Shandong	0.501	0.718	0.697	drs
Henan	0.414	0.438	0.945	drs
Hubei	0.463	0.468	0.990	irs
Hunan	0.494	0.498	0.991	irs
Guangdong	0.916	1.000	0.916	drs
Guangxi	0.444	0.474	0.936	irs
Hainan	1.000	1.000	1.000	–
Chongqing	0.377	0.430	0.878	irs
Sichuan	0.429	0.434	0.989	drs
Guizhou	0.526	0.599	0.879	irs
Yunnan	0.404	0.459	0.881	irs
Shaanxi	0.381	0.413	0.924	irs
Gansu	0.465	0.549	0.847	irs
Qinghai	0.921	1.000	0.921	irs
Ningxia	0.558	0.996	0.560	irs
Xinjiang	0.510	0.606	0.841	irs
Average value of the first stage	0.570	0.644	0.897	
Average value of the third stage	0.582	0.645	0.903	

Note crste = technical efficiency. vrste: pure technical efficiency. Scale: scale efficiency. drs: diminishing returns to scale. –: constant returns to scale. irs: increasing returns to scale. crste = vrste × scale (Data for Tibet are missing; therefore, they were not included in the model calculation.)

scale efficiency of coal utilization has a certain impact on comprehensive technical efficiency. Each province can improve comprehensive technical efficiency through the optimal allocation of resources.

In the view of pure technical efficiency and scale efficiency, the pure technical efficiency of each province is quite different. Hebei, Heilongjiang, Anhui, and other provinces were lower than 0.5, and Tianjin, Shanghai, Jiangsu, Beijing, and other provinces were close to or equal to 1, indicating that when various provinces increased their coal-related inputs, the degree of input efficiency improvement was uneven. In terms of scale efficiency, all provinces were relatively high, indicating that when all provinces increased coal-related inputs, the degree of improvement in input efficiency was roughly the same. From the perspective of returns to scale, the efficiency levels of Beijing, Shanghai, and Hainan reached the optimal efficiency value. Therefore, the returns to scale of these three provinces tended to be stable and positioned at higher levels. The returns to scale in most provinces are increasing; that is, increasing the same proportion of coal input will increase output and income, indicating that improving coal utilization efficiency is beneficial for economic output and environmental protection. The decreasing returns to scale for Hebei, Liaoning, Jiangsu, Zhejiang, Shandong, Henan, Guangdong, and Sichuan indicate that these eight provinces need to improve their coal utilization efficiency. The current input is too high, making it difficult to achieve economic and environmental coordination and reducing coal utilization efficiency.

A three-stage DEA model was constructed based on relevant data on crude oil input and output in various provinces from 2010 to 2019. Due to the lack of data for Shanxi, Chongqing, Guizhou, and Tibet, these four provinces and cities were excluded. The DEAP 2.1 software was used to solve the problem in this study, and the following results were obtained. Without considering factors such as random errors, the average crude oil utilization efficiency of the 27 provinces was 0.764. Using the SFA method to decompose the output slack variable, technical efficiency increased to 0.765, and efficiency improved slightly, indicating that factors such as random errors have an inhibitory impact on crude oil utilization efficiency. By replacing technical efficiency with pure technical efficiency and scale efficiency, pure technical efficiency increased by 0.001, and scale efficiency increased by 0.003. Therefore, the improvement in technical efficiency after adjustment is mainly attributable to increased scale efficiency (Table 9.8).

The comprehensive technical efficiency values of Beijing, Inner Mongolia, Shanghai, Hunan, Guangdong, Hainan, and Qinghai reached the optimal state. The comprehensive technical efficiency levels of Yunnan, Shaanxi, and Gansu are less than 0.5, indicating that the comprehensive technical efficiency of crude oil utilization in these provinces has room for further improvement. Overall, the technical efficiency levels of the 27 provinces remained relatively high. Through in-depth analysis, the reason for the low level of comprehensive technical efficiency in Yunnan, Shaanxi, and Gansu was mainly due to their low pure technical efficiencies of 0.459, 0.468, and 0.549, respectively. That is, there is less R&D investment in each province, including labor and crude oil consumption. In terms of scale efficiency, the scale efficiency of 27 provinces is at a relatively high level, close to or equal to 1, which

Table 9.8 DEA analysis results of crude oil utilization efficiency in 27 provinces

Province	crste	vrste	Scale	
Beijing	1.000	1.000	1.000	–
Tianjin	0.872	0.957	0.911	irs
Hebei	0.731	0.777	0.940	drs
Inner	1.000	1.000	1.000	–
Mongolia	0.506	0.555	0.912	drs
Liaoning	0.546	0.591	0.925	irs
Jilin	0.507	0.530	0.957	irs
Heilongjiang	1.000	1.000	1.000	–
Shanghai	0.855	1.000	0.855	drs
Jiangsu	0.827	0.857	0.966	drs
Zhejiang	0.710	0.716	0.991	irs
Anhui	0.829	0.830	0.998	drs
Fujian	0.678	0.698	0.971	irs
Jiangxi	0.611	0.718	0.850	drs
Shandong	0.883	1.000	0.883	drs
Henan	0.810	0.814	0.995	drs
Hubei	1.000	1.000	1.000	–
Hunan	1.000	1.000	1.000	–
Guangdong	0.836	0.865	0.966	irs
Guangxi	1.000	1.000	1.000	–
Hainan	0.940	1.000	0.940	drs
Sichuan	0.404	0.459	0.881	irs
Yunnan	0.448	0.468	0.958	irs
Shaanxi	0.472	0.549	0.860	irs
Gansu	1.000	1.000	1.000	–
Qinghai	0.662	0.996	0.664	irs
Ningxia	0.510	0.606	0.841	irs
Average value of the first stage	0.764	0.814	0.936	
Average value of the third stage	0.765	0.815	0.939	

Note crste = technical efficiency. vrste: pure technical efficiency. Scale: scale efficiency. drs: diminishing returns to scale. –: constant returns to scale. irs: increasing returns to scale. crste = vrste × scale (The data for Shanxi, Chongqing, Guizhou, and Tibet are missing; therefore, they are not included in the model calculation) indicates that when increasing crude oil-related input, the degree of improvement of input efficiency in each province is relatively consistent. The provinces with lower pure technical efficiency included Yunnan and Shaanxi, consistent with the above analysis. When increasing crude oil-related inputs, the degree of improvement in input efficiency in the relevant provinces needs to be improved. The data obtained in Jiangsu and Ningxia shows that the scale efficiency of crude oil utilization also has a certain impact on comprehensive technical efficiency.

The returns to scale in Hebei, Liaoning, Jiangsu, Zhejiang, Fujian, Shandong, Henan, Hubei, and Sichuan decreased, indicating that the efficiency of crude oil utilization in these nine provinces needs further improvement. The excessive investment makes maintaining effective coordination between the economy and the environment difficult with crude oil. This great environmental pressure has led to an inability to achieve sustainable and stable economic growth, thus reducing the efficiency of crude oil utilization. The returns to scale in Jiangxi, Guangxi, Yunnan, Shaanxi, Gansu, Ningxia, Xinjiang, Tianjin, Jilin, Heilongjiang, and Anhui are increasing, indicating that adding the same proportion of crude oil input factors increases production and income. Improving crude oil utilization efficiency is conducive to economic output and environmental protection. As the crude oil utilization efficiency levels of Beijing, Inner Mongolia, Shanghai, Hunan, Guangdong, Hainan, and Qinghai are already optimal, the returns to scale in these seven provinces tend to be stable and are positioned at higher levels.

9.3.3 *Reasons for the Low Utilization Efficiency of Coal and Oil Resources*

9.3.3.1 Research Ideas

To further analyze the size and influencing factors of coal and oil utilization efficiency in different provinces and to analyze the motivation for the low utilization efficiency of coal and oil resources, based on the above DEA analysis, this study chooses the Malmquist index analysis method to measure the coal and oil utilization efficiency of each province. According to the Malmquist index, fluctuation analyzes the mechanism change process of total factor productivity to further evaluate the overall efficiency and find out the motivation for low utilization efficiency. The Malmquist index decomposition uses 10-year data from 2010 to 2019 for dynamic evaluation.

9.3.3.2 Research Methods

The traditional DEA model, which is only a horizontal analysis, was used to evaluate each province's coal and oil utilization efficiency. With the help of the Malmquist index decomposition method for longitudinal analysis, comprehensive technical efficiency can be decomposed. The degree of influence of other aspects of the efficiency value on comprehensive technical efficiency can be obtained. This method will intuitively show the dynamic trend of efficiency, and the output-oriented Malmquist index model formula is as follows:

$$M(x_{t+1}, y_{t+1}, x_t, y_t) = \left[\frac{D^t(x_{t+1}.y_{t+1})}{D^t(x_t.y_t)} \cdot \frac{D^{t+1}(x_{t+1}.y_{t+1})}{D^{t+1}(x_t.y_t)}\right]^{1/2} \tag{9.6}$$

$D^t(x_t.y_t)$ is the output efficiency in period t expressed by the productivity in period t, M is the factor productivity index, and When $M > 1$, the utilization efficiency of coal and oil increases; when $M < 1$, the efficiency decreases; and when $M = 1$, the efficiency remains unchanged. Transform the Malmquist exponential model to obtain

$$M(x_{t+1}, y_{t+1}, x_t, y_t) = \frac{D^{t+1}(x_{t+1}.y_{t+1})}{D^t(x_t.y_t)}\left[\frac{D^t(x_{t+1}.y_{t+1})}{D^{t+1}(x_{t+1}.y_{t+1})} \cdot \frac{D^t(x_t.y_t)}{D^{t+1}(x_t.y_t)}\right]^{1/2} \tag{9.7}$$

The new Malmquist model demonstrates the combined effect of pure technical efficiency and scale efficiency changes and can evaluate the utilization efficiency of coal and oil more comprehensively and systematically.

9.3.3.3 Analysis of Results

After the DEA horizontal analysis of the sample data, the Malmquist model was used to conduct a longitudinal analysis of the panel data in the previous nine years. The following table shows the results; the nine-year average of efficiency changes replaced the efficiency fluctuations. Technical and technical efficiency changes jointly determine the size of the Malmquist Index.

In Table 9.9, tfpch can be further divided into the products of technological efficiency change (effch) and technological progress (techeh). Technical efficiency changes can be divided into pure technical efficiency (pech) and scale efficiency (sech). When the change rate of total factor productivity is greater than 1, it means that the total factor productivity of coal input increases during the sample period. When the change rate of total factor productivity is less than 1, it indicates a decline in the total factor productivity of coal input. It can be seen from the table that the average rate of change of total factor productivity in each province from 2010 to 2019 was 1.035, indicating that the total factor productivity of each province in these nine years has improved, and the comprehensive utilization efficiency of coal has been at the level of improving efficiency during this period. The annual average value of efficiency change was 0.976, whereas the annual average value of technological change was 1.106, which shows that the improvement in comprehensive coal utilization efficiency in various provinces between 2010 and 2019 was mainly supported by technological progress. The decline in the efficiency of coal mining development has delayed improvements in the coal utilization rate in various provinces. Therefore, it is necessary to improve the technical level of coal mining and production, thereby improving the efficiency of coal utilization. From the perspective of the VRS, the factors influencing efficiency change include pure technical efficiency change and scale efficiency change. The annual average scale efficiency change was 0.989, and the annual average pure technical efficiency change was 0.987. This shows that the loss of coal production efficiency in each province from 2010 to 2019 was the combined effect of scale and pure technical efficiency. For the coal utilization efficiency of each province, it is necessary to pay attention to the change of pure technical efficiency and the size of scale efficiency.

Table 9.10 shows that the Malmquist index of coal efficiency in the 30 provinces increased by 3.5%, mainly due to an increase of 6% in the range of technological progress. Technical efficiency decreased by 2.4%, pure technical efficiency decreased by 1.3%, and scale efficiency decreased by 1.1%. Beijing has the largest Malmquist index with a value of 1.231. The Malmquist indices of most provinces, such as Tianjin, Hebei, Shanxi, and Inner Mongolia, were greater than 1. The total factor productivity of these provinces has increased. The smallest Malmquist index value was for Hainan (0.913), and factor productivity decreased by 8.7%. Technical efficiency, pure technical efficiency, and scale efficiency remained unchanged; however, technological progress fell by 8.7%, and a decline in technological progress entirely caused the decline in total factor productivity. Through comparative analysis among provinces, it can be seen that the Malmquist index of provinces with large changes in technological progress is usually higher, indicating that changes in technological progress have a greater impact on coal utilization efficiency. For example, in Shanxi Province, technical efficiency, pure technical efficiency, and scale efficiency declined by 3.9%, 0.2%, and 3.7%, respectively, but technological progress increased by 6.5%,

Table 9.9 Malmquist index decomposition results of coal utilization efficiency

Year	effch	techch	pech	sech	tfpch
2	0.969	1.089	0.989	0.979	1.055
3	0.979	1.014	0.986	0.994	0.993
4	0.978	1.027	0.984	0.993	1.005
5	1.002	1.002	1.001	1.002	1.004
6	0.974	1.047	0.978	0.996	1.019
7	0.987	1.074	0.989	0.998	1.060
8	1.045	1.058	1.049	0.997	1.106
9	0.885	1.183	0.924	0.957	1.046
Mean	0.976	1.060	0.987	0.989	1.035

Note Calculated using DEP2.1 software. effch: Change in efficiency. techch: Changes in Technology. pech: Change in the PTE. sech: Change in scale efficiency. tfpch: Change in total factor productivity and the final total factor productivity increased by 2.4%. There are also provinces, such as Inner Mongolia, Jilin, Heilongjiang, Zhejiang, and Jiangxi.

Table 9.10 Malmquist index decomposition results of coal utilization efficiencies

Province	effch	techch	pech	sech	tfpch
Beijing	1.000	1.231	1.000	1.000	1.231
Tianjin	0.950	1.071	0.965	0.984	1.017
Hebei	0.960	1.069	0.955	1.005	1.026
Shanxi	0.961	1.065	0.998	0.963	1.024
Inner	0.957	1.068	0.970	0.987	1.023
Mongolia	1.038	1.063	1.056	0.983	1.103
Liaoning	0.949	1.069	0.969	0.980	1.015
Jilin	0.931	1.062	0.983	0.947	0.989
Heilongjiang	1.000	1.064	1.000	1.000	1.064
Shanghai	1.033	1.068	1.000	1.033	1.103
Jiangsu	0.979	1.074	0.996	0.983	1.051
Zhejiang	0.989	1.078	0.983	1.006	1.066
Anhui	0.983	1.078	0.990	0.993	1.060
Fujian	0.981	1.079	0.984	0.997	1.058
Jiangxi	0.973	1.069	0.965	1.009	1.040
Shandong	0.952	1.076	0.977	0.975	1.024
Henan	0.993	1.076	1.005	0.988	1.069
Hubei	0.962	1.083	0.965	0.997	1.042
Hunan	0.944	1.081	1.000	0.944	1.021
Guangdong	0.934	1.083	0.943	0.990	1.012
Guangxi	1.000	0.913	1.000	1.000	0.913
Hainan	1.028	1.077	1.025	1.004	1.107
Chongqing	0.968	1.081	0.996	0.972	1.047
Sichuan	0.925	1.024	0.936	0.988	0.947
Guizhou	0.948	1.053	0.959	0.989	0.998
Yunnan	1.026	1.071	1.026	1.001	1.099
Shaanxi	0.958	0.987	0.999	0.958	0.945
Gansu	1.010	0.948	1.000	1.010	0.958
Qinghai	1.012	1.017	1.000	1.011	1.029
Ningxia	0.960	1.070	0.972	0.987	1.027
Mean	0.976	1.060	0.987	0.989	1.035

Note Calculated using DEP2.1 software. All Malmquist index averages are geometric means. effch: Technical efficiency. techch: Technological progress. pech: Pure technical efficiency. sech: Scale efficiency. tfpch: Total factor productivity

When performing the Malmquist index decomposition on the crude oil sample data, Shanxi, Chongqing, Tibet, and Guizhou data were missing, so the index decomposition for the four provinces was excluded. A longitudinal analysis was carried out on the panel data of the remaining provinces over the past nine years, and the results are shown in Tables 9.11 and 9.12. The Malmquist index decomposition results of crude oil utilization efficiency in the 27 provinces use the average value of efficiency changes over 9 years to replace the degree of efficiency fluctuation.

Table 9.11 shows the Malmquist index decomposition results of crude oil utilization efficiency for each year. The total factor productivity in time periods 3–6 decreased by 2.8%, 3%, and 0.9%, respectively. Total factor productivity rose slightly in other periods, with the largest increase, in total factor productivity, in period 8. When the change rate of total factor productivity is greater than 1, it indicates that the total factor productivity of crude oil input increases during the sample period. When the change rate of total factor productivity is less than 1, the total factor productivity of crude oil input decreases. Overall, an improvement in efficiency occurred in periods 5 and 8, which increased by 3.7% and 6.1%, respectively. Technological progress declined only in Period 5 and improved in all other periods. The pure technical efficiency change is the same as the efficiency change, with improvement in time periods 5 and 8, increasing by 2.8% and 5.2%, respectively. The scale efficiency improved in periods 1, 5, and 8, increasing by 0.4%, 0.9%, and 0.8%, respectively.

The average rate of change in factor productivity for each province from 2010 to 2019 was 1.019 > 1, indicating that the total factor productivity of each province in the past nine years has improved, and the comprehensive utilization of crude oil is at the level of improving efficiency. The annual average value of efficiency change is 0.977 < 1, whereas the annual average value of technological change is 1.043 > 1, which shows that the improvement in the comprehensive utilization efficiency of crude oil in various provinces between 2010 and 2019 depends on technological progress.

Table 9.11 Malmquist index decomposition results of crude oil utilization efficiency

Year	effch	techch	pech	sech	tfpch
2	0.995	1.079	0.992	1.004	1.074
3	0.954	1.019	0.978	0.976	0.972
4	0.962	1.012	0.980	0.982	0.974
5	1.037	0.935	1.028	1.009	0.970
6	0.931	1.064	0.961	0.969	0.991
7	0.987	1.074	0.990	0.997	1.061
8	1.061	1.028	1.052	1.008	1.090
9	0.897	1.145	0.926	0.968	1.027
Mean	0.977	1.043	0.988	0.989	1.019

Note Calculated using DEP2.1 software. effch: Change in Efficiency. techch: Changes in technology. pech: Change in pure technical efficiency. sech: Change in scale efficiency. tfpch: Change in total factor productivity

Table 9.12 Malmquist index decomposition results of crude oil utilization efficiency in 27 provinces

Province	effch	techch	pech	sech	tfpch
Beijing	1.000	1.070	1.000	1.000	1.070
Tianjin	0.950	1.071	0.965	0.984	1.017
Hebei	0.945	1.058	0.938	1.007	0.999
Inner	0.995	1.009	1.000	0.995	1.004
Mongolia	1.038	1.063	1.056	0.983	1.103
Liaoning	0.936	1.094	0.951	0.984	1.024
Jilin	0.928	1.072	0.977	0.949	0.994
Heilongjiang	1.000	1.061	1.000	1.000	1.061
Shanghai	0.981	1.086	1.000	0.981	1.065
Jiangsu	0.956	1.093	0.994	0.961	1.044
Zhejiang	1.020	1.026	1.020	0.999	1.046
Anhui	0.943	1.079	0.952	0.990	1.017
Fujian	0.996	1.026	0.996	0.999	1.021
Jiangxi	0.949	1.078	0.965	0.984	1.023
Shandong	1.016	1.024	1.000	1.016	1.040
Henan	0.985	1.045	0.997	0.988	1.029
Hubei	0.974	1.031	0.974	1.000	1.004
Hunan	0.947	1.068	1.000	0.947	1.012
Guangdong	0.888	1.082	0.893	0.994	0.960
Guangxi	1.000	0.894	1.000	1.000	0.894
Hainan	1.008	0.987	1.000	1.008	0.995
Sichuan	1.024	1.060	1.019	1.005	1.085
Yunnan	1.006	1.080	1.010	0.996	1.086
Shaanxi	0.956	0.999	0.999	0.957	0.955
Gansu	1.000	0.945	1.000	1.000	0.945
Qinghai	0.991	1.017	1.000	0.990	1.007
Ningxia	0.960	1.070	0.972	0.987	1.027
Mean	0.977	1.043	0.988	0.989	1.019

Note Calculated using DEP2.1 software. All Malmquist index averages are geometric. effch: Technical efficiency. techch: Technological progress. pech: Pure technical efficiency. sech: Scale efficiency. tfpch: Total factor productivity

Similar to the motivation for improving the comprehensive utilization efficiency of coal, the impact of efficiency changes can be divided into pure technical efficiency and scale efficiency changes. The annual average change of scale efficiency was 0.989 and the annual average change of pure technical efficiency was 0.988, indicating that the loss of efficiency in oil production in each province from 2010–2019 was affected by both scale efficiency and pure technical efficiency.

Overall, the Malmquist index of crude oil efficiency in the 27 provinces increased by 1.9%, mainly due to a 4.3% increase in technological progress. Pure technical, scale and technical efficiencies decreased by 1.2%, 1.1%, and 2.3%, respectively. Liaoning had the largest Malmquist index, with a value of 1.103. The Malmquist indices of most provinces, such as Beijing, Tianjin, Inner Mongolia, and Liaoning, were greater than 1. The total factor productivity of these provinces has increased slightly. Hainan had the smallest Malmquist index value of 0.894, and its total factor productivity decreased by 10.6%. Pure technical, scale and technical efficiency remained unchanged, but technological progress decreased by 10.6%. Therefore, a decline in total factor productivity is entirely caused by a decline in technological progress. The data for Beijing, Shanghai, Hainan, and Qinghai show that when other efficiencies remain unchanged, changes in technological progress often cause changes in the Malmquist index and total factor productivity. Through a comparative analysis of other provinces, it can also be concluded that the Malmquist indices of provinces with large changes in technological progress tend to be higher, indicating that technological progress significantly impacts crude oil utilization efficiency. For example, the technical, pure technical, and scale efficiencies of Zhejiang Province decreased by 4.4%, 0.6%, and 3.9%, respectively, but technological progress increased by 9.3%, and the final total factor productivity increased by 4.4%. Jilin, Beijing, Fujian, Jiangxi, Shandong, Hubei, and other provinces were included.

9.4 Public Policy

9.4.1 Data Sources and Processing

Based on the research ideas, the energy policies selected in this study were all from national policy documents published by the government between 2005 and August, 2021. By browsing the official websites of the government, such as the National Energy Administration, the Chinese government website, and the National Development and Reform Commission, starting from the energy resource substitution policy and efficiency improvement policy, the keywords are "energy substitution" and "efficiency improvement," respectively. Finally, 1,701 and 644 copies of policy documents were obtained.

However, to ensure the policy's effectiveness and the scientific nature and validity of the follow-up analysis results, the following screenings were conducted: 1. Only keep documents closely related to the research topic. 2. Eliminate repetitive documents that require implementing existing policy documents, such as detailed implementation rules. 3. Eliminate temporary work guidance documents. Finally, 1,016

Table 9.13 The assignment criteria

	Release subject
5	Laws and decisions promulgated by the Standing Committee of the National People's Congress
4	Regulations issued by the State Council, orders of ministries, and commission
3	Interim regulations promulgated by the State Council, regulations and stipulations promulgated by ministries and commissions
2	Opinions, measures, and provisional regulations of various departments
1	Notices, circulars, notices, announcements, local documents

and 527 effective policy documents were obtained. For subsequent quantitative analysis, the collected policy documents were assigned values according to the document's subject, reflecting the policy's implementation. The assignment criteria are as follows (Table 9.13).

9.4.2 *Alternative Policies for Energy Resources*

9.4.2.1 Number and Form of Policies

The number of effective alternative energy and resource policies in China was less than 20 per year from 2005 to 2008 (Fig. 9.7). The same increased from 26 documents in 2008 to 116 in 2016. It decreased to 107 effective documents in 2017 and increased to 127 documents in 2018. The effective documents showed a downward trend from 2018 to August 2021. In 2020, owing to the impact of the pandemic, the number of alternative energy resource policy documents was reduced to 65.

Table 9.14 shows that it is divided into 28 types: measures, reports, standards, ministerial orders, guidelines, and laws. Among them, a few policies are in the form of standards, ministerial orders, guidelines, laws, rules, catalogs, lists, declarations, principles, interim regulations, and regulations, all of which are less than or equal to five. The largest number is notices (452 effective policies) and opinions (108 effective policies), and there are more measures, plans, announcements, and regulations, all more than 50. Other documents, such as reports, regulations, and letters, account for a smaller proportion.

9.4.2.2 Main Body and Strength of Policy Documents

Primary bodies of policy documents. Table 9.15 shows the statistical sources of alternative energy resource policies. Policies for the joint issuance of documents are repeated according to the different issuing subjects. The statistical results are as follows.

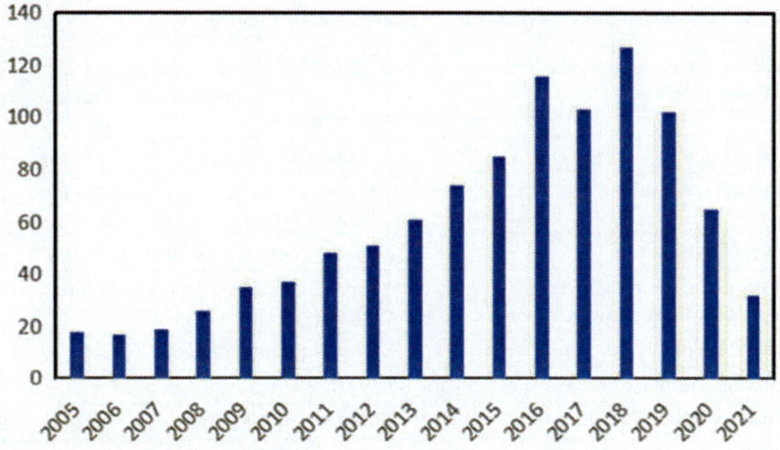

Fig. 9.7 Number of energy alternative policies from 2005 to August, 2021

Table 9.14 The form of policies for alternative energy resources

Form	Copies	Form	Copies	Form	Copies
Method	73	Rule	4	Notice	452
Report	8	Letter	25	Opinion	108
Standard	5	Plan	40	Principle	1
Ministry order	4	Decide	13	Interim measures	27
Guidelines	2	Contents	2	Interim regulation	2
Law	5	List	1	Bylaws	1
Program	58	Statement	2	Guide	8
Announcement	61	Regulations	15	System	7
Regulation	11	Inform	13		
Regularize	59	Notice	7		

As shown in Table 9.15, among the statistical issuing bodies, the National Energy Administration issued 614 documents, ranking first. The National Development and Reform Commission ranked second, issuing 205 documents. The State Council and the Ministry of Finance issued 107 and 81 documents, respectively. The State Economic and Trade Commission, State Archives Bureau, National Bureau of Statistics, and State-owned Assets Supervision and Administration Commission issued fewer documents.

Strengths of the policy documents. According to the statistics of the issuing entities and the standard of issuing papers in Table 9.13, strengths are assigned to the policies of different issuing entities of the energy and resource alternative policies in China. Among them, the highest strength is assigned a value of 5, which implies that the policy is the most powerful, with the highest degree of attention. The lowest value is

Table 9.15 Statistics of policy issuers for alternative energy resources

Policy issuers	Copies	Policy issuers	Copies
National Energy Board	614	National Development and Reform Commission	205
State Department	107	Ministry of Finance	81
Ministry of Agriculture	23	Ministry of Industry and Information Technology	45
Ministry of Construction	18	State Administration of Work Safety	11
Certification and Accreditation Administration Commission	11	State Coal Mine Safety Supervision Bureau	11
Administration of Quality Supervision	11	State Administration of Taxation	9
State Oceanic Administration	9	transportation Department	9
State Forestry Administration	7	Ministry of Education	7
National People's Representative Meeting	6	Ministry of Water Resources	6
General Administration of Customs	5	Ministry of Environmental Protection	6
Ministry of Science and Technology	5	Ministry of Commerce	5
China International Fusion Energy Program Implementation Center	5	State Organs Affairs Administration	4
General Office of the National Health and Family Planning Commission	3	Ministry of Land and Resources	3
State Economic and Trade Commission	2	State Archives	2
National Bureau of Statistics	2	State-owned Assets Supervision and Administration Commission	2
National Energy Board	614	National Development and Reform Commission	205

1, which means that the policy strength is the weakest, the implementation effect is average and receives the least attention. The assignments to alternative energy and resource policies obtain policy strength, as shown in Fig. 9.8.

From 2015 to 2016, China paid the most attention to energy and resource substitution, and policy implementation was the most vigorous (Fig. 9.8). In 2017 and 2018, these intensities were 169 and 178, respectively. The overall policy strength from 2005 to 2008 was relatively small. The policy strength from 2006 to 2015 showed a steady upward trend, indicating that China's emphasis on energy resource substitution is gradually increasing. From 2016 to 2019, this was the focus of policy capture. In 2020, attention decreased due to the impact of the COVID-19 pandemic. However, replacing traditional fossil fuels with new energy sources is an irresistible trend.

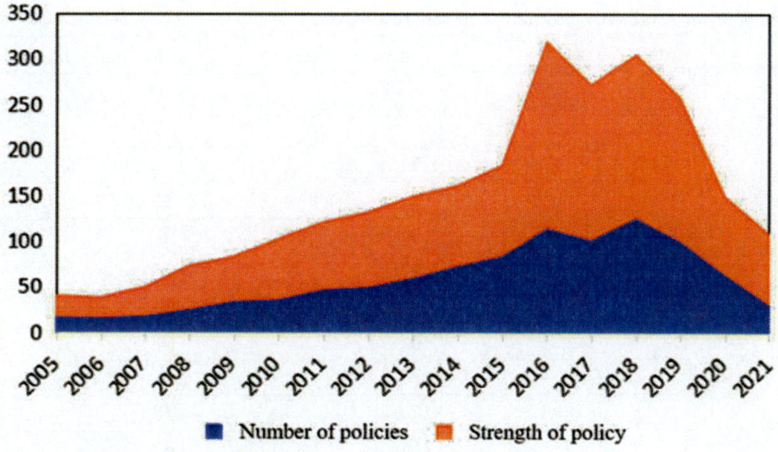

Fig. 9.8 Policy strength of alternative energy resources

9.4.2.3 Analysis of Policy Tools

This study divides policy tools into three categories, according to their roles: stimuli, guarantees, and guidance (Dong et al., 2021). Stimulative policy tools are mainly used to stimulate market changes by releasing and implementing policies to achieve expected results faster. However, the effect achieved by such means deviates from expectations because there are many uncertain factors in the market. Guaranteed policies, such as granting new energy subsidies, aim to ensure the development and production of relevant alternative energy sources to achieve the expected results. This method can directly achieve the expected purpose. The guiding policy promotes the development of alternative energy resources by indirectly acting on the market supply and demand. According to the division of policy tools, the policy tools of different years are as follows.

Table 9.16 shows that the most common types of policy tools are guidance, safeguard, and stimulus tools. From 2005 to 2008, the government used fewer of the three policy tools, indicating that it was still focusing on resource consumption to promote economic development during this period. Sustainable development had not yet been fully implemented in selecting energy consumption varieties. From 2009 to 2014, the number of policy-guided tools gradually increased, the substitution of energy resources was gradually emphasized, and innovative tools and methods began. From 2015 to 2019, the guiding policy tools in China were fully implemented, and the number of policy tools was relatively stable each year, which indicates that the government has carried out concentrated policy development in these years to promote the development of alternative energy resources.

Table 9.16 Policy tools for alternative energy resources

Year	Stimulative policies	Guarantee policies	Guiding policies
2005	2	6	10
2006	3	7	7
2007	4	7	8
2008	5	6	15
2009	6	9	20
2010	5	9	23
2011	8	9	31
2012	7	6	38
2013	7	13	41
2014	9	17	48
2015	7	16	62
2016	9	21	86
2017	5	21	77
2018	11	25	91
2019	7	13	82
2020	9	13	33
2021	3	8	21

9.4.3 *Efficiency Improvement Policy*

9.4.3.1 Number and Form of Policies

Figure 9.9 shows that the number of efficiency improvement policies in China fluctuated from 2005 to August, 2021. From 2005 to 2016, it was in a state of rising volatility, which declined in 2007, 2013, and 2014. In 2015 and 2016, the number of policies rapidly increased, reaching 87 in 2016. After 2016, efficiency improvement policies declined to 46 in 2017. In 2018, the number of effective policies was adjusted back to 47; however, 2019 and 2020 showed a downward trend, which may be due to the epidemic's impact. As of August, 2021, this number did not show an upward trend.

Table 9.17 shows the release forms of the efficiency improvement policies. Overall, the largest number was noticed, and the total number of effective policies was 236. Following this method, the effective number was 48, with over 20 plans, announcements, plans, and opinions. A few policies are reports, standards, ministerial orders, guidelines, laws, rules, catalogs, lists, declarations, interim measures, interim regulations, charters, guidelines, and systems, all of which have fewer than or equal to five copies. Other documents, such as regulations and decisions, accounted for a smaller proportion.

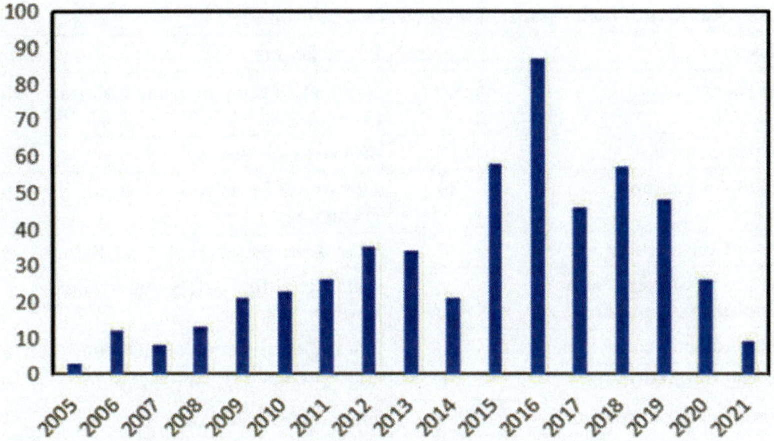

Fig. 9.9 Number of efficiency improvement policies from 2005 to August 2021

Table 9.17 The form of efficiency improvement policies

Form	Copies	Form	Copies	Form	Copies
Method	48	Rule	2	Notice	236
Report	4	Letter	13	Opinion	26
Standard	3	Plan	21	Principle	13
Ministry order	2	Decide	7	Interim measures	1
Guidelines	1	Contents	1	Interim regulation	1
Law	3	List	0	Bylaws	4
Program	35	Statement	1	Guide	4
Announcement	31	Regulations	8	System	1
Regulation	6	Inform	7		
Regularize	31	Notice	4		

9.4.3.2 Main Body and Strength of Policy Documents

Primary bodies of policy documents. Table 9.18 shows the issuers of statistical efficiency improvement policies. The policies of joint issuers were repeatedly counted according to the issuers. The statistical results are as follows.

The National Energy Administration issued the largest number of documents (369), followed by the National Development and Reform Commission (98) (Table 9.18). The State Council has issued 87 documents. The State Organ Affairs Administration and the General Office of the National Health and Family Planning Commission have not issued similar documents. The China International Fusion Energy Program Implementation Center, Ministry of Land and Resources, and State Archives Administration have published fewer papers.

Table 9.18 Statistics of policy issuers for efficiency improvement

Policy issuers	Copies	Policy issuers	Copies
National Energy Board	369	National Development and Reform Commission	98
State Department	87	Ministry of Finance	32
Ministry of Agriculture	11	Ministry of Industry and Information Technology	26
Ministry of Construction	13	State Administration of Work Safety	8
Certification and Accreditation Administration Commission	8	State Coal Mine Safety Supervision Bureau	8
Administration of Quality Supervision	8	State Administration of Taxation	6
State Oceanic Administration	6	transportation Department	6
State Forestry Administration	4	Ministry of Education	4
National People's Representative Meeting	5	Ministry of Water Resources	5
General Administration of Customs	3	Ministry of Environmental Protection	3
Ministry of Science and Technology	5	Ministry of Commerce	3
China International Fusion Energy Program Implementation Center	1	State Organs Affairs Administration	0
General Office of the National Health and Family Planning Commission	0	Ministry of Land and Resources	1
State Economic and Trade Commission	1	State Archives	1
National Bureau of Statistics	1	State-owned Assets Supervision and Administration Commission	1
National Energy Board	369	National Development and Reform Commission	98

Strengths of the policy documents. According to the same policy intensity calculation standard as above, the statistical results for the policy intensity of China's efficiency improvement policies from 2005 to August 2021 were obtained, as shown in Fig. 9.10.

The intensity of efficiency improvement policies and the number of policies showed similar trends (Fig. 9.10). Compared with alternative energy and resource policies, the intensity of efficiency improvement policies reached a maximum in 2016. In 2016, China paid unprecedented attention to efficiency improvements and energy resource substitution. Since 2015, the intensity of efficiency improvement policies has increased. In the early period, from 2005 to 2009, China paid less attention to efficiency improvement; however, from 2007 to 2013, China paid more attention to efficiency improvement each year. Combined with the growth in the number of policies, it can be seen that the Chinese government's emphasis on energy efficiency improvement has been greatly enhanced.

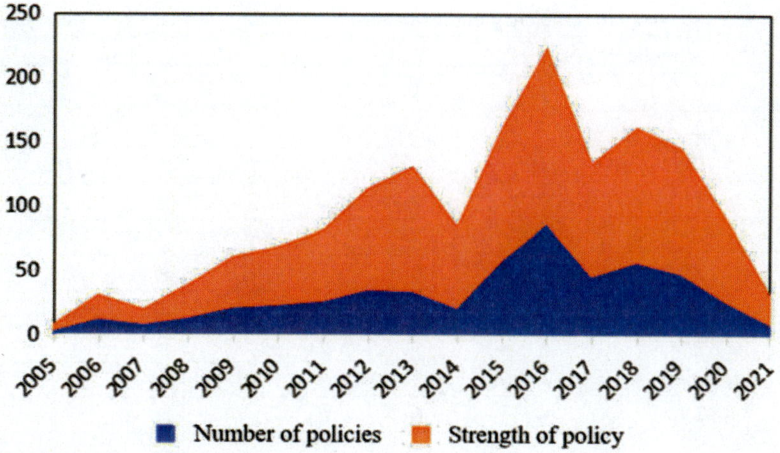

Fig. 9.10 Efficiency improvement policy strength

From 2013 to 2018, efficiency improvement policies' intensity fluctuated significantly, and performance was extremely unstable, indicating that there were still certain difficulties and obstacles in exploring efficiency improvement paths. For example, the efficiency of departmental collaboration needs to be improved, and a strictly unified version of the implementation policy has not yet been formulated. Regarding follow-up road exploration, combining the actual situation and rhythmically improving efficiency are necessary.

9.4.3.3 Analysis of Policy Tools

Based on this classification method, we obtained the number of efficiency improvement policy tools in China. The classification results are presented in Table 9.19. Comparing the division table of alternative energy resource policy tools with efficiency improvement policy tools, most policy tools are still guided,with the number of policy-guided tools for efficiency improvement showing a fluctuating upward trend. The peak number was 67 in 2016, followed by 2015. Regarding guarantee tools, the peak was still in 2016 with 13 copies, followed by 2019 with 12 copies. For stimulus tools, the most were also in 2016, with 7. In 2016, the Chinese government attached significant importance to efficiency improvements and related changes. In terms of the overall number of the three tools, the number of efficiency improvements from 2005 to 2008 was the lowest, with zero stimulus tools in 2005 and 2007, and only one assurance tool. However, both tools show an increasing volatility trend, indicating that the government's policy formulation and implementation stability improved to a certain extent.

Table 9.19 Policy tools for efficiency improvement

Year	Stimulative policies	Guarantee policies	Guiding policies
2005	0	1	2
2006	2	3	7
2007	0	1	7
2008	1	2	10
2009	2	3	16
2010	2	6	15
2011	3	7	16
2012	5	7	23
2013	2	9	23
2014	2	6	13
2015	5	9	44
2016	7	13	67
2017	4	11	31
2018	3	11	43
2019	4	12	32
2020	3	7	16
2021	1	2	6

9.5 Summary

The first subsection examines the current state of coal and oil resources, and explains the total amount and spatial distribution of coal and oil resources. The results showed that, compared with the proven coal reserves of various countries in the world in 2020 (which refers to the proven reserves at the end of 2020) and the reserve-production ratio, China's coal energy reserves are in the middle, with 143,197 million tons. The crude oil energy reserves are relatively low at 25 billion barrels, less than 25% of that of neighboring Russia. Raw coal resources are the most abundant resources in China in terms of regional distribution. The central region has the highest number of basic coal reserves. In 2007 and 2017, it exceeded the sum of the basic coal reserves in the Eastern and Western regions, followed by the Western and Eastern regions. The economy has a certain impact on energy consumption; however, its impact is not decisive.

The second section analyzes the utilization efficiency of coal and oil resources. First, we analyze the problems associated with using coal and oil resources. The natural environment in coal-rich areas is relatively harsh, and the conditions for resource extraction are poor. The oil safety risk factor is increasing daily, and the utilization efficiency needs to be improved. The use of coal and oil negatively affects the environment. Second, we use the DEA model to evaluate the utilization efficiency of coal and oil and explain the meaning of the utilization efficiency of coal and oil resources. Based on the principles of rationality, representativeness, feasibility, and scientific nature, and by focusing on the combination of input and output, the evaluation index system for the utilization efficiency of coal and petroleum resources is established from the three aspects of "financial resources," "human resources," and "material resources." The DEA analysis results of coal utilization efficiency in 30 provinces and the DEA analysis results of crude oil utilization efficiency in 27 provinces were obtained. The results show that increasing returns to scale is the main trend in the 30 provinces, mainly reflected in the increase in output and income with the same coal input. This indicates that the improvement of coal utilization efficiency is beneficial to economic output and environmental protection. The decreasing returns to scale for Hebei, Liaoning, Jiangsu, Zhejiang, Shandong, Henan, Guangdong, and Sichuan indicate that these eight provinces need to improve their coal utilization efficiency further. The overall technical efficiency of crude oil production in the 27 provinces was relatively high. The scale efficiency of the 27 provinces is at a relatively high level, close to or equal to 1, which indicates that crude oil-related input is relatively close to efficiency improvement ability. In Jiangsu and Ningxia, the scale efficiency of crude oil utilization also has a certain impact on comprehensive technical efficiency. Each province can improve its comprehensive technical efficiency through the optimal allocation of resources.

Finally, we analyzed the motivation behind the low utilization efficiency of coal and oil resources. Based on the DEA analysis, we measured each province's time-series changes in coal and oil utilization efficiency. We then analyzed the time-series changes in total factor productivity, accurately evaluated the overall efficiency, and determined the reasons for low utilization efficiency. The Malmquist index decomposition results for 30 coal utilization efficiencies, and the Malmquist index decomposition results for crude oil utilization efficiencies in 27 provinces were obtained. The Malmquist index of coal efficiency in the 30 provinces improved by 3.5%, mainly because of greater technological progress. The average value of the total factor productivity change rate of the provinces in the nine years from 2010 to 2019 was $1.035 > 1$, indicating that the total factor productivity of the provinces in these nine years improved, and the comprehensive utilization efficiency of coal was conducive to improving the efficiency level during this period—improvement in the comprehensive utilization efficiency of coal under technological progress. Additionally, the improvement in the coal utilization rate in various provinces was due to a decline in the efficiency of coal development. Therefore, it is necessary to improve the technical level of coal mining and production to improve the efficiency of coal utilization. The improvement in the comprehensive utilization efficiency of crude oil in various provinces between 2010 and 2019 was mainly supported by technological progress. Changes in pure technical and scale efficiency were the main factors affecting efficiency changes. This is similar to the improvement in the comprehensive utilization efficiency of coal. The Malmquist index of crude oil efficiency in the 27 provinces increased by 1.9%, mainly because of a 4.3% increase in technological progress. There is an inherent relationship between the Malmquist index, technological progress, and crude oil utilization efficiency. Technological changes affect crude oil utilization efficiency by affecting the Malmquist index. Large changes in technological progress often lead to a higher Malmquist index; that is, technological progress is the leading factor in efficiency.

The third subsection analyzes China's public policies, divided into energy and resource substitution, and efficiency improvement policies. They were analyzed from four aspects: the number and form of policies, the subject and strength of policy documents, and policy tools. In terms of energy and resource alternative energy and resource policies, China had fewer than 20 effective policies every year from 2005 to 2008. After 2008, the number of alternative energy and resource policies increased. Regarding policy issuance, a few policy forms were standards, ministerial orders, guidelines, laws, etc. The largest numbers were notifications, followed by opinions. The National Energy Administration has issued the largest number of documents, with as many as 614 documents, followed by the State Economic and Trade Commission, the State Archives Bureau, etc. From 2015 to 2016, China paid the most attention to substituting energy and resources, and its policy implementation was the most vigorous. It peaked in 2016 with an intensity of 203. Policy intensity from 2006 to 2015 showed a steady upward trend. We divided the policy tools into three categories: stimulus, guarantee, and guidance. Most of these tools are guidance tools. From 2015 to 2019, the guidance policy tools were fully implemented in China. From 2005 to 2008, the government used fewer of these three policy tools. The number of policy-guided tools gradually increased. Among efficiency improvement policies, the number of efficiency improvement policies in China from 2005 to August 2021 has fluctuated. Judging from the overall number, the largest number was of notices, and the total number of effective policies was 236. Among the statistics on issuing subjects, the National Energy Administration issued 614 documents, which ranked first. The National Development and Reform Commission issued 205 documents, ranking second. The intensity of efficiency improvement policies and the number of policies show similar trends. Compared with alternative energy and resource policies, the intensity of efficiency improvement policies reached a maximum in 2016. From 2013 to 2018, the intensity of efficiency improvement policies fluctuated greatly, indicating that there are still some difficulties and obstacles in exploring efficiency improvement paths. For example, the efficiency of departmental collaboration needs to be improved, and most policy tools remain guiding tools. From 2005 to 2008, the number of efficiency improvements was the smallest among the three tools, with zero stimulus tools and one assurance tool in 2005 and 2007, respectively; however, both tools showed an upward trend.

References

Dong, C. G., Qi, Y., & Nemet, G. (2021). A government approach to address coal overcapacity in China. *Journal of Cleaner Production, 278*, 123417.

Liu, L. C., Cheng, L., Zhao, L. T., Cao, Y., & Wang, C. (2020). Investigating the significant variation of coal consumption in China in 2002–2017. *Energy, 207*, 118307.

Wang, S. B., Zhao, C., Liu, H. B., & Tian, X. L. (2021). Exploring the spatial spillover effects of low-grade coal consumption and influencing factors in China. *Resources Policy, 70*, 101906.

Wu, K. (2014). China's energy security: Oil and gas. *Energy Policy, 73*, 4–11.
Zhang, C., Zhao, Z. W., Wang, Q. W., & Xu, B. (2022). Holistic governance strategy to reduce carbon intensity. *Technological Forecasting and Social Change, 179*, 121600.

Chapter 10
Renewable and Common Resources: Marine Fishery Resources

10.1 Introduction

With the advances of globalization, the world economy has developed rapidly, resulting in population growth and a series of problems, such as the shortage of resources and environmental degradation. Among these, the shortage of several important land-based non-renewable resources is a significant problem. Countries worldwide have been focusing on ocean energy as a new pillar of economic development. In May 2001, the United Nations (UN) State Party document clearly stated that "the twenty-first century is the maritime century," suggesting that this is an era of marine economic development. Consequently, numerous countries have seized this new opportunity for development, formulating marine development strategies and developing the marine economy. For example, the marine economy has become an important part of the American economy.

As a global maritime power, the United States (US) is the first country to have utilized marine resources, and its degree of marine resource development is also the highest. The US has adopted a series of marine development policies to ensure the coordinated development of the marine economy and environment, including a blueprint for the twenty-first century, marine action plan, national marine policy on oceans and coasts, the presidential executive order related to the Great Lakes, and the marine policy executive order on promoting maritime economy and environmental interests. Similarly, several other countries have adopted various policies for marine development. For example, the United Kingdom (UK) has adopted the marine science strategy 2010–2025 and the 2050 maritime policy to promote the development of emerging marine industries and make the UK a global maritime power. Japan has also adopted a series of related policies and plans to develop its marine economy, including the Ocean and Japan: Marine Policy Suggestions for the 21st Century, the Basic Marine Plan, the Marine Development Promotion Plan, and the Development Plan for Marine Science and Technology.

X. Deng et al., *Environmental and Natural Resources Economics*,
https://doi.org/10.1007/978-981-99-9923-1_10

After the Reform and Opening-up in China, the development of the marine economy has become an important engine for promoting economic development and improving people's living standards. According to the China Marine Economic Development Report 2020, China's marine GDP reached 8.9 trillion yuan in 2019, up 6.2% year on year, accounting for 17.1% of the coastal GDP and contributing 9.1% of the national economic growth and 0.6% to the GDP growth in that year. The report to the 18th CPC National Congress clearly stated the strategic goal of building China as a "maritime power." The report to the 19th CPC National Congress emphasizes the decision of "adhering to the land and sea planning and accelerating the building of a maritime power," which provides policy guidance for China to develop the marine economy and build itself into a maritime power.

The ocean accounts for over 70% of the earth's surface area and is rich in non-renewable and renewable resources such as oil, natural gas, and fresh water. The continuous provision of new resources supports the sustainable development of humankind. Data has shown that the ocean has a wide variety of organisms, including more than 200,000 species. More than 80% of the world's organisms live in the ocean, making it a natural protein warehouse. The ocean is 1,000 times habitable to humans than all the arable land on land. As of 1998, more than 60% of the world's population lives near the ocean, no more than 100 km away. According to the estimates of the UN Agenda 21, this proportion will reach more than 75% by 2020. Therefore, the future of the ocean is bound to become a new opportunity for the development of coastal countries and regions. For mankind, the ocean, as the origin of life, is the ultimate space on earth for its survival and development. It can even be said that the ocean is the way out for future civilization. Marine resources include marine water resources, marine chemical resources, marine biological resources, and marine mineral resources. The forms of human exploitation and utilization of marine resources are also diversified, such as seawater desalination, extraction of sea salt, and fishing grounds, among which marine fishery is the most important form of the utilization of marine resources.

As an important part of the marine economy, the marine fishery not only provides food for human beings but also plays an inestimable role in regulating the climate of the marine ecosystem and purifying the water quality. According to the 2020 report of FAO, marine fishery produce averaged 91.4 million tons in 1996–2005, 89.8 million tons in 2006–2015, 89.6 million tons, and 96.4 million tons in 2018, up 5.4% from the average of the previous three years. With social development and a growing population, human demand is increasing, and the progress of fishing technology makes the development of marine fisheries possible. However, this has also led to a series of environmental problems, resource decline, population depletion, and fishing and ecological environment deterioration. These are a serious threat to the diversity of marine life and the sustainable development of the marine ecosystem. FAO statistics show that the proportion of the bio-sustainable fish population decreased from 90% in 1974 to 65.8% in 2017 (down 1.1% since 2015), whereas the global bio-sustainable fishing population proportion increased from 10% in 1974 to 34.2% in 2017. This data reflects the excessiveness of exploitation and the resultant resource decline. The sustainable development of marine fishery is facing a crisis. Thus, implementing a scientific and effective management system for marine fishery and fishing and realizing the importance of the sustainable development of marine fishery resources play a vital role in protecting marine biodiversity, the marine ecosystem, and the marine ecological environment.

Since the United Nations Convention on the Law of the Sea in 1982, the impact of marine fisheries on marine ecosystems and marine biodiversity has received widespread attention in the international community, which has proposed a series of regulations and policy systems for sustainable marine fisheries. The United Nations Convention on the Law of the Sea emphasizes protecting marine fishery resources and the need for restricted fishing. The International Code of Conduct for Responsible Fisheries stipulates the responsibilities of countries worldwide for fishing activities such as fishing and breeding. A series of conventions, such as the Conservation and Management of Chinese and Western Pacific highly migratory fish resources convention, the South Pacific high seas fishery resources conservation and management convention, the North Pacific high seas fisheries resources conservation and management convention, the North Pacific anadromous fish population protection convention, and the southern bluefin tuna conservation convention, have been conducted for specific oceans, and specific population protection and management regulations. These international conventions and policies have laid the foundation for developing sustainable marine fishery by providing guidance and guarantee for protecting marine life and ecosystems from a legal perspective.

As a major country practicing marine fishery, China has rapidly progressed since the Reform and Opening-up. Since 1989, China has ranked first in the world in total aquatic produce and marine breeding. In 2019, the total aquatic produce of China reached 64.8036 million tons, which comprised 32.825 million tons of seawater produce, accounting for more than 50% of the total output. With the strategy of making China a "maritime power," its marine fishery has entered a new period of development. According to the China Fishery Statistical Yearbook 2020, the output value of China's fishery in 2019 was 1,293.4 billion yuan, of which the total output value of the mariculture and marine fishing industry was 569.1 billion yuan, accounting for about 40% of the total output value. The per capita net income of fishermen nationwide was 21,108.29 yuan in the same year. These data reflect the importance of marine fishery in China's economic development. However, with the problems of overfishing and over-aquaculture, the sustainable utilization of marine fishery resources and the sustainable economic development of China has been seriously affected.

To solve these problems, relevant departments in China have adopted a series of laws, regulations, and policies. These include marine fishery development in 1986 and the fishing license management regulations through the implementation of the fishing license system, limiting fishing access, controlling the development of fishery resources, the 1987 "double control" system, the 2002 "switch" system, which limits the number of fishing vessels. The target of "zero growth" of total fishing was proposed in 2000, and the fishing limit system was proposed in the Fisheries Law. These measures were designed to control the development of marine fisheries from the perspective of total fishing management, based on the principle that the fishing volume is lower than the growth volume of fishery resources. In 2016, the Ministry of Agriculture issued the 13th Five-Year Plan for National Fishery Development (2016–2020); the total marine fishing volume should be limited to 10 million tons by 2020. Since 2013, relevant departments have produced documents on the sustainable development of fisheries and the creation of marine ranches to promote the sustainable development of marine fishery resources. In 2017, China began to create a national marine pasture demonstration zone to promote fishery transformation and upgrading. The promulgation and implementation of these policies, laws, and regulations play an important role in promoting the sustainable utilization of marine fishery resources and providing an institutional guarantee for its management. However, the basic situation of China's marine fishery resources is still very severe. Though the total amount of resources is large, the per capita share is low, and the resource utilization efficiency is poor. The development and utilization of marine fishery resources scientifically, reasonably, and sustainably are the focus and challenge in developing marine resources, which we must resolve.

This chapter first describes the entire situation and distribution of marine fishery resources in the world and China. Then, it compares the current situation of Chinese marine fishery resources with the global situation, analyzes the utilization efficiency of the Slack Based Measure (SBM) and Malmquist-Luenberger (ML) index, and finally proposes the public policy of fishing cost, tax, and personal transfer quota.

10.2 Current Situation of Marine Fishery Resources

10.2.1 Current Status of Global Marine Fishery Resources

10.2.1.1 Global Marine Fishery Production Situations

With the growth of the global population and the improvement of people's living standards, the human demand for food has risen sharply. However, the types and quantities of food available for human consumption are limited. Thus, the fishery has become an important source of human food. Figure 10.1 shows the state of global fishery production since 1950, displaying a significant upward trend. In 1950, the total global fishery production was 19.89 million tons. In 2019, the total global fishery production reached 213.69 million tons, more than tenfold compared with the 1950 production, fully reflecting the increasing demand for fisheries. Fishery production mainly takes place in two types of waters—marine waters and inland waters. In the early stage of fishery development, marine fishery accounted for nearly 90% of fishery production. However, with the increase in marine fishery production, the marine biodiversity and marine ecological environment have been damaged. To protect marine resources, inland waters fishery production has been emphasized, and the proportion of marine fishery in fishery production has gradually been reduced. In 2010, the proportion of marine fishery among total fishery production fell to about 70%. The two kinds of

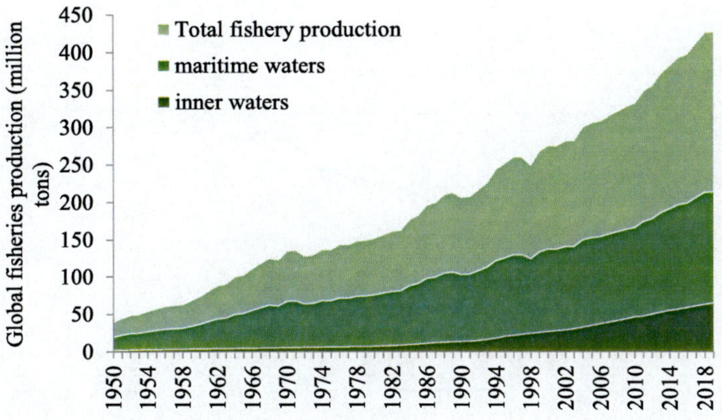

Fig. 10.1 Global fishery production in 1950–2019. *Data source* The FAO Fishstat Plus Database

fishery production also present an obvious upward trend. Marine fishery production rose faster after 1950; the late upward trend slowed, and the inland waters fishery production rose relatively slowly, growing faster since the 1990s. This is because we have realized the importance of the sustainable development of marine fishery and adopted a series of policies to control it.

Fishery production statistics mainly include fishing and aquaculture. Figure 10.2 shows the output of global fishing and aquaculture since 1950. In the 1950s, fishery production mainly relied on fishing, which was not large in proportion. In the early stage of fishery development, though fishery production increased rapidly, the growth rate of aquaculture production was relatively slow. The fishing volume relatively stagnated in the late 1980s, and aquaculture production began to increase rapidly, becoming the main driving force for the increase in total fishery production. The development of marine fishery is analyzed from the two aspects of fishing and aquaculture, respectively.

10.2.1.2 Global Marine Fisheries and Fishing Conditions

Global marine fishery and fishing production

As an important part of fishery production, fishery fishing plays an inestimable role. Figure 10.3 shows the fishing production and total production trends in the global inland and marine waters from 1950 to 2019. The overall global fishing production has increased since 1950 (Fig. 10.3). In 1950, it was about 19.25 million tons, and about 93.59 million tons in 2019, which is about five times that of 1950. Thus, global fishery production developed rapidly in its early stages. In the late 1980s, the development of global fishery production was relatively stagnant, showing small fluctuations, and the output did not increase. Similar to fishery production, the proportion of marine fishery output far exceeded that of fishery fishing in inland waters. In 1950,

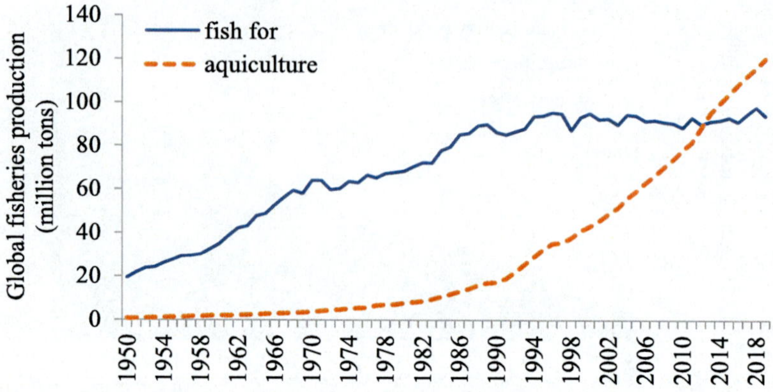

Fig. 10.2 Global fisheries fishing and aquaculture production from 1950–2019. *Data source* The FAO Fishstat Plus Database

the marine fishery output was about 17.32 million tons, accounting for about 90% of the total output. By 2019, the proportion of marine fishery output reached 81.5 million tons, accounting for about 87% of the total output. Compared with 1950, the proportion of marine fishery production in the total output decreased slightly; however, though the decline was small, marine fishing continues to be the most significant component of fishing. The development trends of both marine fishing and inland waters fishing production are similar to the growth trend of global fishing production, and both became relatively stagnant after the 1980s.

Thus, the contribution of marine fishery fishing in fishery production is much greater than that of fishery fishing in inland waters. The following is a specific analysis of marine fishery fishing. Figure 10.4 shows the state of marine fishing production during 2010–2019, globally. Marine fishery production has changed little

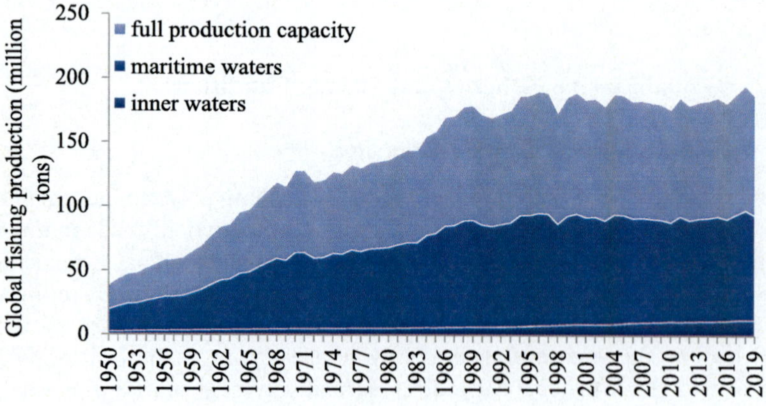

Fig. 10.3 Global fishery production from 1950 to 2019. *Data source* The FAO Fishstat Plus Database

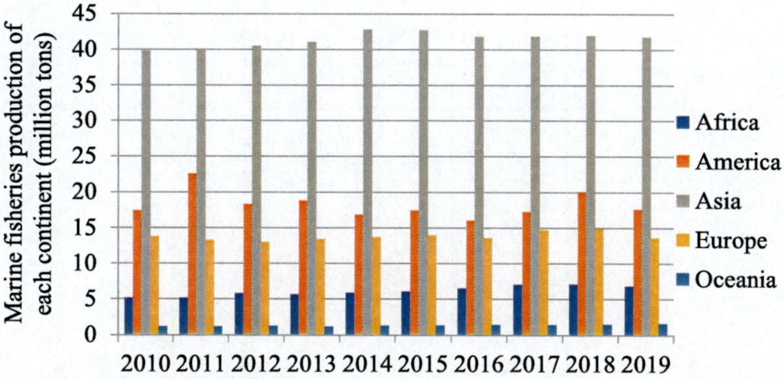

Fig. 10.4 Production of marine fisheries based on each continent from 2010–2019. *Data source* The FAO Fishstat Plus Database

since 2010. In 2010, the marine fishery output was 68.82 million tons, 41.77 million tons, and 136.41 million tons, respectively. A comparison of the production of all continents shows that Asian marine fisheries produce far more than any other continents, followed by America, and Europe, whereas the lowest production comes from Oceania.

Major varieties of global marine fishery and fishing

In recent years, although the global trend of marine fishery production has become relatively stable and the production of the marine fishery has changed little, the production of a single variety has changed. As the oceans are rich in marine resources, marine fishery is not only the fishing of fish, shrimp, and crab, but also the fishing of aquatic plants and animals, such as algae, seaweed, kelp, frogs, turtles, and mollusks such as oysters and scallops, as well as aquatic animal products such as coral and pearls. Table 10.1 shows the output of major global marine fishery species during 2010–2019. Marine fish is the most important marine fishing species in the world, which far exceeds other species in terms of production, accounting for more than 80% of marine fishery production. This is followed by mollusks and crustaceans, with freshwater fish and various aquatic animal products coming at last. In fact, the largest contribution to the freshwater-catch is by inland waters rather than the ocean.

Marine fish is the main species of marine fishing and has a diverse classification. The categories include flatfish, sole, cod, hake, haddock, sardine, anchovies, shark, ray, tuna, and other coastal and ocean fish; each category has further specific species division. Figure 10.5 shows the species and production of marine fish in the global marine fisheries caught in 2010–2019. As shown in Fig. 10.5, sardines and anchovies have the largest proportion, followed by various pelagic fish, including halibut, sole, shark, and ray, and silver shark accounting for a small proportion.

Table 10.1 Major varieties of global marine fisheries. Unit: Million tons

Year	2010	2011	2012	2013	2014	2015	2016	2017	2018	2019
Hydrophyte	107.07	113.70	114.43	130.27	120.53	107.49	111.00	112.22	95.31	108.33
Crustacean	540.31	559.32	568.85	586.27	609.69	611.19	596.87	603.56	599.33	564.12
Freshwater fishes	1.75	1.99	2.06	2.21	2.25	2.20	2.38	2.22	2.06	2.23
Fish with fan-shaped veins	139.43	152.02	135.61	162.19	137.61	149.60	140.14	151.83	162.31	154.48
Marine fishes	6278.1	6698.4	6362.8	6421.0	6382.0	6512.0	6461.4	6710.4	7048.7	6650.1
Various aquatic animal products	1.37	1.38	1.42	1.48	1.59	1.42	1.37	1.43	1.45	1.06
All kinds of aquatic animals	39.90	53.30	50.00	55.96	64.51	63.55	55.41	55.55	53.09	55.02
Mollusca	626.27	646.65	653.97	654.22	734.97	712.67	572.50	601.46	595.61	614.95

Data source The FAO Fishstat Plus Database

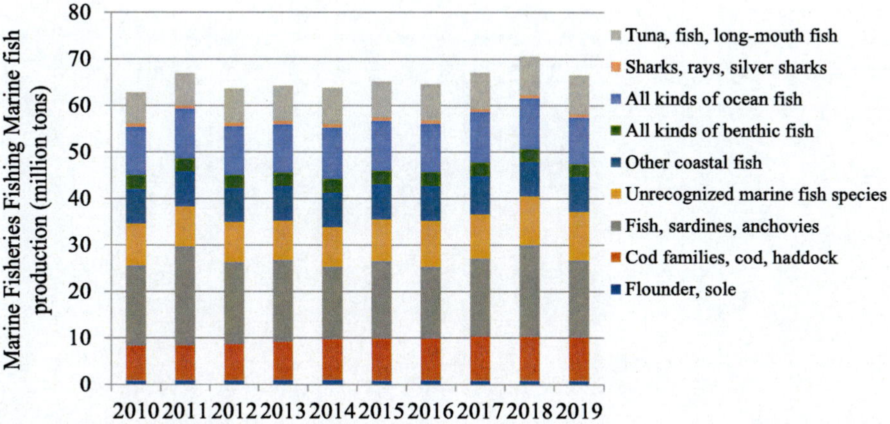

Fig. 10.5 Marine fish production from global marine fisheries in 2010–2019. *Data source* The FAO Fishstat Plus Database

Major fishing areas of global marine fishing

The Pacific Ocean is the largest and deepest ocean in the world, and its phytoplankton, seabed plants, fish, and other animals are more abundant than in other oceans. Table 10.2 shows the production of marine fisheries in major global fishing areas from 2010 to 2019. The Pacific marine fishing production is far more than in other oceans, accounting for more than half of all fishing production. The Northwest Pacific is the world's largest marine fishing production, accounting for about a quarter of the global marine fishing production, ranked second in the Midwest Pacific. The production in the Southeast Pacific and Northeast Atlantic is relatively high. Compared with the marine fishing production in each fishing area in 2010, eight fishing areas decreased in 2019, with the Southwest Pacific falling the most, followed by the Northwest Pacific, with a decline of more than 20%.

Major countries of global marine fishing production

Asia has the largest marine fishery output. However, as a major fishery producer, China's contribution cannot be ignored. Table 10.3 shows the marine fishery production of 25 major global producers during 2010–2019. China's marine fishery production ranks first globally, surpassing other countries by far. It is followed by Indonesia, the US, and the Russian Federation. Eleven of the twenty-five major production countries show a growing trend of marine fishing production compared with 2010. The biggest increase has been recorded by Vietnam and Ecuador, with an increase of more than 50%. Fourteen countries, including South Korea, Norway, and the Philippines, have been showing a downward trend. The most significant decrease has been recorded by Denmark, Japan, and Chile, by more than 20%.

Table 10.2 Global marine fishery production capacity in 2010–2019. Unit: 10,000 tons

Year	2010	2011	2012	2013	2014	2015	2016	2017	2018	2019
Northeast Atlantic Ocean	895.8	831.7	834.7	873.5	893.0	936.9	861.0	960.7	959.8	841.0
Northwest Atlantic Ocean	215.2	205.6	201.0	188.3	185.3	187.0	183.2	177.2	170.7	171.1
Southeast Atlantic Ocean	137.4	133.1	166.7	138.7	157.7	169.1	171.4	170.3	159.6	137.3
Southwest Atlantic Ocean	176.4	173.6	187.9	196.6	240.3	241.5	155.0	181.8	176.3	173.5
Middle East Atlantic Ocean	451.9	445.2	426.4	432.7	453.7	445.5	489.7	540.7	547.6	541.7
Midwest Atlantic Ocean	111.1	146.4	134.1	123.0	119.5	142.2	156.1	148.1	151.3	139.6
Mediterranean Sea and Black Sea	142.8	144.0	128.8	125.4	112.8	132.6	125.7	135.8	129.9	139.0
East Indian Ocean	597.5	617.2	616.7	626.3	636.4	639.8	644.1	697.6	676.1	686.4
West Indian Ocean	429.2	426.1	454.2	459.5	482.6	471.1	504.1	545.4	555.0	549.0
Northeast Pacific Ocean	242.7	294.9	285.1	320.7	316.3	317.3	311.1	337.7	309.1	316.8
Northwest Pacific Ocean	2086.4	2121.3	2109.5	2110.2	2159.2	2145.3	2127.6	2053.0	2050.6	1965.3
Southeast Pacific Ocean	817.8	1272.8	871.9	905.4	732.8	806.6	666.9	764.9	1060.7	823.7
Southwest Pacific Ocean	58.6	58.2	60.2	58.3	57.6	55.1	47.6	47.3	45.0	45.5
Central East Pacific Ocean	195.6	192.8	202.4	206.8	191.0	167.2	165.0	174.6	171.4	187.4
Midwest Pacific Ocean	1152.9	1143.4	1193.0	1223.3	1284.9	1277.5	1304.8	1276.5	1361.9	1395.0
The Arctic and Antarctic Regions	23.0	20.6	16.8	24.9	30.0	25.5	27.7	27.1	33.1	38.2

Data source The FAO Fishstat Plus Database

Table 10.3 Marine fishery production capacity in 2010–2019. Unit: 10,000 tons

Year	2010	2011	2012	2013	2014	2015	2016	2017	2018	2019
Chile	304.8	346.7	300.8	228.9	259.3	213.2	182.7	233.5	237.0	237.7
China	1292.9	1321.3	1340.6	1358.9	1433.6	1465.1	1401.6	1339.0	1286.7	1232.9
Denmark	82.8	71.6	50.3	66.8	74.5	86.9	67.0	90.4	78.9	62.9
Ecuador	40.0	50.6	51.2	51.3	66.2	64.3	71.5	62.3	59.7	60.8
Iceland	108.1	116.5	146.7	138.0	109.5	133.6	108.5	119.8	128.0	92.3
India	327.2	327.5	343.2	343.9	374.4	351.6	373.5	395.8	363.6	369.0
Indonesia	504.6	539.1	547.2	573.0	608.2	626.7	615.2	635.5	675.3	700.3
Japan	414.9	384.0	373.0	370.4	370.0	346.5	325.8	325.7	331.4	320.9
Korea	21.2	21.4	20.8	20.4	22.1	21.5	20.4	20.4	20.4	20.4
Malaysia	143.3	137.7	147.6	148.7	146.2	149.0	157.9	146.9	145.7	146.0
Mexico	141.2	145.9	147.4	151.1	140.6	132.8	132.5	146.9	147.6	142.5
Morocco	113.6	97.0	116.7	126.0	136.0	136.9	145.7	138.7	137.1	146.0
Burma	117.6	117.8	113.2	108.1	111.8	110.7	118.6	126.8	114.8	106.4
Norway	283.8	244.2	228.2	224.1	244.9	245.1	220.0	255.7	266.4	247.2
Peru	426.2	821.8	481.9	584.9	357.4	480.6	380.7	415.7	718.9	483.3
The Philippines	231.9	203.0	201.2	200.2	203.6	195.2	186.8	172.7	189.0	190.2
Russian Federation	381.3	401.2	407.5	410.4	401.1	417.9	448.1	460.3	484.8	472.9
Spain	96.9	100.5	92.5	98.4	105.9	97.2	91.1	94.8	92.3	87.7
Taiwan, China	85.3	90.6	90.7	92.5	106.9	98.9	75.2	74.9	80.4	75.0
Thailand	160.1	161.0	150.0	161.5	148.8	131.7	134.3	130.0	139.3	141.1
Britain	61.0	59.9	62.9	63.0	75.5	70.5	70.3	72.5	70.0	61.9
America	429.4	515.8	499.0	507.3	496.7	502.3	488.6	502.2	473.5	478.5
Vietnam	206.7	223.2	236.0	246.4	257.7	271.4	292.9	315.3	319.0	328.5

Data source The FAO Fishstat Plus Database

10.2.1.3 Global Marine Aquaculture Situation

Global marine aquaculture production

Aquaculture refers to breeding aquatic life, including fish, mollusks, crustaceans, and aquatic plants. In the late 1980s, though the global marine fishery fishing development was relatively stagnant, aquaculture developed rapidly. This chapter focuses on the rapid developmental stage of aquaculture. Figure 10.6 shows the global aquaculture production during 1985–2019. The total global aquaculture production showed a significant upward trend after 1985, increasing from 11.8 million tons in 1985 to 120.1 million tons in 2019. The total output of marine aquaculture and inland aquaculture also showed an obvious upward trend. In 1985, the total output of aquaculture in inland waters and marine waters was 4.7 million tons and 7.1 million tons,

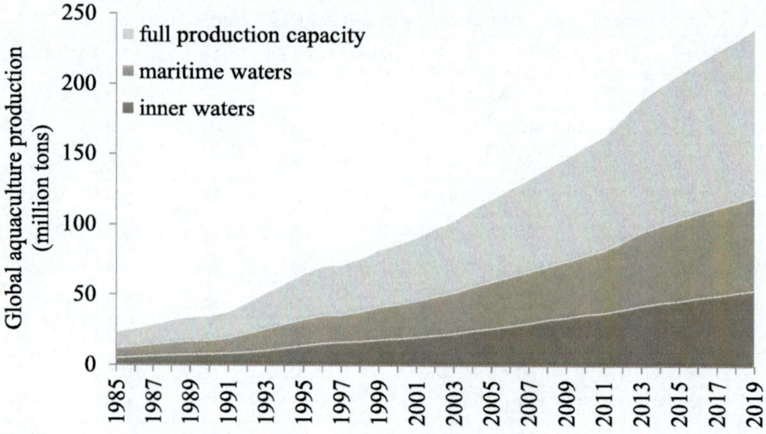

Fig. 10.6 Global aquaculture production from 1985 to 2019. *Data source* The FAO Fishstat Plus Database

respectively. In 2019, the total output of each aquaculture was 53.36 million tons and 66.74 million tons, respectively. Compared with 1985, although aquaculture output increased in 2019, the proportion of marine aquaculture output in the total output decreased from 60% in 1985 to 56% in 2019, and the proportion of aquaculture in inland waters increased.

The global output value of marine aquaculture

Different from the growth trend of aquaculture production, the output value of aquaculture grew relatively slowly in the early stage of development in 1985 (Fig. 10.7). In the twenty-first century, aquaculture's output value increased rapidly, reflecting the increasingly important substitution role of aquaculture in human food. In addition, in the early stage of development, the output value of marine aquaculture was higher than that of inland aquaculture. However, after 2007, the output value of inland aquaculture exceeded that of marine aquaculture, indicating that the contribution of inland aquaculture to global aquaculture is increasing.

Major varieties of global marine aquaculture

Table 10.4 shows the main varieties of global marine aquaculture during 2010–2019. Marine aquaculture and marine fishing production are quite different. Marine fish's highest production occurs in marine fishing; however, in marine aquaculture, aquatic life has the highest production varieties, followed by mollusks. Both have production reaching more than ten million tons, and the trend is that of a rise. Although the yield of various aquatic animal products and freshwater fish is still the two lowest yield varieties, it is significantly higher than that of marine fisheries.

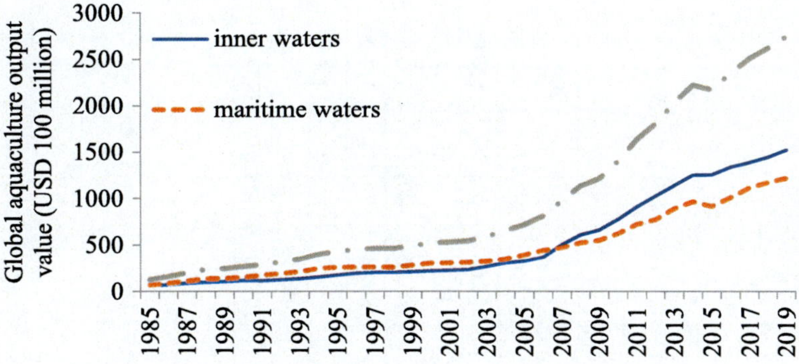

Fig. 10.7 Global aquaculture output value in 1985–2019. *Data source* The FAO Fishstat Plus Database

Major fishing areas of global marine aquaculture

Table 10.5 shows that between 2010 and 2019, marine aquaculture production, similar to marine fishery fishing main fisheries development. The Pacific is the world's largest marine aquaculture production ocean, especially the Northwest Pacific Ocean. Aquaculture production is predominant in the Midwest Pacific, far more than other fisheries. The marine aquaculture output in the Northeast Atlantic Ocean, the Eastern Indian Ocean, and the Southeast Pacific Ocean are also relatively high compared with other fishery areas, reaching more than one million tons. The marine aquaculture outputs in the Southeast Atlantic Ocean and the Central and Eastern Atlantic Ocean are relatively low.

Major producers of marine aquaculture

Asia is not only the largest marine fishery fishing producing continent but also the largest marine aquaculture output continent. Table 10.6 shows the marine aquaculture output of the twenty major global producers during 2010–2019. Among many aquaculture-practicing countries, China's marine aquaculture has ranked first in the world over the years. Its contribution to Asian marine aquaculture output cannot be ignored. Similarly, in Indonesia, marine aquaculture production has reached more than 10 million tons since 2013. Chile, North Korea, Norway, the Philippines, and Vietnam have also recorded a relatively high amount of marine aquaculture production, with an annual output of more than one million tons in the past five years.

Table 10.4 Global marine aquaculture major variety, 2010–2019. Unit: ten thousand tons

Year	2010	2011	2012	2013	2014	2015	2016	2017	2018	2019
Hydrophyte	2008.06	2170.00	2459.55	2791.94	2898.63	3099.30	3157.63	3254.01	3336.71	3467.91
Crustacean	324.70	359.45	370.90	386.96	425.74	450.67	481.33	550.62	578.34	630.06
Freshwater fishes	22.84	20.99	20.50	27.70	34.43	34.74	47.76	32.32	31.87	28.88
Fish with fan-shaped veins	268.14	308.19	351.26	352.55	378.13	384.41	376.08	386.76	409.54	455.55
Marine fishes	180.84	196.21	207.79	215.39	225.26	235.69	248.48	280.64	291.56	311.29
Various aquatic animal products	6.03	5.13	1.99	3.17	4.64	3.93	5.26	6.15	2.61	2.62
All kinds of aquatic animals	37.37	28.55	31.90	34.98	35.63	36.84	39.92	40.99	39.01	42.63
Mollusks	1346.24	1355.19	1408.26	1464.75	1518.65	1557.72	1658.91	1708.26	1729.27	1734.98

Data source The FAO Fishstat Plus Database

Table 10.5 Marine aquaculture output of major global fishery areas in 2010–2019. Unit: 10,000 tons

Year	2010	2011	2012	2013	2014	2015	2016	2017	2018	2019
Northeast Atlantic Ocean	174.77	189.72	208.00	197.50	213.24	217.21	212.46	216.71	222.45	235.00
Northwest Atlantic Ocean	12.49	12.16	13.86	12.53	11.02	12.65	13.25	12.57	12.22	12.62
Southeast Atlantic Ocean	0.47	0.44	0.49	0.53	0.60	0.55	0.67	0.51	0.62	0.76
Southwest Atlantic Ocean	8.60	8.86	9.93	8.48	8.79	9.24	7.37	6.28	6.07	7.03
Middle East Atlantic Ocean	0.58	0.75	0.55	0.67	0.83	0.88	0.89	0.78	0.86	0.91
Midwest Atlantic Ocean	15.55	11.23	14.62	15.11	15.16	13.08	15.37	15.98	17.55	19.52
Mediterranean Sea and Black Sea	51.75	50.88	51.90	53.54	55.75	59.54	64.65	76.85	82.72	88.71
East Indian Ocean	57.41	83.86	87.57	85.76	95.07	106.41	109.17	125.55	127.33	139.71
West Indian Ocean	29.02	18.02	19.69	15.91	20.90	25.83	22.17	25.30	26.98	28.68
Northeast Pacific Ocean	11.39	12.54	12.74	11.06	10.70	13.69	13.48	13.01	12.95	13.30
Northwest Pacific Ocean	2821.18	2875.56	3116.78	3270.40	3365.09	3497.46	3709.78	3932.79	4057.37	4238.00
Southeast Pacific Ocean	97.20	125.67	133.98	139.09	157.53	150.54	151.30	171.61	188.82	218.22
Southwest Pacific Ocean	11.53	12.07	10.27	10.02	11.24	9.37	11.16	11.90	10.65	11.70
Central East Pacific Ocean	16.73	18.96	18.70	15.25	18.85	21.98	21.48	25.08	26.32	27.30
Midwest Pacific Ocean	885.53	1023.01	1153.03	1441.59	1536.34	1664.89	1662.19	1624.81	1625.99	1632.45

Data source The FAO Fishstat Plus Database

Table 10.6 Marine aquaculture output of major global producers in 2010–2019. Unit: 10,000 tons

Year	2010	2011	2012	2013	2014	2015	2016	2017	2018	2019
Bangladesh	12.52	14.51	15.08	16.18	17.44	18.07	20.59	21.23	21.46	21.79
Chile	67.70	92.81	101.60	100.13	115.87	105.71	104.92	121.66	128.59	140.51
China	2506.91	2572.33	2793.17	2948.50	3035.20	3150.24	3347.34	3524.08	3647.40	3823.70
Ecuador	22.33	26.00	28.11	30.40	34.00	40.31	42.24	46.01	56.01	68.01
Egypt	11.71	11.10	12.38	11.54	12.27	14.04	15.79	24.27	26.60	27.22
India	14.17	32.25	36.61	39.97	48.43	60.64	62.59	75.70	81.06	90.32
Indonesia	494.46	621.67	749.79	1072.52	1150.24	1268.82	1278.55	1234.42	1222.57	1206.54
Japan	111.13	86.87	103.95	99.71	98.76	106.90	103.25	98.61	100.49	91.22
Korea	50.56	50.56	50.56	50.67	56.25	56.25	61.35	61.55	66.57	66.57
Democratic People's Republic of Korea	135.50	147.75	148.89	151.52	154.67	165.26	183.32	230.68	224.98	237.20
Malaysia	42.56	40.43	47.06	39.73	41.38	39.43	30.40	32.44	29.02	30.72
Norway	101.97	114.38	132.10	124.78	133.24	138.08	132.61	130.85	135.50	145.28
The Philippines	223.77	230.12	223.19	205.46	203.83	204.50	190.10	192.27	198.18	203.73
Spain	23.44	25.48	24.74	20.74	26.67	27.32	26.60	29.37	33.69	29.05
Taiwan, China	16.71	17.76	19.77	19.62	19.46	18.14	13.66	16.76	16.49	16.12
Thailand	85.37	81.71	81.78	56.18	48.27	50.88	54.47	48.07	49.38	54.75
Turkey	8.92	8.84	10.12	11.08	12.61	13.75	14.87	16.95	20.65	25.51
Britain	18.84	18.71	19.23	19.21	20.32	20.10	18.39	21.18	17.60	21.05
America	19.61	15.60	19.12	18.33	18.39	18.26	19.39	19.03	20.37	23.61
Vietnam	64.73	72.36	71.77	82.18	106.45	112.12	116.70	129.03	136.55	147.17

Data source The FAO Fishstat Plus Database

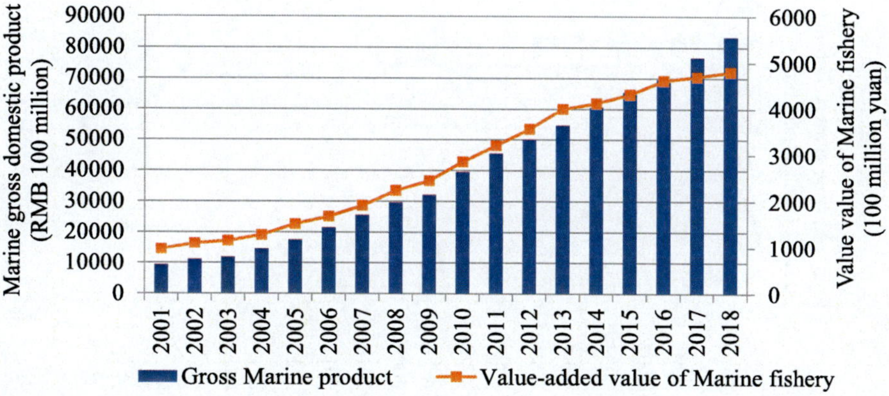

Fig. 10.8 China's marine economy output value in 2001–2018. *Data source* China Marine Economic Statistical Yearbook

10.2.2 Status Quo of Marine Fishery Resources in China

10.2.2.1 Economic Output Value of China's Marine Fishery

China has a vast marine area, stretching for about 473 km^2, which plays an important role in national production. Figure 10.8 shows that China's marine GDP has been rising in recent years, accounting for nearly 10% of the national GDP. The proportion of the tertiary industry in marine production is increasing, far more than the first and secondary industries, reaching about 60%. Marine-related industries are diverse, including marine fishery, marine oil and gas industry, marine salt industry, and marine transportation industry, among others. The increasing added value of marine fishery since 2001 rose from 96.6 billion yuan in 2001 to 480 billion yuan in 2018; however, the added proportional value of marine fishery in marine GDP declined from 10% in 2001 to 6% in 2018.

10.2.2.2 Fishing Situation in Marine Fisheries

Marine fishery and fishing output in China

As a major fishing country, China has recorded the largest marine fishery output in the world in recent years, far more than other countries. Figure 10.9 shows its marine fishery fishing output from 2001 to 2018. However, after entering the twenty-first century, marine fishery production has shown an overall downward trend. In 2001, the marine fishery production was about 14.41 million tons; in 2018, it was 10.44 tons, which is a decrease of 3.97 million tons since 2001. This is due to the "zero growth" target of fishing in 2000. China has implemented a series of marine fishery policies to achieve the goal of sustainable development of marine fisheries.

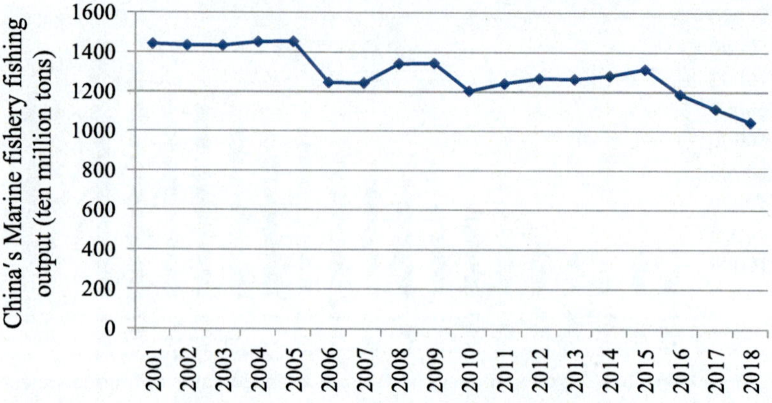

Fig. 10.9 Production of Chinese marine fishery in 2001–2018. *Data source* China Fishery Statistics Yearbook

With the increasing importance of the ocean's role in China's economic development and the implementation of the "ocean power" strategy, China's coastal provinces have grasped the strategic opportunities of national marine development to formulate marine economic development planning with provincial characteristics. For example, Liaoning and Shandong adopted the strategy of "sea Liaoning," and the Zhejiang and Guangdong provinces proposed to transform themselves into strong marine economy provinces. Table 10.7 shows the marine fishery production in coastal China from 2010 to 2018. The marine fishery production in eight of the eleven coastal provinces showed a downward trend. Only three provinces are on the rise. Marine fishery production in Shandong province decreased from 2.3509 million tons in 2010 to 1.7023 million tons in 2018. Liaoning province dropped from 1,000.74 million tons in 2001 to 524,400 tons in 2018, with the most decline.

Major varieties of marine fishery and fishing in China

Although the marine fishery output of various varieties fluctuates, the overall marine fishery output in China has been showing a downward trend. Figure 10.10 shows the varieties that comprise the total Chinese marine fishery production, which include fish, crustaceans, shellfish, algae, and cephalopods. The other varieties of total production have barely changed; the fish production in marine fishery production is the highest, accounting for about 70%; at about 0.2%, algae account for the lowest proportion.

Major waters of China's marine fishing industry

China's main sea areas include the Bohai Sea, the Yellow Sea, the East China Sea, and the South China Sea; marine fishing yields vary across these regions. Figure 10.11 shows the changing trend of marine fishery production in China's four major waters from 2001 to 2018. Barring the case of 2001, marine fishery production in all four major waters shows an obvious downward trend. The Bohai Sea fell from 1.37 million

Table 10.7 Marine fishery production in the coastal areas of China from 2010 to 2018. Unit: 10,000 tons

Year	2010	2011	2012	2013	2014	2015	2016	2017	2018
Liaoning	100.74	106.16	107.93	107.93	107.60	110.79	108.15	55.20	52.44
Hebei	25.33	25.18	25.26	23.05	23.96	25.04	24.78	23.40	21.23
Tianjin	1.58	1.71	1.65	5.34	4.55	4.71	4.52	2.75	2.70
Shandong	235.09	238.44	236.33	231.52	229.72	228.23	229.22	174.96	170.23
Jiangsu	57.04	56.81	56.61	55.38	54.80	55.43	54.89	53.03	47.52
Shanghai	2.15	2.15	2.04	1.96	1.99	1.70	1.69	1.48	1.37
Zhejiang	282.10	303.02	316.02	319.20	324.27	336.70	347.06	309.33	287.39
Fujian	190.85	191.66	192.72	193.73	197.51	200.39	203.86	174.32	170.12
Guangdong	142.96	145.26	151.05	149.08	149.37	150.51	148.05	144.14	127.16
Guangxi	66.30	66.53	66.66	65.06	65.06	65.20	65.29	61.08	55.91
Hainan	99.47	105.03	110.93	112.13	122.01	136.07	140.75	112.73	108.39

Data source China Fishery Statistics Yearbook

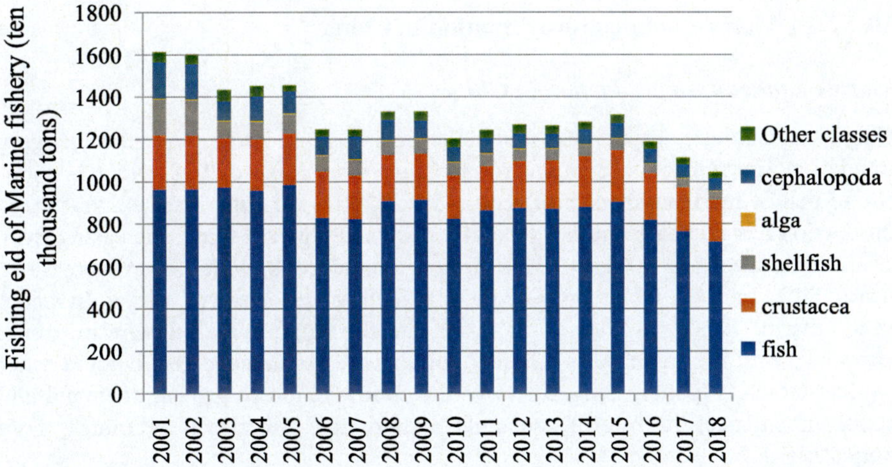

Fig. 10.10 China marine fishery fishing variety in 2001–2018. *Data source* China Fishery Statistics Yearbook

tons in 2001 to 790,000 tons in 2018. The Yellow Sea decreased from 3.22 million tons in 2001 to 2.39 million tons in 2018. The East China Sea fell from 5.4 million tons in 2001 to 4.17 million tons in 2018. The South China Sea dropped from 3.58 million tons in 2001 to 3.1 million tons in 2018. The proportions of the decline are about 42%, 26%, 23%, and 14%, respectively. The biggest drop in production was in the Yellow Sea, and the biggest drop was in the Bohai Sea.

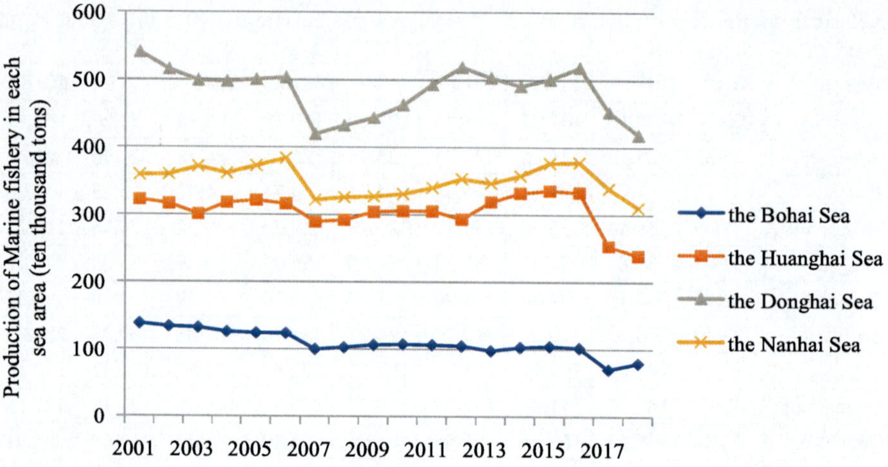

Fig. 10.11 Marine fishery production in the major waters of China from 2001 to 2018. *Data source* China Fishery Statistics Yearbook

10.2.2.3 Marine Aquaculture Situation in China

Marine aquaculture production in China

China's aquaculture output ranks first in the world. Figure 10.12 shows the changing trend of marine aquaculture production in China from 2001 to 2018, which is opposite to that of marine fishery production. Since 2001, although marine aquaculture production has fluctuated, it has generally shown an upward trend, increasing from 11.32 million tons in 2001 to 20.31 million tons in 2018, which is an increase of nearly 80%. In 2000, China proposed a total fishing "zero growth" target. In recent years, marine fishery production in China has shown an overall downward trend. However, with the increase in China's population, the demand for fish and other aquatic products is increasing. To solve the contradiction of growing demand and supply imbalance, the role of aquaculture alternative fisheries is becoming more noticeable.

Table 10.8 shows the output value of marine aquaculture in China's coastal areas from 2010 to 2018. As the statistical data of marine aquaculture output for Shanghai after 2010 was not found, it has not been listed in Table 10.8; only the data of the ten other provinces and cities have been considered. Although the output of marine aquaculture in the other nine provinces, barring Tianjin, has fluctuated since 2010, the overall trend shows an upward curve (Table 10.8). The output of marine aquaculture in Liaoning, Hebei, Shandong, Jiangsu, Zhejiang, Fujian, Guangdong, Guangxi, and Hainan increased from 231.47, 32.93, 396.26, 78.52, 82.57, 303.90, 249.07, 87.74, and 184,200 tons in 2010 to 286.36, 48.98, 521.09, 91.83, 120.90, 478.83, 316.73, 136.32, and 2942 million tons in 2018, respectively. The increases were 24%, 49%, 31%, 17%, 46%, 58%, 27%, 55%, and 60%, respectively. Among them, Shandong

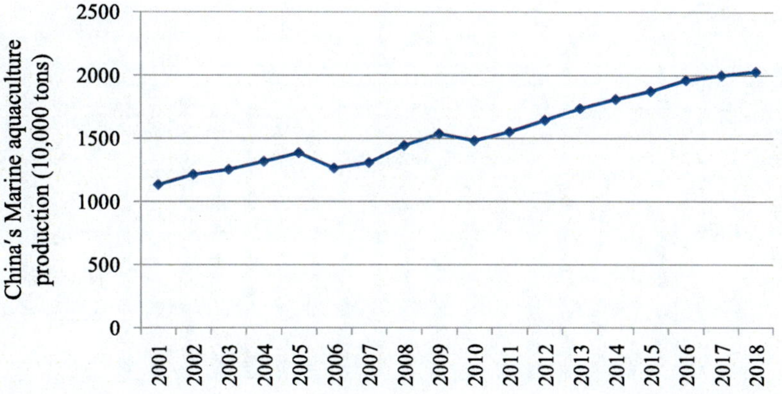

Fig. 10.12 Productivity of Chinese marine aquaculture in 2001–2018. *Data source* China Fishery Statistics Yearbook

province has the highest marine aquaculture output, followed by Fujian, Guangdong, Liaoning, Guangxi, and Zhejiang provinces, with an output of more than one million tons.

The main varieties of marine aquaculture in China

Similar to marine fishery fishing varieties, the main varieties of marine aquaculture also include algae, shellfish, crustaceans, and fish. However, the breeding yield of different varieties significantly differs from that of marine fishery fishing. Though fish is the most productive in China's marine fisheries, shellfish is the most productive in marine aquaculture. Compared with 2001, though the output of marine aquaculture has increased, the proportion in the total output varies. The highest yield of shellfish

Table 10.8 Marine aquaculture output in the coastal areas of China from 2010–2018. Unit: 10,000 tons

Year	2010	2011	2012	2013	2014	2015	2016	2017	2018
Liaoning	231.47	243.52	263.56	282.76	289.05	294.20	310.27	308.14	286.36
Hebei	32.93	31.15	38.21	45.23	49.20	50.65	51.14	52.92	48.98
Tianjin	1.42	1.33	1.43	1.23	1.16	1.05	1.13	0.92	0.77
Shandong	396.26	413.48	436.24	456.64	479.91	499.57	512.78	519.08	521.09
Jiangsu	78.52	84.24	90.50	93.87	93.59	89.35	90.42	93.08	91.83
Zhejiang	82.57	84.49	86.14	87.17	89.79	93.34	101.77	116.26	120.90
Fujian	303.90	316.15	332.66	354.90	379.43	404.13	432.38	445.32	478.83
Guangdong	249.07	265.57	275.74	287.00	294.40	303.22	313.81	302.91	316.73
Guangxi	87.74	92.38	97.73	105.65	109.10	114.22	121.45	129.94	136.32
Hainan	18.42	19.01	21.61	24.81	27.01	25.90	27.97	32.15	29.42

Data source China Fishery Statistics Yearbook

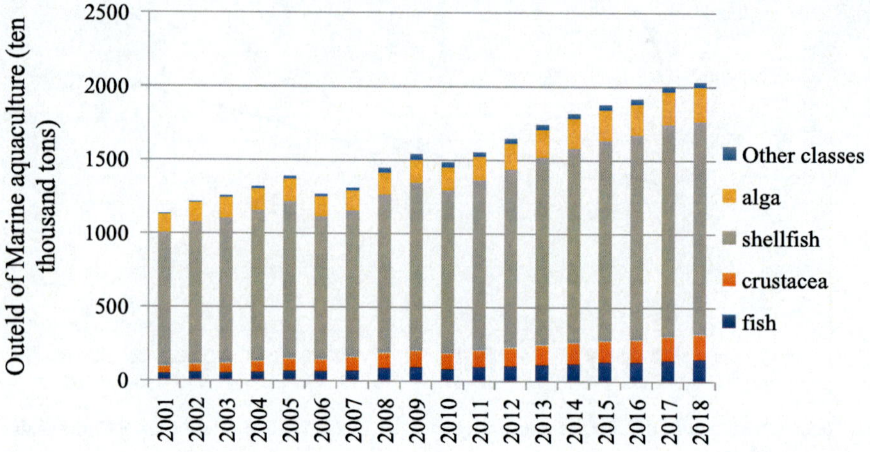

Fig. 10.13 Marine aquaculture variety in China in 2001–2018. *Data source* China Fishery Statistics Yearbook

increased from tons in 2001 to 14.44 million tons in 2018, an increase of nearly 60%. However, the proportion of shellfish in the total aquaculture production decreased from 81% in 2001 to 71% in 2018, whereas the proportion of other varieties in the total production increased (Fig. 10.13).

10.3 Analysis of the Development and Utilization Efficiency of Marine Fishery

10.3.1 Problems Existing in the Utilization of Marine Fishery Resources

Fishery development is putting tremendous pressure on global marine ecosystems (Cury et al., 2011). About 30% of fish stocks have been overexploited, and about 17% of fish species are either endangered or extinct (Kleisner et al., 2013; Rosenberg et al., 2018). Human food security depends on the sustainable utilization of natural resources (Godfray et al., 2010). Marine fisheries, as one of the important food sources, are facing the serious risk of unsustainable utilization. With the growing demand for fish and food, the development of fishing gear and fleets (Pauly & Zeller, 2016), and the expansion of marine fishery industry space (Swartz et al., 2010; Tickler et al., 2018), the importance of protecting the marine ecological environment and promoting the sustainable utilization of marine fishery resources to establish fishery management systems have achieved international consensus (Danielsen & Agnarsson, 2018; Tickler et al., 2018; Zou & Huntington, 2018).

China is the world's leading fishing country. Although aquaculture is now dominant in China's fishing production, the marine fishing fishery remains an important part of the overall supply (Wang & Shi, 2015; Zhang, 2016; Zhang et al., 2020). In recent years, the over-exploitation and depletion of Chinese offshore fishery resources have severely restricted the development of China's fishery (Ding et al., 2020; Xue, 2006; Yang et al., 2016; Zhao & Shen, 2016). In addition, marine aquaculture occupies marine areas and may have a significant negative impact on the marine environment (Grefsrud et al., 2018; Lorenzen et al., 2012). For example, undesirable chemicals and fish wastes generated by drug use or other inputs can pollute the marine environment and even affect the growth of wild marine fish stocks. Since the twenty-first century, the fishery development policy in China has resulted in great changes in fishery and aquaculture practices. Fishery-related policies aim to control fishing capacity and resource utilization intensity, and aquaculture policy has shifted from focusing on quantity growth to improving quality. The national development concept of China has shifted from emphasizing economic growth to emphasizing ecological protection. Fishery resource protection has become the top priority in China's marine fishery management. However, marine fishery resources are still in a challenging situation. Although the resources are abundant, the per capita share is low, and the resource utilization efficiency is poor. The problems existing in the utilization process of marine fishery resources are discussed below.

10.3.1.1 Overfishing of Marine Fishery Resources and Low Proportion of Ocean Fishing

In 2018, marine fishery production amounted to 10.44 million tons. According to the survey data, China's best offshore fishing production was 9 million tons in 1994, when the best fishery resource utilization was reached, followed by a historical peak in 1999 (Yang et al., 2016). Thereafter, the marine fishery production in China has been showing a downward trend; however, the total fishing production is still higher than the topmost recorded level. Marine fishery resources are still in a state of overfishing, causing the actual fishing production to exceed the potential fishing production beyond the carrying capacity of the marine environment. Long-term overfishing has seriously affected marine ecosystems, resulting in the gradual decline of fishery resources and even the imminent extinction of some fish populations. Moreover, with the increase of marine fishery practitioners, the progress of fishing technology, and the improvement of fishing mechanization degree, the fishing capacity of fishing boats have been enhanced, further intensifying the fishing intensity and fishing intensity. In addition, from the perspective of the composition of the marine fishing fleet, the overall number of small and medium-sized fishing vessels is large, and the number and engine power of ocean-going fishing vessels occupy a small proportion of the number and total power of the fishing vessels. This indicates that the fishing intensity of ocean-going fishing is small, and the fishing proportion is unreasonable.

10.3.1.2 Insufficient Utilization of Marine Aquatic Products for Processing, with Primary Products and Rough Processing being the Mainstay of Aquatic Products Processing

The processing of marine aquatic products includes frozen aquatic products, fish millet products, dried pickled products, algae processing products, tank products, aquatic feed, pearls, etc. The processing of marine aquatic products involves several aspects, such as the lack of high-value-added products, the low utilization efficiency of processing waste, and the waste caused by retrograde technology. In recent years, though the overall processing of aquatic products has shown an upward trend, the processed products are mainly primary products and rough processing. The aquatic-frozen products account for a high proportion at more than 85%. Some marine fishery resources with medicinal value are less developed and utilized, such as oysters, tortoises, seaweed, etc. In addition, many valuable ingredients can be extracted from processed waste, such as proteins, highly unsaturated fatty acids, polysaccharides, etc. However, at least 3.9 million tons of processed waste are not effectively utilized each year.

10.3.1.3 Destruction of Marine Life Habitat and Reduction of Marine Biodiversity

Marine fishing methods include trawling nets, purse seine nets, barbed nets, piling nets, and fishing rods. For a long time, sea fishing in China mainly comprised trawled fishing, often targeting marine animals on or near the seabed. Though trawling fishing can capture much marine life, it has a serious negative impact on benthic communities. Bottom trawls, especially double trawls, have been widely used in shallow waters across China, which is a major threat to seabed ecosystems and causes reduced benthic biodiversity. In addition, some coastal projects, such as dam construction and artificial land reclamation, have also caused damage to the habitat of marine life, destroying biodiversity. The survival of mangroves and other marine plants has become seriously threatened, and many marine lives have become almost endangered.

10.3.1.4 Intensification of Marine Ecological and Environmental Pollution and the Restricted Development and Utilization of Fishery Resources

Due to the influence of human activities and the pollution caused by industrial activities, the ecosystem of the coastal areas is more fragile and more vulnerable to pollution. China discharges pollutants into the sea in various ways annually, including "three wastes" emissions, marine ship drainage, waste dumping, and oil exploration, development, and utilization. The sea pollutants are mainly inorganic nitrogen, active phosphate, and petroleum hydrocarbons. The spread of pollution seriously affects the marine ecosystem, leads to ecological imbalance, and reduces biodiversity and biological resources. In addition, in areas where mariculture is practiced, aquaculture biological diseases are seriously increasing due to the intensification of offshore eutrophication and the deterioration of water quality in aquaculture areas.

10.3.2 Efficiency Calculation and Analysis

10.3.2.1 Efficiency Measurement Method

SBM ultra-efficiency Data Envelopment Analysis (DEA) model

Data Envelopment Analysis (DEA) is a non-parametric method that computes multiple inputs and outputs and relatively evaluates their efficiency (Charnes et al., 1978). When analyzing the DEA results of multiple input–output indicators, there are often multiple decision units (Decision Making Unit, DMU) with an efficiency value of 1, which makes it impossible to distinguish between the efficiencies. To address this problem, Andersen and Petersen (1993) proposed the ultra-efficiency DEA model. Furthermore, Tone (2001) proposed the SBM ultra-efficiency DEA model.

Suppose the marine fishery resources utilization efficiency has n decision unit, recorded as DMU_j $(j = 1, 2, \ldots, n)$. Each DMU includes an input vector, expected output vector, and unexpected output vector, respectively as x_i $(i = 1, 2, \ldots, m)$, y_r $(r = 1, 2, \ldots, q_1)$, and b_h $(h = 1, 2, \ldots, q_2)$. $x_j^t = (x_{1j}^t, x_{2j}^t, \ldots, x_{mj}^t) \in R_+^m$ is the input index of the decision unit j in period t, and the expected output index is $y_j^t = (y_{1j}^t, y_{2j}^t, \ldots, y_{q_1 j}^t) \in R_+^{q_1}$, whereas the undesirable output index is $b_j^t = (b_{1j}^t, b_{2j}^t, \ldots, b_{q_2 j}^t) \in R_+^{q_2}$. Defining matrix as $x = (x_1^t, x_2^t, \ldots, x_n^t) \in R_+^{m \times n}$, $y = (y_1^t, y_2^t, \ldots, y_n^t) \in R_+^{q_1 \times n}$, $b = (b_1^t, b_2^t, \ldots, b_n^t) \in R_+^{q_2 \times n}$, under the condition of variable scale income, production may be set as:

$$P(x) = \{(y, b) : x\ (y, b)\} \tag{10.1}$$

According to the model defined by Cooper et al. (2007), the SBM model is:

$$\rho^{*} = \min \frac{1 - \frac{1}{m}\sum_{i=1}^{m}\frac{s_i^x}{x_{ik}^t}}{1 + \frac{1}{q_1+q_2}\left(\sum_{r=1}^{q_1}\frac{s_r^y}{y_{rk}^t} + \sum_{h=1}^{q_2}\frac{s_h^b}{b_{hk}^t}\right)}$$

$$\text{s.t.}\begin{cases}\sum_{j=1}^{n}\lambda_j^t x_{ij}^t + s_i^x = x_{ik}^t \\ \sum_{j=1}^{n}\lambda_j^t y_{rj}^t - s_r^y = y_{rk}^t \\ \sum_{j=1}^{n}\lambda_j^t b_{hj}^t + s_h^b = b_{hk}^t \\ \sum_{j=1}^{n}\lambda_j^t = 1 \\ \lambda_j^t, s_i^x, s_r^y, s_h^b \geq 0 \\ i = 1, 2, \ldots, m;\ j = 1, 2, \ldots, n;\ r = 1, 2, \ldots, q_1;\ h = 1, 2, \ldots, q_2\end{cases} \tag{10.2}$$

where (n,m,q_1,q_2) represents the number of decision units, input, expected output, and undesired outputs, and $(x_{ik}^t,y_{rk}^t,b_{hk}^t)$, $(x_{ij}^t,y_{rj}^t,b_{hj}^t)$ represents the input, expected output, and unexpected output in the decision unit k and j in period t, respectively. λ_j^t represents the weight of input and output in the decision unit j in period t, and (s_i^x,s_r^y,s_h^b) represents the relaxation variables of input, expected output, and undesired output.

The non-radial SBM super-efficiency DEA model considering the unexpected output is:

$$\rho = \min \frac{1 + \frac{1}{m}\sum_{i=1}^{m}\frac{s_i^x}{x_{ik}^t}}{1 - \frac{1}{q_1+q_2}\left(\sum_{r=1}^{q_1}\frac{s_r^y}{y_{rk}^t} + \sum_{h=1}^{q_2}\frac{s_h^b}{b_{hk}^t}\right)}$$

$$\text{s.t.}\begin{cases}\sum_{j=1,j\neq k}^{n}\lambda_j^t x_{ij}^t - s_i^x \leq x_{ik}^t \\ \sum_{j=1,j\neq k}^{n}\lambda_j^t y_{rj}^t + s_r^y \geq y_{rk}^t \\ \sum_{j=1,j\neq k}^{n}\lambda_j^t b_{hj}^t - s_h^b \leq b_{hj}^t \\ 1 - \frac{1}{q_1+q_2}\left(\sum_{r=1}^{q_1}\frac{s_r^y}{y_{rk}^t} + \sum_{h=1}^{q_2}\frac{s_h^b}{b_{hk}^t}\right) > 0 \\ \sum_{j=1,j\neq k}^{n}\lambda_j^t = 1 \\ \lambda_j^t, s_i^x, s_r^y, s_h^b \geq 0 \\ i = 1, 2, \ldots, m;\ j = 1, 2, \ldots, n;\ r = 1, 2, \ldots, q_1;\ h = 1, 2, \ldots, q_2\end{cases} \tag{10.3}$$

The Malmquist-Luenberger index

As the decision unit data is the panel data, the total factor productivity analysis was performed using the ML index (Färe et al., 1992). According to Chung et al. (1997), the directional distance function is defined as:

$$\vec{D}_0(x,y,b;g) = \sup\{\beta: (y,b) + \beta g \in P(x)\} \tag{10.4}$$

where $g = (g_y, -g_b)$ is the direction vector and β is the distance function value. Solving the directional distance function of the decision unit j in the period t through the linear programming model is:

$$\vec{D}_0^t(x_j^t, y_j^t, b_j^t; y_j^t, -b_j^t) = \max\beta$$

$$\text{s.t.}\begin{cases} \sum_{j=1}^{n} \lambda_j^t y_{rj}^t \geq (1+\beta) y_{rj}^t \\ \sum_{j=1}^{n} \lambda_j^t b_{hj}^t \leq (1-\beta) b_{hj}^t \\ \sum_{j=1}^{n} \lambda_j^t x_{ij}^t \leq x_{ij}^t \\ \lambda_j^t \geq 0 \\ i = 1, 2, \ldots, m;\ j = 1, 2, \ldots, n;\ r = 1, 2, \ldots, q_1;\ h = 1, 2, \ldots, q_2 \end{cases} \tag{10.5}$$

The results obtained by the model have been presented in (10.5). According to the ML index defined by Chung et al. (1997), the total factor productivity index of marine fishery resource utilization from the period t to the adjacent period $t + 1$ is:

$$ML_t^{t+1} = \left[\frac{1 + \vec{D}_0^t(x^t, y^t, b^t; y^t, -b^t)}{1 + \vec{D}_0^t(x^{t+1}, y^{t+1}, b^{t+1}; y^{t+1}, -b^{t+1})} \cdot \frac{1 + \vec{D}_0^{t+1}(x^t, y^t, b^t; y^t, -b^t)}{1 + \vec{D}_0^{t+1}(x^{t+1}, y^{t+1}, b^{t+1}; y^{t+1}, -b^{t+1})}\right]^{\frac{1}{2}} \tag{10.6}$$

If $ML_t^{t+1} > 0$, the total factor productivity of marine fishery resources increases; however, if $ML_t^{t+1} = 1$, the productivity is unchanged.

10.3.2.2 Selection of Indicators and Data Sources

To comprehensively measure the utilization efficiency of marine fishery resources, this chapter selects the utilization data of marine fishery resources in eleven coastal areas of China from 2003 to 2018 for empirical research, including input indicators, expected output indicators, and unexpected output indicators. The evaluation index system is shown in Table 10.9. The data are mainly derived from China Statistical

Table 10.9 Evaluation index system for utilization efficiency of marine fishery resources

Indicator category	Level 1 indicators	Secondary indicators
Investment index	Capital	Marine fishery capital stock (RMB 100 million Yuan)
	Labor force	Marine fishery practitioners (persons)
Expect output	Economic benefits	Economic output value of marine fishery (ten thousand yuan)
Undesired output	Environmental pollution	Industrial wastewater is directly discharged into mass (million tons)

Yearbook, China Marine Statistical Yearbook, China Fishery Statistical Yearbook, and China Environmental Statistical Yearbook.

The input index selects the marine fishery practitioners as the labor input index and the marine fishery capital stock as the capital input index. The calculation of the marine fishery capital stock is based on He et al. (2014), as follows:

$$\begin{aligned}&\text{Marine resources utilization capital stock}\\&\quad = (\text{Marine fishery output value/regional GDP}) \times \text{regional capital stock}\end{aligned}$$

The regional capital stock is calculated by perpetual inventory and distribution (Zhang, 2004). The formula is:

$$K_{it} = I_{it} \big/ P_{it} + (1 - \delta_{it})K_{i,t-1} \tag{10.8}$$

Among them, K_{it} is the actual capital stock for the provincial i in year t, and $K_{i,t-1}$ is the actual capital stock for the provincial i in year $t-1$. I_{it} is new fixed assets for the provincial i in year t assessed using the fixed assets investment price index. δ_{it} is the capital depreciation rate for the provincial i in year t, choosing 9.6% (Zhang, 2004). The capital stock utilization for the base year (2003) is calculated as $K_{i0} = I_{i0} \big/ 10\%$.

The expected output index considers the benefits of marine fishery production. Marine fishery includes fishing in the ocean and collecting and breeding aquatic animals and plants for aquatic production activities, which can be divided into marine fishing and mariculture. Thus, the value of marine fishery economic output includes marine fishing and mariculture value. Considering the pollution of the marine environment in coastal areas, industrial wastewater is discharged directly into massive quantities as an undesired output index.

10.3.2.3 Calculation Results and Analysis

Efficiency analysis of eleven coastal provinces and cities

According to the sample data, the utilization efficiency of marine fishery resources in eleven coastal areas from 2003 to 2018 was calculated by MaxDEA under the condition of variable scale remuneration. The results are shown in Table 10.10. The average efficiency of marine fishery resources in China was about 0.6528 in 2018. The average efficiency in Tianjin was the highest, reaching 0.9246, followed by Guangdong Province, with an average efficiency of 0.8930, and the average efficiency of the marine fishery resources of Hebei was the lowest at only 0.3979. In addition, the utilization efficiency value of marine fishery resources in the eleven coastal regions fluctuated greatly. Compared with 2003, the efficiency value decreased in Guangxi, Hainan, Hebei, and Liaoning from 0.6200, 0.4847, 0.4676, and 0.7399 in 2003 to 0.4505, 0.4626, 0.3019, and 0.3088. The overall utilization efficiency value of marine fishery resources in other regions showed an upward trend.

Efficiency analysis of the three major coastal economic zones

Based on the spatial division scale of the land economic zone, the coastal areas in China are divided into the Bohai Rim Economic Zone, the Yangtze River Delta Economic Zone, and the Pan-Pearl River Delta Economic Zone. The Bohai Rim Economic Zone includes Liaoning, Hebei, Tianjin, and Shandong. The Yangtze River Delta Economic Zone includes Jiangsu, Shanghai, and Zhejiang. The Pan-Pearl River Delta Economic Zone includes Fujian, Guangdong, Guangxi, and Hainan. The average utilization efficiency of marine fishery resources is calculated based on the utilization efficiency of marine fishery resources in coastal areas, compared with the overall mean value of marine fishery resources utilization efficiency in coastal areas. As can be seen from Fig. 10.14, the utilization efficiency of marine fishery resources in the three major economic zones fluctuated from 2003 to 2018 around the average overall efficiency, with a small gap. Specifically, in the initial stage, the utilization efficiency gap of marine fishery resources in the three economic zones was not noticeable. After 2007, the utilization efficiency of marine fishery resources in the Yangtze River Delta Economic Zone increased rapidly, exceeding the overall average, and decreased in 2009. Compared with 2003, the utilization efficiency of marine fishery resources in the three economic zones showed an upward trend, among which the average efficiency of the Pan-Pearl River Delta Economic Zone was the highest, and that of the Bohai Rim Economic Zone was the lowest; however, the overall gap is not large.

Dynamic analysis of the ML index

To study the dynamic evolution and the intrinsic driving mechanism of the efficiency of marine fishery resource utilization in the coastal areas of China, the ML index was used (Table 10.11). The average ML index of marine fishery resources utilization in the eleven coastal provinces and cities in China from 2003 to 2018 was 1.0267, indicating that the utilization efficiency of marine fishery resources in

Table 10.10 Utilization efficiency of marine fishery resources in eleven coastal provinces and cities in 2003–2018

Year	Fujian	Guangdong	Guangxi	Hainan	Hebei	Jiangsu	Liaoning	Shandong	Shanghai	Tianjin	Zhejiang
2003	0.4771	0.5988	0.6200	0.4847	0.4676	0.4709	0.7399	0.5504	0.6506	0.6298	0.4712
2004	0.4936	0.7274	0.5581	0.5379	0.4474	0.4679	0.4818	0.5371	0.6358	0.6668	0.4603
2005	0.5223	0.6623	0.6346	0.6039	0.4545	0.6417	0.5088	0.6374	0.6103	0.5747	0.4686
2006	0.5548	1.0217	0.8352	0.6858	0.5344	0.6866	0.5132	0.6580	0.8486	0.7185	0.5318
2007	0.5518	1.0136	0.8303	0.8048	0.5103	0.7397	0.5404	0.6736	1.0787	1.0744	0.5862
2008	0.5431	0.8320	0.5512	0.6456	0.4529	1.0954	0.5056	0.6262	1.0214	0.8142	0.5052
2009	0.5637	1.0163	0.6023	0.6384	0.4253	0.7001	0.4237	0.6930	0.7029	1.0157	0.5004
2010	0.5526	1.0147	0.5902	0.7399	0.4177	0.7990	0.4595	0.5835	0.5643	0.9178	0.4967
2011	0.4698	0.7885	0.4192	0.6019	0.3253	0.6588	0.4258	0.4378	0.5791	1.2449	0.4231
2012	0.5441	1.0032	0.4677	0.6246	0.3487	0.7070	0.4128	0.7104	0.6273	1.0308	0.4599
2013	0.5923	0.8134	0.4388	0.5328	0.3601	1.0109	0.4230	0.7636	0.6220	1.0203	0.4352
2014	0.7484	1.0171	0.4398	0.4975	0.3466	0.9213	0.4310	1.0038	0.6081	1.0076	0.4463
2015	1.0043	0.9879	0.4373	0.4935	0.3206	0.7762	0.4096	1.0157	0.6238	1.0317	0.4371
2016	1.0177	0.8403	0.4529	0.5014	0.3345	0.8051	0.3863	1.0306	0.6671	1.0198	0.3926
2017	0.8034	0.9249	0.4438	0.4666	0.3181	0.8417	0.3160	1.0031	0.7647	0.9801	0.4444
2018	1.0085	1.0267	0.4505	0.4626	0.3019	1.1459	0.3088	1.0149	1.0286	1.0461	0.5248
Mean	0.6530	0.8930	0.5482	0.5826	0.3979	0.7793	0.4554	0.7462	0.7271	0.9246	0.4740

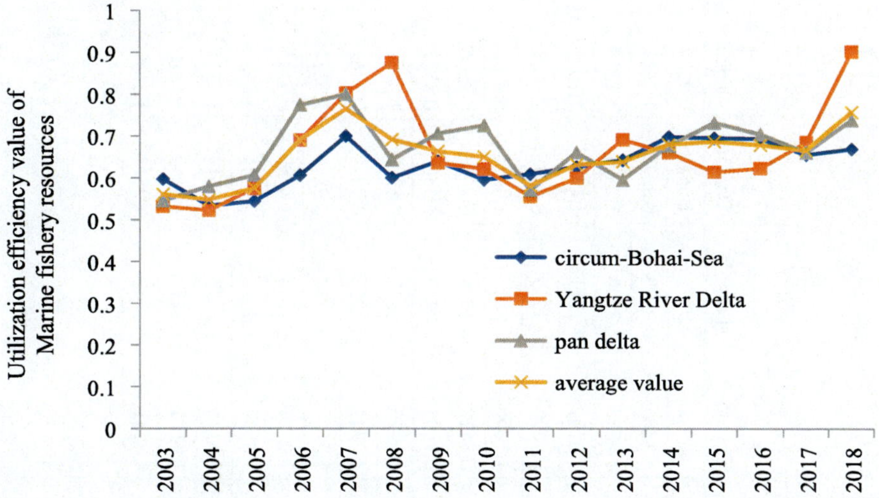

Fig. 10.14 Utilization efficiency of marine fishery resources in the three major coastal economic zones from 2003 to 2018

China's coastal areas showed an upward trend, with an average annual growth rate of 2.67%. Comparing the mean values of the ML indices across the regions, the average ML index of Fujian, Guangdong, Hainan, Jiangsu, Shandong, Shanghai, Tianjin, and Zhejiang from 2003 to 2018 was 1.0608, 1.0563, 1.0040, 1.0866, 1.0581, 1.0468, 1.0518, and 1.0119, respectively. The total utilization of marine fishery resources in these regions is on the rise, with annual growth of 6.08, 5.63, 0.40, 8.66, 5.81, 4.68, 5.18, and 1.19%. The mean ML index of Guangxi, Hebei, and Liaoning were 0.9917, 0.9751, and 0.9504, respectively. The total utilization of marine fishery resources in these regions shows a downward trend. The average annual decline was 0.83%, 2.49%, and 4.96%, respectively. From the perspective of time change, the ML indexes of Guangdong, Hainan, and Hebei showed a downward trend, respectively, from 1.2148, 1.1098, and 0.9569 in 2003–2004 to 1.1101, 0.9915, and 0.9490 in 2017–2018. However, other provinces have shown an upward trend. With the improvement of science and technology, efficiency should have an upward trend. This phenomenon should be paid great attention to by the relevant departments, and effective measures must be taken.

10.3.3 Motives for Inefficient Utilization of Marine Fishery Resources

As mentioned above, although China has rich marine fishery resources, it is facing a series of problems in the utilization process, including overfishing, insufficient

Table 10.11 ML index of marine fishery resources utilization efficiency in eleven coastal provinces and cities from 2003–2018

Year	Fujian	Guangdong	Guangxi	Hainan	Hebei	Jiangsu	Liaoning	Shandong	Shanghai	Tianjin	Zhejiang
2003–2004	1.0346	1.2148	0.9002	1.1098	0.9569	0.9938	0.6511	0.9758	0.9773	1.0587	0.9769
2004–2005	1.0581	0.9105	1.1371	1.1228	1.0159	1.3713	1.0560	1.1866	0.9599	0.8620	1.0181
2005–2006	1.0621	1.5428	1.3161	1.1357	1.1756	1.0699	1.0087	1.0325	1.3905	1.2501	1.1348
2006–2007	0.9947	0.9921	0.9942	1.1735	0.9549	1.0774	1.0531	1.0237	1.2711	1.4953	1.1022
2007–2008	0.9842	0.8209	0.6639	0.8022	0.8876	1.4809	0.9355	0.9296	0.9469	0.7578	0.8619
2008–2009	1.0379	1.2215	1.0926	0.9888	0.9390	0.6391	0.8381	1.1066	0.6882	1.2475	0.9905
2009–2010	0.9804	0.9984	0.9799	1.1589	0.9821	1.1413	1.0843	0.8420	0.8028	0.9036	0.9926
2010–2011	0.8501	0.7771	0.7103	0.8135	0.7788	0.8246	0.9268	0.7503	1.0262	1.3565	0.8517
2011–2012	1.1582	1.2723	1.1158	1.0378	1.0721	1.0731	0.9693	1.6225	1.0832	0.8280	1.0871
2012–2013	1.0887	0.8108	0.9382	0.8530	1.0326	1.4299	1.0248	1.0749	0.9916	0.9899	0.9462
2013–2014	1.2635	1.2504	1.0022	0.9338	0.9624	0.9114	1.0190	1.3146	0.9777	0.9876	1.0254
2014–2015	1.3420	0.9713	0.9942	0.9921	0.9250	0.8425	0.9503	1.0118	1.0258	1.0239	0.9795
2015–2016	1.0133	0.8506	1.0358	1.0160	1.0433	1.0372	0.9431	1.0147	1.0693	0.9884	0.8981
2016–2017	0.7895	1.1006	0.9799	0.9305	0.9510	1.0455	0.8181	0.9734	1.1463	0.9611	1.1321
2017–2018	1.2552	1.1101	1.0152	0.9915	0.9490	1.3614	0.9772	1.0117	1.3451	1.0673	1.1810
Mean	1.0608	1.0563	0.9917	1.0040	0.9751	1.0866	0.9504	1.0581	1.0468	1.0518	1.0119

product processing and utilization, and ecological environment damage. The insufficient utilization of marine fishery resources and low utilization efficiency is not conducive to the sustainable development of marine fishery resources. The motivation for marine fishery resource utilization efficiency is discussed below.

10.3.3.1 Poorly Defined Property Rights of Marine Fishery Resources

The state owns sea areas, and marine fishery resources are state-owned assets. They are neither owned by an individual nor are they quasi-public goods. Therefore, marine fishery resources are the resources available to all people. The public nature of marine fishery resources encourages fishermen to continuously promote their development and use marine fishery resources to seek benefits so that their utilization exceeds their total amount. This forms a "tragedy of common" which is not conducive to the long-term development of marine fishery resources. Fishery management includes planning, water use, and fishery resource development, which different departments manage. The government fishery administrative departments generally hold the ownership of fishery resources on behalf of the state; these departments have the right to license and supervise the utilization of marine fishery resources, which also results in the unclear definition of marine fishery resources property rights, not conducive to the sustainable utilization of marine fishery resources and its utilization efficiency.

10.3.3.2 Unreasonable Structure of the Marine Fishery Industry

The fishery industry is divided into primary, secondary, and tertiary industries, among which the primary industry is mainly fishing and aquaculture. The secondary industry includes fishery construction, and the tertiary industry includes fishery circulation and service industry. Though the marine fishery industry is mainly concentrated in the primary and secondary industries, the utilization efficiency of resources and energy in marine fishery fishing and breeding is generally low, which impacts the utilization efficiency of marine fishery resources. Moreover, the processing of China's marine fishery products mainly includes frozen and rough processing products. The product processing level is low, and the processing of marine fishery products is not fully utilized, which is not conducive to the improvement of the utilization efficiency of marine fishery resources. In addition, the primary and secondary industries produce more pollutants than the tertiary marine fishery industry. The discharge of pollutants into the sea causes pollution in the marine water environment, resulting in even more undesired output, which is not conducive to improving the utilization efficiency of marine fishery resources.

10.3.3.3 Inadequate Modernization Level of Marine Fisheries

The development of marine fishery cannot be separated from the support of advanced technology. Although China's marine fishery has achieved rapid development in recent decades, there is still a big gap between China's marine fishery compared with developed countries. In countries with developed marine fishery practices, the stage of realizing the modernization of marine fishery has almost been reached. However, marine fishery in China is still in the middle stage of development, and the modernization of marine fishery still faces a series of challenges. For instance, marine fishery science and technology power is relatively scattered, resource investment cannot be shared, and the phenomenon of repeated investment is more serious. The low utilization rate of scientific research equipment seriously restricts the improvement of marine fishery science and technological innovations. Basic marine research is relatively weak, and the research on resource conservation and ecological restoration in mariculture areas started late, which cannot provide excellent technical support for mariculture. Despite the large number, marine fishing boats are usually small and medium in size, and ocean-going fishing activities are limited. In addition, the aquatic products processing technology is not advanced, which has seriously affected the industrialization of fishery production and is not conducive to the improvement of the utilization efficiency of marine fishery resources.

10.4 Public Policy

10.4.1 Fishing Cost

Fishing costs include purchasing or maintaining fishing boats, oil, and nets, which the fishermen bear. To ensure the sustainable development of marine fishery, several countries have implemented institutional policies to reduce fishing production. However, these systems increased the cost of fishing. Effective policy is essential not only to achieve effective production but also to ensure that the yield is achieved at the lowest cost. Although the implementation of the policy brings about a decline in the catch volume, the reasons for the effectiveness of the policy cannot be explained. In Fig. 10.15, TC1 is the cost curve corresponding to the effective catch, and TC2 is the cost curve corresponding to the system policy. The net income of the fishermen after the policy implementation is the longitudinal distance between the income curve and the home of the cost curve. When a series of policies, such as fees and limited fishing time, is implemented, the fishing cost is improved to a certain extent. For example, when fishing time is limited, to increase the catch in a short time, the cost is increased to buy larger fishing boats, or the method of fishing is improved to ensure that the fishermen can catch more fish quickly. As the increased cost reduces the net income of the fishermen, the sustainable utilization of marine fishery resources does not protect them.

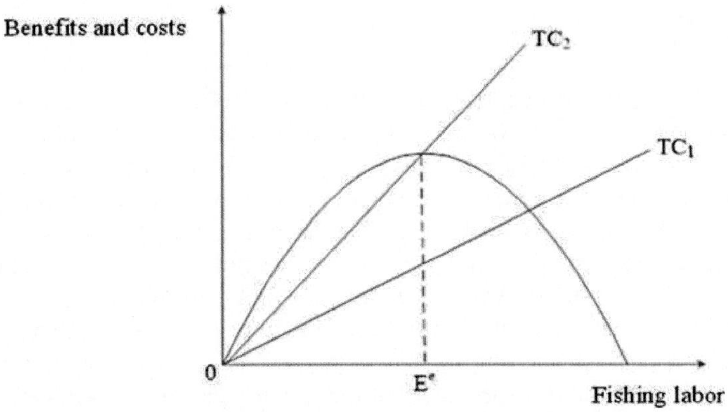

Fig. 10.15 Relationship between higher catch cost and higher catch volume

Since the Reform and Opening-up, China has formulated a series of fishing management systems to protect the fishery resources, such as the fishing permit system, fishing moratorium system, fishing quota system, fishery resources proliferation protection system, etc. Though some of these systems have not been strictly implemented, they raise the fishing cost. These systems also have their problems. For example, under the fishing permit system, fishing practitioners hope to control the total catch at the best production through fishing type, place, fishing gear quantity, and production. However, the system has no limit on total fishing and fishing time. Fishermen can increase fishing by extending operation time. Companies can indirectly apply for certificates through market channels and increase the number of fishery practitioners, which is not conducive to the sustainable utilization of marine fishery resources. Fishing systems stipulate the fishing area and fishing period based on the law of fishery resources reproduction, growth, and development. Although the system gives time for breeding, after the end of the fishing, a large number of fishermen concentrate on fishing and its improvement. The effect of fishing will soon be offset after larger fishing, but it is not conducive to sustainable fishing. The fishery resource proliferation and protection fee system stipulates the expenses fishery producers bear and increases fishery production costs. However, implementing this system is also very difficult due to the large number of fishermen, the scattered distribution, the unclear definition of household registration, and the unclear legal positioning. Therefore, simply relying on increasing the cost of fishing cannot reduce the amount of marine fisheries caught, especially when the fishermen's income is not high. Although improving the cost of marine fishing can encourage people to act upon the development and utilization of marine fishery resources for cost accounting, the most fundamental need is to pay attention to fishermen's income problem. Helping fishermen improve their income can truly guarantee the sustainable utilization of marine fishery resources.

10.4.2 Subsidies

Fishery subsidies are the financial contributions and subsidies for income or prices provided by the government or any public body to fishery producers or operators. They are of four types: direct government payments, loans or physical transfers of goods (such as subsidized fuel, ice, or bait) or services (such as training) to fishermen or fishery enterprises, direct benefits from goods, services, or activities (e.g., port infrastructure, marketing, and promotion programs), tax rebates, and other tax breaks to fishermen or fishery enterprises, and direct planned expenditures for fishery conservation and management. The impact of fishery subsidies on fish population sustainability is of three main types: beneficial subsidies for fish population growth, harmful impact subsidies for fisheries, and ambiguous subsidies. Beneficial subsidies include support for fisheries management plans and services, fisheries research and development, and the establishment of marine reserves. Harmful subsidies distort economic incentives that may otherwise be unprofitable, including fuel subsidies, ships, fishing ports, tax exemptions, and vague subsidies for uncertain beneficial or harmful effects on fish population growth, such as fishermen assistance programs, ship repurchase programs, and rural fishermen community development programs.

Since the 1990s, controlling the negative effects of fisheries subsidies has attracted the attention of scientists, policymakers, and managers worldwide. In 1992, the FAO noted that providing subsidies to the fisheries sector could lead to the depletion of fish resources. In 2001, the World Trade Organization (WTO) began strengthening the discipline in fishery subsidies. Since its accession to the WTO, China has followed the FAO code of action for responsible fisheries and the WTO subsidy and countervailing agreement. In recent years, a series of subsidies have been implemented in fishery, such as fishery diesel subsidies, fishery management subsidies, tax breaks, fishing boat subsidies, fishery insurance subsidies, and ocean fishery development subsidies. The scale of the Chinese fishery subsidy policy is expanding, and policy subsidies are increasing. Marine fishery subsidies gradually formed with indirect subsidies, direct subsidies for complementary subsidy structure. These subsidies help reduce fishermen's production costs and improve basic fishery conditions. To a certain extent, fishermen's income and effective maintenance of fishery resources are ensured to stabilize fishery production, which plays an important role in marine fishery modernization. However, fishery subsidy policy also has problems, including unreasonable subsidy structure, nonstandard subsidy accounting system, fishery insurance, and financial subsidy mechanism. Future fishery subsidies in China must be reformed to adjust to the structure of the fishery subsidies, standardize the fishery subsidy accounting system, and optimize the fishery insurance subsidy scheme.

10.4.3 Individual Transferable Quota

An Individual Transferable Quota (ITQ) is a tenure right that allows owners to decide when and how to use their quota units; however, this excludes the full right to resources or the ability to decide how much resources can be harvested. ITQs are usually considered an appropriate way to solve the management of the fisheries sector. The public nature of fisheries prompted fishermen to catch as many fish populations as possible, leading to overfishing. ITQs ensure the longest use of fishing time, with the total share of the fishing season allocated to individual fishermen, enterprises, and vessels. To allocate fishing rights to individual fishermen, ITQs, though associated with additional social, environmental, and management costs, can also help fishermen purchase others' fishing rights, thus reducing overall fishing capacity, improving efficiency, and reducing excess capacity in the fisheries sector. ITQs can also reduce the cost of fishing as fishermen no longer compete with other fishermen to ensure a certain share of the catch, which can save fishermen many costs over time and increase economic profits by avoiding increased capture. Since the first adoption of ITQs in Iceland and New Zealand in the 1980s, several countries have adopted ITQs as their management system. Canada introduced ITQs in 1983 due to overcapacity, resource depletion, fishing competition, and the safety of fishermen. ITQs were introduced in Australia in 1984 due to population depletion and fishing competition. The US adopted ITQs in 1990 over resource exhaustion, overcapacity, and fisherman conflicts resulting from competitive fishing. Denmark adopted ITQs in 2003 due to overcapacity and overfishing. Argentina introduced ITQs in 2010 due to overcapacity and inventory depletion. Other countries, such as Chile, the UK, Peru, the Netherlands, and Sweden, have also introduced the ITQ system due to overcapacity, inventory depletion, and overfishing.

After 1989, China's aquatic production has ranked first in the world. The fishery achievements made at the cost of fishery resources exhaustion make China face severe domestic pressure. At the same time, the entry into force of the United Nations Convention on the Law of the Sea in the international community also makes the world's fishery resources face exhaustion pressure. To meet the dual pressure of many international and domestic sources, China stipulated in the Fisheries Law that "the state determines the total fishing volume of fishery resources according to the principle that the fishing volume of fishery resources is lower than the increased volume, and implements the fishing quota system." Although quota fishing has been incorporated in the Fishery Law, it has not yet been truly implemented in China's fishery management practice. Thus, advanced fishery management policies such as ITQs have not yet been implemented nationwide in China. To ensure the sustainable utilization of national fishery resources, China should learn from the experience of developed foreign countries and gradually implement the ITQ system with Chinese characteristics in the concrete practice of fishery management. Before implementing the ITQ management system, some prerequisites must be met. First, sufficient monitoring, control, and supervision are needed so that fishermen cannot catch without ITQs. If fishermen violate the regulations, the administration can restrict their behavior through strong law enforcement. Moreover, as implementing the ITQs system requires determining the total recoverable amount of fishery resources, the management costs are often higher than the equivalent investment of management fees for controlling fisheries. Second, at the fisheries level, if ITQs are to be implemented effectively, it must be ensured that the total recoverable amount of fishery resources is binding on fishermen's fishing activities. Third, sufficient flexibility in coordinating quotas and incentives is needed to enable fishermen to balance their actual catches with quota holdings. Fourth, the ITQ system needs to fully cover the main target species, without which fishermen may catch species outside the provisions of the ITQ system to replace them, thus threatening the sustainability of other fisheries. Finally, the annual changes in biomass must be taken into account.

10.5 Summary

With rich marine fishery resources, China is one of the most important marine and fishery countries in the world. The development and utilization of marine fishery resources promote the economic and social development of marine fishery and coastal fishing areas and meet Chinese urban and rural residents' diversified dietary and nutritional needs. Initially, China's fishery development showed the pattern of "fishery first and aquaculture second." By the mid and late 1980s, to protect offshore fishery resources and meet the growing demand for high-quality protein, the fishery development strategy was adjusted to "aquaculture as the main, fishery as the auxiliary." In the process of the exploitation and utilization of China's marine resources, a series of problems have emerged, including overfishing, resource depletion, and ecological environment deterioration. To solve these problems, the Chinese government has promulgated and implemented a series of policies, mainly focusing on the three core elements of the fishery, fishermen, and ecological resources. In addition, China has established a basic institutional system to achieve the policy objectives set by each element. However, these three core elements are interdependent, which means that failing to implement policies serving one goal may hinder realizing other goals. There are internal contradictions between the multiple goals of this institutional system, which may also be the main reason why China's marine fishery governance failed to get rid of the "tragedy of commons." Although China faces many challenges in formulating effective marine fishery policies, marine fishery management policies in China have been continuously improved by drawing on the successful experience of other countries in marine fishery management to realize the sustainable development of marine fishery resources and build maritime power.

References

Andersen, P., & Petersen, N. C. (1993). A procedure for ranking efficient units in data envelopment analysis. *Management science, 39*(10), 1261–1264.

Charnes, A., Cooper, W. W., & Rhodes, E. (1978). Measuring the efficiency of decision making units. *European journal of operational research, 2*(6), 429–444.

Chung, Y. H., Färe, R., & Grosskopf, S. (1997). Productivity and undesirable outputs: a directional distance function approach. *Journal of Environmental Management, 51*(3), 229–240.

Cooper, W. W., Seiford, L. M., & Tone, K. (2007). Data envelopment analysis: a comprehensive text with models, applications, references and DEA-solver software (Vol. 2). New York: Springer.

Cury, P. M., Boyd, I. L., Bonhommeau, S., Anker-Nilssen, T., Crawford, R. J., Furness, R. W., ... & Sydeman, W. J. (2011). Global seabird response to forage fish depletion-one-third for the birds. *Science, 334*(6063), 1703–1706.

Danielsen, R., & Agnarsson, S. (2018). Fisheries policy in the Faroe Islands: Managing for failure?. *Marine Policy, 94*, 204–214.

Ding, Q., Shan, X., & Jin, X. (2020). Ecological footprint and vulnerability of marine capture fisheries in China. *Acta Oceanologica Sinica, 39*(4), 100–109.

Färe, R., Grosskopf, S., Lindgren, B., & Roos, P. (1992). Productivity changes in Swedish pharamacies 1980–1989: A non-parametric Malmquist approach. *Journal of productivity Analysis, 3*(1), 85–101.

Godfray, H. C. J., Beddington, J. R., Crute, I. R., Haddad, L., Lawrence, D., Muir, J. F., ... & Toulmin, C. (2010). Food security: the challenge of feeding 9 billion people. *science, 327*(5967), 812–818.

Grefsrud, E. S., Glover, K., Grøsvik, B. E., Husa, V., Karlsen, Ø., Kristiansen, T., ... & Svasand, T. (2018). Risikorapport norsk fiskeoppdrett 2018.

Kleisner, K., Zeller, D., Froese, R., & Pauly, D. (2013). Using global catch data for inferences on the world's marine fisheries. *Fish and Fisheries, 14*(3), 293–311.

Lorenzen, K., Beveridge, M. C., & Mangel, M. (2012). Cultured fish: integrative biology and management of domestication and interactions with wild fish. *Biological Reviews, 87*(3), 639–660.

Pauly, D., & Zeller, D. (2016). Global Atlas of Marine Fisheries: A critical appraisal of catches and ecosystem impacts. Island Press.

Rosenberg, A. A., Kleisner, K. M., Afflerbach, J., Anderson, S. C., Dickey?Collas, M., Cooper, A. B., ... & Ye, Y. (2018). Applying a new ensemble approach to estimating stock status of marine fisheries around the world. *Conservation Letters, 11*(1), e12363.

Swartz, W., Sala, E., Tracey, S., Watson, R., & Pauly, D. (2010). The spatial expansion and ecological footprint of fisheries (1950 to present). *PloS one, 5*(12), e15143.

Tickler, D., Meeuwig, J. J., Palomares, M. L., Pauly, D., & Zeller, D. (2018). Far from home: Distance patterns of global fishing fleets. *Science advances, 4*(8), eaar3279.

Tone, K. (2001). A Slacks-Based Measure of Super-Efficiency in Data Envelopment Analysis. European *Journal of Operational Research, 143*(1), 32–41.

Wang, Z. K., & Shi, J. C. (2015). Research on China's marine fishery resources protection and optimal allocation. *Chin Fish Econ, 33*, 66–73.

Xue, G. J. (2006). China's distant water fisheries and its response to flag state responsibilities. *Marine Policy, 30*(6), 651–658.

Yang, Z., Li, S., Chen, B., Kang, H., & Huang, M. (2016). China's aquatic product processing industry: Policy evolution and economic performance. *Trends in Food Science & Technology, 58*, 149–154.

Zhang, H. (2016). Chinese fishermen in disputed waters: Not quite a "people's war". *Marine Policy, 68*, 65–73.

Zhang, W., Liu, M., Sadovy de Mitcheson, Y., Cao, L., Leadbitter, D., Newton, R., ... & Zhou, W. (2020). Fishing for feed in China: Facts, impacts and implications. *Fish and Fisheries, 21*(1), 47–62.

Zhao, W., & Shen, H. (2016). A statistical analysis of China's fisheries in the 12th five-year period. *Aquaculture and Fisheries, 1*, 41–49.

Zou, L., & Huntington, H. P. (2018). Implications of the Convention on the Conservation and Management of Pollock Resources in the Central Bering Sea for the management of fisheries in the Central Arctic Ocean. *Marine Policy, 88*, 132–138.